Birkhäuser

Frontiers in Mathematics

Ştefan Cobzaş

Functional Analysis in Asymmetric Normed Spaces

Birkhäuser

Ştefan Cobzaş
Department of Mathematics
Babeş-Bolyai University
Cluj-Napoca
Romania

ISSN 1660-8046 ISSN 1660-8054 (electronic)
ISBN 978-3-0348-0477-6 ISBN 978-3-0348-0478-3 (eBook)
DOI 10.1007/978-3-0348-0478-3
Springer Basel Heidelberg New York Dordrecht London

Library of Congress Control Number: 2012950766

Printed on acid-free paper

Springer Basel is part of Springer Science+Business Media (www.springer.com)

Contents

Introduction

The main goal of this book is to present the basic results on asymmetric normed spaces. Since the basic topological tools come from quasi-metric spaces and quasi-uniform spaces, the first chapter contains a thorough presentation of some fundamental results from the theory of these spaces. The focus is on those which are most used in functional analysis – completeness, compactness and Baire category. For a good presentation of the general theory of quasi-uniform and quasi-metric spaces, a well-established and thickly developed branch of general topology, one can consult the classical monograph by Fletcher and Lindgren [80] and some subsequent survey papers by Künzi (see the bibliography at the end of the book). The survey paper [45] may be viewed as a skeleton of this book.

A quasi-metric is a function ρ on $X \times X$ satisfying all the axioms of a metric with the exception of the symmetry: it is possible that $\rho(y,x) \neq \rho(x,y)$ for some $x, y \in X$. In this case $\bar{\rho}(x,y) = \rho(y,x)$ is another quasi-metric on X, called the conjugate of ρ, and $\rho^s = \rho \vee \bar{\rho}$ is a metric on X. Asymmetric metric spaces are called quasi-metric spaces. The term quasi-metric was proposed as early as 1931 by Wilson [239], see also [1]. Quasi-metric spaces were considered also by Niemytzki [168] in connection with the axioms defining a metric space and metrizability. In [27], [186] they are called oriented metric spaces and in [187] spaces with weak metric.

This apparently innocent modification of the axioms of a metric space drastically changes the whole theory, mainly with respect to completeness, compactness and total boundedness. There are a lot of completeness notions in quasi-metric and quasi-uniform spaces, all agreeing with the usual notion of completeness in the case of metric or uniform spaces, each of them having its advantages and weaknesses.

Also, concerning compactness, the situation is totally different in quasi-metric spaces – for instance, sequential compactness does not agree with compactness, in contrast to the case of metric spaces. In spite of these peculiarities there are a lot of positive results relating compactness with various kinds of completeness and total boundedness. Baire category also needs a special treatment, including some bitopological results.

Quasi-uniform spaces form a natural extension of both quasi-metric spaces and uniform spaces. A quasi-uniformity is a family $\mathcal{U}$ of subsets of $X \times X$, called entourages, satisfying all the axioms of a uniformity excepting symmetry: one does not suppose that $\mathcal{U}$ has a base formed of symmetric entourages. Again, $\mathcal{U}^{-1} = \{U^{-1} : U \in \mathcal{U}\}$ is another quasi-uniformity on X, called the conjugate of $\mathcal{U}$, and $\mathcal{U}^s = \mathcal{U} \vee \mathcal{U}^{-1}$ is a uniformity. The notions of completeness can be transposed from quasi-metric spaces to quasi-uniform spaces by replacing sequences with nets or filters. Again the focus is on the relations between compactness, completeness and total boundedness within this framework.

On a quasi-metric space (X, ρ) there are two natural topologies generated by the quasi-metric ρ and its conjugate $\bar{\rho}$, respectively by the quasi-uniformity $\mathcal{U}$ and its conjugate $\mathcal{U}^{-1}$, making quasi-metric and quasi-uniform spaces bitopological spaces. For this reason the first chapter of the book contains a quite detailed introduction to bitopological spaces, including Urysohn and Tietze type theorems for semi-continuous functions on bitopological spaces, compactness and Baire category.

Following the advice of Einar Hille [105] that "a functional analyst is an analyst, first and foremost, and not a degenerate species of a topologist", after this detour in topology we turn to functional analysis. Functional analysis in the asymmetric case, meaning the study of asymmetric normed spaces, asymmetric locally convex spaces and of operators acting between them, with emphasis on linear functionals and dual spaces, is treated in the second chapter.

An asymmetric norm is a positive definite sublinear functional p on a real vector space X. Since the possibility that $p(x) = p(-x)$ for some $x \in X$ is not excluded, $\bar{p}(x) = p(-x)$, $x \in X$, is another asymmetric norm on X called the conjugate of p, and $p^s = p \vee \bar{p}$ is a norm on X. The topological notion are considered with respect to the attached metric $\rho_p(x, y) = p(y - x)$, $x, y \in X$. Any asymmetric norm can be obtained as the Minkowski gauge functional of an absorbing convex subset of X. Asymmetric locally convex spaces are defined as vector spaces equipped with a topology generated by a family of asymmetric seminorms.

Of great importance is the asymmetric norm u on $\mathbb{R}$ given by $u(t) = t^+$, $t \in \mathbb{R}$, with conjugate $\bar{u}(t) = t^-$ and $u^s = |\cdot|$. The topology generated by u is called the upper topology of $\mathbb{R}$, while that generated by its conjugate $\bar{u}$, the lower topology. If (T, τ) is a topological space, then a real-valued function f on T is upper semi-continuous as a function from T to $(\mathbb{R}, |\cdot|)$ if and only if it is continuous from T to $(\mathbb{R}, u)$. Similarly, f is $(\tau, |\cdot|)$-lower semi-continuous if and only if it is $(\tau, \bar{u})$-continuous.

The main differences with respect to the classical functional analysis (meaning analysis over the fields $\mathbb{R}$ or $\mathbb{C}$) come from the fact that the asymmetric norm p does not generate a vector topology on X: the addition is continuous with respect to the product topology on X, but the multiplication by scalars is continuous only when restricted to $(0; \infty) \times X$. Also, for each fixed x, the function $f : \mathbb{R} \to X$ given by $f(t) = tx$, $t \in \mathbb{R}$, is continuous. The dual space of an asymmetric normed space (X, p), denoted by $X_p^\flat$, formed by all linear and $|\cdot|$-upper semi-continuous functions, or, equivalently, linear continuous functionals from (X, p) to $(\mathbb{R}, u)$, is not a linear space but merely a cone contained in the dual space $X^* = (X, p^s)^*$ of the associated normed space (X, p^s). The situation is similar for the set of all continuous linear operators between two asymmetric normed spaces, as well as for asymmetric locally convex spaces.

In spite of these differences, many results from classical functional analysis have their counterparts in the asymmetric case, by taking care of the interplay between the asymmetric norm p and its conjugate $\bar{p}$. Among the positive results we mention: Hahn-Banach type theorems and separation results for convex sets, Krein-Milman type theorems, analogs of the fundamental principles – open mapping and closed graph theorems – an analog of the Schauder theorem on the compactness of a conjugate mapping. Applications are given to best approximation problems and, as relevant examples, one considers normed lattices equipped with asymmetric norms and spaces of semi-Lipschitz functions on quasi-metric spaces.

It is difficult to localize the first moment when asymmetric norms were used, but it goes back as early as 1968 in a paper by Duffin and Karlovitz (1968) [70], who proposed the term asymmetric norm. Krein and Nudelman (1973) [129] used also asymmetric norms in their study of some extremal problems related to the Markov moment problem. Remark that the relevance of sublinear functionals for some problems of convex analysis and of mathematical analysis was emphasized also by H. König in the 1970s. A systematic study of the properties of asymmetric normed spaces started with the papers of S. Romaguera, from the Polytechnic University of Valencia, and his collaborators from the same university and from other universities in Spain: Alegre, Ferrer, García-Raffi, Sánchez Pérez, Sánchez Álvarez, Sanchis, Valero (see the bibliography). Besides its intrinsic interest, their study was motivated also by applications in Computer Science, namely to the complexity analysis of programs, results obtained in cooperation with Professor Schellekens from the National University of Ireland.

Containing very recent results, some of them appearing for the first time in print, in the focus of current research, on quasi-metric, quasi-uniform, asymmetric normed and asymmetric locally convex spaces, the book can be used as a reference by researchers in this domain. Due to the detailed exposition of the subject, the book can be also used as an introductory text for newcomers.

Acknowledgement. The author expresses his gratitude to the staff of Birkhäuser-Springer, particularly to the Editors Anna Mätzener and Sylvia Lotrovsky, for the professional work and excellent cooperation during the publication process, and to Ute McCrory from Springer DE for support.

I want to mention also that this research was supported by
Grant CNCSIS 2261, ID 543.

Notation. We present here, for the convenience of the reader, some symbols that are used throughout the text, which could differ from the standard ones. Other notations are standard or explained in the text, some of them being included in the index at the end of the book.

- $\mathbb{N} = \{1, 2, \dots\}$ – the set of natural numbers (positive integers);
- $[a; b]$, $(a; b)$, $(a; b]$, $[a; b)$ – intervals;
- (a, b) – an ordered pair;
- $B_\rho[x, r] = \{y \in X : \rho(x, y) \leq r\}$ – a closed ball in a quasi-metric space (X, ρ);
- $B_\rho(x, r) = \{y \in X : \rho(x, y) < r\}$ – an open ball;
- $\rho_p(x, y) = p(y - x)$ – the quasi-metric associated to an asymmetric norm p;
- $B_p = \{x \in X : p(x) \leq 1\}$ – the closed unit ball of an asymmetric normed space (X, p);
- $B'_p = \{x \in X : p(x) < 1\}$ – the open unit ball;
- $S_p = \{x \in X : p(x) = 1\}$ – the unit sphere;
- u is the standard asymmetric norm $u(t) = t^+$ on $\mathbb{R}$.

Cluj-Napoca, July 2012 *Ştefan Cobzaş*

Chapter 1

Quasi-metric and Quasi-uniform Spaces

The first chapter of the book is concerned with the topological properties of quasi-metric, quasi-uniform, asymmetric normed and asymmetric locally convex spaces. A quasi-metric on a set X is a positive function ρ on $X \times X$ satisfying all the axioms of a metric excepting symmetry: it is possible that $\rho(y, x) \neq \rho(x, y)$ for some $x, y \in X$. Similarly, a quasi-uniformity is a family $\mathcal{U}$ of subsets of $X \times X$ satisfying all the requirements of a uniformity excepting symmetry: $U \in \mathcal{U}$ does not imply that $U^{-1} = \{(y, x) : (x, y) \in U\}$ belongs to $\mathcal{U}$. The lack of symmetry in the definition of quasi-metric spaces and of quasi-uniform spaces causes a lot of troubles, mainly concerning completeness, compactness and total boundedness in such spaces. There are several notions of completeness in quasi-metric and quasi-uniform spaces, all agreeing with the usual notion of completeness in the case of metric or uniform spaces, each of them having its advantages and weaknesses. Also, countable compactness, sequential compactness and compactness do not agree in quasi-metric spaces, in contrast to the metric case. In spite of these differences, there are a lot of positive results relating compactness, completeness and total boundedness in quasi-metric and in quasi-uniform spaces which are presented in this chapter. A quasi-metric ρ and its conjugate $\bar{\rho}(x, y) = \rho(y, x),\ x, y \in X$, generate in the usual way two topologies τ_ρ and $\tau_{\bar{\rho}}$, that makes X a bitopological space, with a similar situation for quasi-uniform spaces. For this reason we have included in this chapter a quite detailed study of bitopological spaces including pairwise separation properties, Urysohn and Tietze type theorems for semi-continuous functions and Baire category.

1.1 Topological properties of quasi-metric and quasi-uniform spaces

In this section we shall present the basic topological properties of quasi-metric and quasi-uniform spaces, with emphasis on asymmetric normed spaces and asymmetric locally convex spaces. Since quasi-metric and quasi-uniform spaces are particular cases of bitopological spaces, a quite detailed presentation of bitopological spaces is also given, including compactness, normality, regularity, a Tikhonov type theorem on the existence of semi-continuous functions and Tietze-Urysohn type theorems on the extension of semi-continuous functions.

1.1.1 Quasi-metric spaces and asymmetric normed spaces

A *quasi-semimetric* on a set X is a mapping $\rho : X \times X \to [0;\infty)$ satisfying the following conditions:

$$\begin{aligned} &\text{(QM1)} \qquad \rho(x,y) \ge 0, \quad \textit{and} \quad \rho(x,x) = 0; \\ &\text{(QM2)} \qquad \rho(x,z) \le \rho(x,y) + \rho(y,z)\ , \end{aligned}$$

for all $x, y, z \in X$. If, further,

$$\text{(QM3)} \qquad \rho(x,y) = \rho(y,x) = 0 \Rightarrow x = y\ ,$$

for all $x, y \in X$, then ρ is called a *quasi-metric*. The pair (X, ρ) is called a *quasi-semimetric space*, respectively a *quasi-metric space*. The conjugate of the quasi-semimetric ρ is the quasi-semimetric $\bar{\rho}(x,y) = \rho(y,x)$, $x, y \in X$. The mapping $\rho^s(x,y) = \max\{\rho(x,y), \bar{\rho}(x,y)\}$, $x, y \in X$, is a semimetric on X which is a metric if and only if ρ is a quasi-metric. Sometimes one works with *extended* quasi-semimetrics, meaning that the quasi-semimetric ρ can take the value $+\infty$ for some $x, y \in X$. The following inequalities hold for these quasi-semimetrics for all $x, y \in X$:

$$\rho(x,y) \le \rho^s(x,y) \quad \text{and} \quad \bar{\rho}(x,y) \le \rho^s(x,y)\ . \tag{1.1.1}$$

An *asymmetric norm* on a real vector space X is a functional $p : X \to [0,\infty)$ satisfying the conditions

$$\begin{aligned} &\text{(AN1)} \qquad p(x) = p(-x) = 0 \Rightarrow x = 0; \\ &\text{(AN2)} \qquad p(\alpha x) = \alpha p(x); \\ &\text{(AN3)} \qquad p(x+y) \le p(x) + p(y)\ , \end{aligned}$$

for all $x, y \in X$ and $\alpha \ge 0$.

If p satisfies only the conditions (AN2) and (AN3), then it is called an *asymmetric seminorm*. The pair (X, p) is called an *asymmetric normed* (respectively *seminormed*) *space*. Again, in some instances, the value $+\infty$ will be allowed for

p in which case we shall call it an *extended asymmetric norm* (or seminorm). An asymmetric seminorm p defines a quasi-semimetric ρ_p on X through the formula

$$\rho_p(x, y) = p(y - x),\ x, y \in X . \tag{1.1.2}$$

Defining the conjugate asymmetric seminorm $\bar{p}$ and the seminorm p^s by

$$\bar{p}(x) = p(-x) \quad \text{and} \quad p^s(x) = \max\{p(x), p(-x)\} , \tag{1.1.3}$$

for $x \in X$, the inequalities (1.1.1) become

$$p(x) \le p^s(x) \quad \text{and} \quad \bar{p}(x) \le p^s(x) , \tag{1.1.4}$$

for all $x \in X$. Obviously, p^s is a norm when p is an asymmetric norm and (X, p^s) is a normed space.

The conjugates of ρ and p are denoted also by ρ^{-1} and p^{-1}, a notation that we shall use occasionally.

If (X, ρ) is a quasi-semimetric space, then for $x \in X$ and $r > 0$ we define the balls in X by the formulae

$$B_\rho(x, r) = \{y \in X : \rho(x, y) < r\} - \text{ the open ball, and}$$
$$B_\rho[x, r] = \{y \in X : \rho(x, y) \le r\} - \text{ the closed ball.}$$

In the case of an asymmetric seminormed space (X, p) the balls are given by

$$B_p(x, r) = \{y \in X : p(y-x) < r\}, \text{ respectively } B_p[x, r] = \{y \in X : p(y-x) \le r\} .$$

The closed unit ball of X is $B_p = B_p[0, 1]$ and the open unit ball is $B'_p = B_p(0, 1)$. In this case the following formulae hold true:

$$B_p[x, r] = x + rB_p \quad \text{and} \quad B_p(x, r) = x + rB'_p , \tag{1.1.5}$$

that is, any of the unit balls of X completely determines its quasi-metric structure. If necessary, these balls will be denoted by $B_{p,X}$ and $B'_{p,X}$, respectively.

The conjugate $\bar{p}$ of p is defined by $p(x) = p(-x)$, $x \in X$, and the associate seminorm is $p^s(x) = \max\{p(x), \bar{p}(x)\}$, $x \in X$. The seminorm p is an asymmetric norm if and only if p^s is a norm on X. Sometimes an asymmetric norm will be denoted by the symbol $\| \cdot |$, a notation proposed by Krein and Nudelman, [129, Ch. IX, §5], in their book on the theory of moments.

Remark 1.1.1. Since the terms "quasi-norm", "quasi-normed space" and "quasi-Banach space" are already "registered trademarks" (see, for instance, the survey by Kalton [106]), we can not use these terms to designate an asymmetric norm, an asymmetric normed space or an asymmetric biBanach space. A *quasi-normed space* is a vector space X equipped with a functional $\| \cdot \| : X \to [0; \infty)$, satisfying all the axioms of a norm, excepting the triangle inequality which is replaced by

$$\|x + y\| \le C(\|x\| + \|y\|),\ x, y \in X ,$$

for some constant $C \geq 1$. It is obvious that for $C = 1$ the functional $\|\cdot\|$ is a norm. The reverse situation is also encountered: in [232] a quasi-metric space is a metric space (X, ρ) in which the triangle inequality is replaced by $\rho(x,z) \leq C\,(\rho(x,y) + \rho(y,z))$, for some $C > 0$.

Note also that the alternative terms "quasi-pseudometric" is used by many topologists instead of "quasi-semimetric". We have preferred the term semimetric to be in concordance with the notion of seminorm – in this way an asymmetric seminorm induces a quasi-semimetric.

1.1.2 The topology of a quasi-semimetric space

The topology $\tau(\rho)$ of a quasi-semimetric space (X, ρ) can be defined starting from the family $\mathcal{V}_\rho(x)$ of neighborhoods of an arbitrary point $x \in X$:

$$
\begin{aligned}
V \in \mathcal{V}_\rho(x) &\iff \exists r > 0 \text{ such that } B_\rho(x, r) \subset V \\
&\iff \exists r' > 0 \text{ such that } B_\rho[x, r'] \subset V\,.
\end{aligned}
$$

To see the equivalence in the above definition, we can take, for instance, $r' = r/2$.

A set $G \subset X$ is $\tau(\rho)$-open if and only if for every $x \in G$ there exists $r = r_x > 0$ such that $B_\rho(x, r) \subset G$. Sometimes we shall say that V is a ρ-neighborhood of x or that the set G is ρ-open.

The convergence of a sequence (x_n) to x with respect to $\tau(\rho)$, called ρ-convergence and denoted by $x_n \xrightarrow{\rho} x$, can be characterized in the following way:

$$
x_n \xrightarrow{\rho} x \iff \rho(x, x_n) \to 0\,. \tag{1.1.6}
$$

Also

$$
x_n \xrightarrow{\bar{\rho}} x \iff \bar{\rho}(x, x_n) \to 0 \iff \rho(x_n, x) \to 0\,. \tag{1.1.7}
$$

The following proposition contains some simple properties of convergent sequences.

Proposition 1.1.2. *Let (x_n) be a sequence in a quasi-semimetric space (X, ρ).*

1. *If (x_n) is τ_ρ-convergent to x and $\tau_{\bar{\rho}}$-convergent to y, then $\rho(x, y) = 0$.*
2. *If (x_n) is τ_ρ-convergent to x and $\rho(y, x) = 0$, then (x_n) is also τ_ρ-convergent to y.*

Proof. 1. Letting $n \to \infty$ in the inequality $\rho(x, y) \leq \rho(x, x_n) + \rho(x_n, y)$, one obtains $\rho(x, y) = 0$.

2. Follows from the relations $\rho(y, x_n) \leq \rho(y, x) + \rho(x, x_n) = \rho(x, x_n) \to 0$ as $n \to \infty$. □

Using the conjugate quasi-semimetric $\bar{\rho}$ one obtains another topology $\tau(\bar{\rho})$. A third one is the topology $\tau(\rho^s)$ generated by the semimetric ρ^s. Sometimes, (see, for instance, Menucci [151] and Collins and Zimmer [50]) the balls with respect to ρ are called *forward balls* and the topology $\tau(\rho)$ is called the *forward topology*, while the balls with respect to $\bar{\rho}$ are called *backward balls* and the topology $\tau(\bar{\rho})$ the *backward topology*. We shall use sometimes the alternative notation τ_ρ, $\tau_{\bar{\rho}}$, τ_{ρ^s} to designate these topologies.

As a space with two topologies, τ_ρ and $\tau_{\bar{\rho}}$, a quasi-semimetric space can be viewed as a bitopological space in the sense of Kelly [111] (see also the book [72]) and so, all the results valid for bitopological spaces apply to a quasi-semimetric space. A *bitopological space* is simply a set T endowed with two topologies τ and ν. A bitopological space is denoted by (T, τ, ν).

The following example is very important in what follows.

Example 1.1.3. On the field $\mathbb{R}$ of real numbers consider the asymmetric norm $u(\alpha) = \alpha^+ := \max\{\alpha, 0\}$. Then, for $\alpha \in \mathbb{R}$, $\bar{u}(\alpha) = \alpha^- := \max\{-\alpha, 0\}$ and $u^s(\alpha) = |\alpha|$. The topology $\tau(u)$ generated by u is called the *upper topology* of $\mathbb{R}$, while the topology $\tau(\bar{u})$ generated by $\bar{u}$ is called the *lower topology* of $\mathbb{R}$. A basis of open $\tau(u)$-neighborhoods of a point $\alpha \in \mathbb{R}$ is formed of the intervals $(-\infty; \alpha + \varepsilon)$, $\varepsilon > 0$. A basis of open $\tau(\bar{u})$-neighborhoods is formed of the intervals $(\alpha - \varepsilon; \infty)$, $\varepsilon > 0$.

In this space the addition is continuous from $(\mathbb{R} \times \mathbb{R}, \tau_u \times \tau_u)$ to $(\mathbb{R}, \tau_u)$, but the multiplication need not be continuous at every point $(\alpha, \beta) \in \mathbb{R} \times \mathbb{R}$.

The continuity property can be directly verified. To see the last assertion, let $V = (-\infty; \alpha\beta + \varepsilon)$, be a τ_u-neighborhood of $\alpha\beta$, for some $\varepsilon > 0$. Since the τ_u-neighborhoods of α and β contain $-n$, for $n \in \mathbb{N}$ sufficiently large, it follows that $n^2 = (-n)(-n)$ does not belong to V, for n large enough.

Remark 1.1.4 (communicated by M.D. Mabula). In the asymmetric normed space $(\mathbb{R}, u)$ the sequence $\alpha_n = (-1)^n$, $n \in \mathbb{N}$, is u-convergent to 1 and $\bar{u}$-convergent to -1, but it is not convergent in $(\mathbb{R}, |\cdot|)$.

For a sequence (x_n) in a quasi-semimetric space (X, ρ), denote by $L((x_n))$ the set of all ρ-limits of the sequence (x_n), that is

$$L_\rho((x_n)) = \{x \in X : \lim_n \rho(x, x_n) = 0\}. \tag{1.1.8}$$

The following proposition gives a characterization of this set in $(\mathbb{R}, u)$.

Proposition 1.1.5 ([51]). *Let (α_n) be a sequence of real numbers. Then*

$$L_u((\alpha_n)) = \begin{cases} [\limsup_n \alpha_n; \infty) & \text{if } \limsup_n \alpha_n \in \mathbb{R}, \\ \mathbb{R} & \text{if } \limsup_n \alpha_n = -\infty. \end{cases} \tag{1.1.9}$$

If $\limsup_n \alpha_n = \infty$, then the sequence (α_n) is not u-convergent.

Similarly,

$$L_u((\alpha_n)) = \begin{cases} (-\infty; \liminf_n \alpha_n] & \text{if } \liminf_n \alpha_n \in \mathbb{R}, \\ \mathbb{R} & \text{if } \limsup_n \alpha_n = \infty. \end{cases} \tag{1.1.10}$$

If $\limsup_n \alpha_n = -\infty$, *then the sequence* (α_n) *is not* $\bar{u}$*-convergent.*

Another important topological example is the so-called Sorgenfrey topology on $\mathbb{R}$.

Example 1.1.6 (The Sorgenfrey line). For $x, y \in \mathbb{R}$ define a quasi-metric ρ by $\rho(x,y) = y - x$, if $x \leq y$ and $\rho(x,y) = 1$ if $x > y$. A basis of open τ_ρ-open neighborhoods of a point $x \in \mathbb{R}$ is formed by the family $[x; x+\varepsilon)$, $0 < \varepsilon < 1$. The family of intervals $(x - \varepsilon; x]$, $0 < \varepsilon < 1$, forms a basis of open $\tau_{\bar{\rho}}$-open neighborhoods of x. Obviously, the topologies τ_ρ and $\tau_{\bar{\rho}}$ are Hausdorff and $\rho^s(x,y) = 1$ for $x \neq y$, so that $\tau(\rho^s)$ is the discrete topology of $\mathbb{R}$.

We shall present, for the convenience of the reader, the separation axioms. A topological space (T, τ) is called

- T_0 if for any pair s, t of distinct points in T, at least one of them has a neighborhood not containing the other;
- T_1 if for any pair s, t of distinct points in T, each of them has a neighborhood not containing the other (this is equivalent to the fact that the set $\{t\}$ is closed for every $t \in T$);
- *Hausdorff* or T_2 if for any pair s, t of distinct points in T, there exist neighborhoods U of s and V of t such that $U \cap V = \emptyset$;
- *regular* if for each $t \in T$ and each closed subset S of T, not containing t, there are disjoint open subsets U, V of T such that $t \in U$ and $S \subset V$. In other words a point and a closed set not containing it can be separated by open sets. This is equivalent to the fact that every point in T has a neighborhood base formed of closed sets. If T is regular and T_1, then it is called a T_3 space.
- *completely regular*, or *Tikhonov*, or $T_{3\frac{1}{2}}$, if for every $t \in T$ and every closed subset S of T not containing t there is a continuous function $f : T \to [0; 1]$ such that $f(t) = 1$ and $f(s) = 0$ for each $s \in S$.
- *normal* if any pair S_1, S_2 of disjoint closed sets can be separated by open sets, that is there exist two disjoint open sets $G_1 \supset S_1$ and $G_2 \supset S_2$. A normal T_1 space is called a T_4 space.

In general, we shall say that a bitopological space (T, τ, ν) has a property P if both the topologies τ and ν have the property P.

Now we introduce, following Kelly [111], some separation properties specific to a bitopological space (T, τ, ν).

The bitopological space (T, τ, ν) is called *pairwise Hausdorff* if for each pair of distinct points $s, t \in T$ there exists a τ-neighborhood U of s and a ν-neighborhood

V of t such that $U \cap V = \emptyset$. It is obvious that if T is pairwise Hausdorff, then both of the topologies τ and ν are T_1.

Remark 1.1.7. Taking into account the symmetry ($x \neq y \iff y \neq x$) it follows that a bitopological space (T, τ, ν) is pairwise Hausdorff if and only if for every pair of distinct points x, y from X the following condition holds:

$$\begin{aligned} &(\exists U \in \tau,\ \exists V \in \nu,\ x \in U\ \wedge y \in V\ \wedge U \cap V = \emptyset) \\ \wedge &(\exists U_1 \in \tau,\ \exists V_1 \in \nu,\ y \in U_1\ \wedge\ x \in V_1\ \wedge\ U_1 \cap V_1 = \emptyset)\ . \end{aligned} \tag{1.1.11}$$

The topology τ is called *regular with respect to* ν if every $t \in T$ has a τ-neighborhood base formed of ν-closed sets or, equivalently, if for every $t \in T$ and every τ-closed subset S of T not containing t, there exist a τ-open set U and a ν-open set V such that $t \in U$, $S \subset V$ and $U \cap V = \emptyset$.

One says that the bitopological space (T, τ, ν) is *pairwise regular* if τ is regular with respect to ν and ν is regular with respect to τ.

The bitopological space (T, τ, ν) is called *pairwise normal* if given a τ-closed subset A of T and a ν-closed subset B of T with $A \cap B = \emptyset$, there exist a ν-open subset U of T and a τ-open subset V of T such that $A \subset U$, $B \subset V$, and $U \cap V = \emptyset$. Equivalently, the bitopological space (T, τ, ν) is pairwise normal if given an τ-closed set C and a ν-open set D with $C \subset D$ there exist a ν-open set G and a τ-closed set F such that

$$C \subset G \subset F \subset D\,, \tag{1.1.12}$$

or, equivalently, there exists a ν-open set G such that

$$C \subset G \subset \overline{G}^{\tau} \subset D\,, \tag{1.1.13}$$

where $\overline{G}^{\tau}$ denotes the closure of G with respect to τ. A bitopological space (T, τ, ν) is called *quasi-semimetrizable* if there exists a quasi-semimetric ρ on T such that $\tau = \tau_\rho$ and $\nu = \tau_{\bar{\rho}}$. If ρ is a semimetric, then $\tau = \nu$.

The following topological properties are true for quasi-semimetric spaces. We use the abbreviation lsc for lower semicontinuous and usc for upper semicontinuous.

Proposition 1.1.8. *If (X, ρ) is a quasi-semimetric space, then*

1. *Any ball $B_\rho(x, r)$ is $\tau(\rho)$-open and a ball $B_\rho[x, r]$ is $\tau(\bar{\rho})$-closed. The ball $B_\rho[x, r]$ need not be $\tau(\rho)$-closed.*

   Also, the following inclusions hold:

   $$B_{\rho^s}(x, r) \subset B_\rho(x, r) \ \ \textit{and}\ \ B_{\rho^s}(x, r) \subset B_{\bar{\rho}}(x, r)\ ,$$

   with similar inclusions for the closed balls.
2. *The topology $\tau(\rho^s)$ is finer than the topologies $\tau(\rho)$ and $\tau(\bar{\rho})$. This means that:*

- *any $\tau(\rho)$-open (closed) set is $\tau(\rho^s)$-open (closed); similar results hold for the topology $\tau(\bar{\rho})$;*
- *the identity mappings from $(X, \tau(\rho^s))$ to $(X, \tau(\rho))$ and to $(X, \tau(\bar{\rho}))$ are continuous;*
- *a sequence (x_n) in X is $\tau(\rho^s)$-convergent to $x \in X$ if and only if it is $\tau(\rho)$-convergent and $\tau(\bar{\rho})$-convergent to x.*

3. *If ρ is a quasi-metric, then the topologies $\tau(\rho)$ and $\tau(\bar{\rho})$ are T_0, but not necessarily T_1 (and so nor T_2, in contrast to the case of metric spaces). The topology $\tau(\rho)$ is T_1 if and only if $\rho(x,y) > 0$ whenever $x \neq y$. In this case, $\tau(\bar{\rho})$ is also T_1 and, as a bitopological space, X is pairwise Hausdorff.*
4. *For every fixed $x \in X$, the mapping $\rho(x,\cdot) : X \to (\mathbb{R}, |\cdot|)$ is τ_ρ-usc and $\tau_{\bar{\rho}}$-lsc. For every fixed $y \in X$, the mapping $\rho(\cdot,y) : X \to (\mathbb{R}, |\cdot|)$ is τ_ρ-lsc and $\tau_{\bar{\rho}}$-usc.*
5. ([145]) *The mapping $\rho(x,\cdot) : X \to (\mathbb{R}, |\cdot|)$ is τ_ρ-continuous at $x \in X$, if and only if τ_ρ-cl$(B_\rho(x,r)) \subset B_\rho[x,r]$ for all $r > 0$.*

Similar results hold for an asymmetric seminorm p, its conjugate $\bar{p}$ and the associated seminorm p^s.

Proof. 1. For $y \in B_\rho(x,r)$ we have $B_\rho(x,r') \subset B_\rho(x,r)$, where $r' := r - \rho(x,y) > 0$. Also, if $y \in B_\rho[x,r]$ and $r' := \rho(x,y) - r > 0$, then $B_{\bar{\rho}}(x,r') \cap B_\rho[x,r] = \emptyset$, or, equivalently, $B_{\bar{\rho}}(x,r') \subset X \setminus B_\rho[x,r]$. Indeed, if $z \in B_{\bar{\rho}}(x,r') \cap B_\rho[x,r]$, then

$$\rho(x,y) \le \rho(x,z) + \rho(z,y) = \rho(x,z) + \bar{\rho}(y,z) < r + r' = \rho(x,y) ,$$

a contradiction.

The inclusions from 1 follows from the inequalities (1.1.1) and, in their turn, they imply the assertions from the second point of the proposition.

The assertions from 2 are obvious.

3. If x, y are distinct points in the quasi-metric space (X,ρ), then $\max\{\rho(x,y), \rho(y,x)\} > 0$. If $\rho(x,y) > 0$, then $y \notin B_\rho(x,r)$, where $r = \rho(x,y)$. Similarly, if $\rho(y,x) > 0$, then $x \notin B_\rho(y,r')$, where $r' = \rho(y,x)$. Consequently, $\tau(\rho)$ is T_0 and $\tau(\bar{\rho})$ as well.

Suppose that $\rho(x,y) > 0$ for every $x \neq y$. Then $y \notin B_\rho(x, \rho(x,y))$. Since $\rho(y,x) > 0$ too, $x \notin B_\rho(y, \rho(y,x))$, showing that the topology τ_ρ is T_1. Similarly $\tau_{\bar{\rho}}$ is T_1.

Also, $B_\rho(x,r) \cap B_{\bar{\rho}}(y,r) = \emptyset$, where $r > 0$ is given by $2r := \rho(x,y) > 0$. Indeed, if $z \in B_\rho(x,r) \cap B_{\bar{\rho}}(y,r)$, then

$$\rho(x,y) \le \rho(x,z) + \rho(z,y) < r + r = \rho(x,y) ,$$

a contradiction which shows that the bitopological space $(X, \tau_\rho, \tau_{\bar{\rho}})$ is pairwise Hausdorff.

It is easy to check that if τ_ρ is T_1, then $\rho(x,y) > 0$ for every pair of distinct elements $x, y \in X$.

4. To prove that $\rho(x,\cdot)$ is τ_ρ-usc and $\tau_{\bar\rho}$-lsc, we have to show that the set $\{y \in X : \rho(x,y) < \alpha\}$ is τ_ρ-open and $\{y \in X : \rho(x,y) > \alpha\}$ is $\tau_{\bar\rho}$-open, for every $\alpha \in \mathbb{R}$, properties that are easy to check.

Indeed, for $y \in X$ such that $\rho(x,y) < \alpha$, let $r := \alpha - \rho(x,y) > 0$. If $z \in X$ is such that $\rho(y,z) < r$, then

$$\rho(x,z) \le \rho(x,y) + \rho(y,z) < \rho(x,y) + r = \alpha,$$

showing that $B_\rho(y,r) \subset \{y \in X : \rho(x,y) < \alpha\}$.

Similarly, for $y \in X$ with $\rho(x,y) > \alpha$ take $r := \rho(x,y) - \alpha > 0$. If$z \in X$ satisfies $\rho(z,y) = \bar\rho(y,z) < r$, then

$$\rho(x,y) \le \rho(x,z) + \rho(z,y) < \rho(x,z) + r\ ,$$

so that $\rho(x,z) > \rho(x,y) - r = \alpha$. Consequently, $B_{\bar\rho}(y,r) \subset \{y \in X : \rho(x,y) > \alpha\}$.

5. Suppose that $\rho(x,\cdot)$ is continuous on X. If $y \in \tau_\rho\text{-cl}(B_\rho(x,r))$, then there exists a sequence (y_n) in $B_\rho(x,r)$ such that $y_n \xrightarrow{\rho} y$ as $n \to \infty$. The continuity of $\rho(x,\cdot)$ implies $\rho(x,y) = \lim_{n\to\infty} \rho(x,y_n) \le r$, that is $y \in B_\rho[x,r]$.

To prove the converse, suppose that there exists $x \in X$ such that $\rho(x,\cdot)$ is discontinuous at some $y \in X$. Then there exist $r > 0$ and a sequence (y_n) in X such that $y_n \xrightarrow{\rho} y$ and $\rho(x,y_n) \in (-\infty, \rho(x,y) - r) \cup (\rho(x,y) + r, \infty)$ for all $n \in \mathbb{N}$. If there exists an infinity of $n \in \mathbb{N}$ such that $\rho(x,y_n) > \rho(x,y) + r$, then, by the τ_ρ-usc of the function $\rho(x,\cdot)$, $\rho(x,y) + r \le \limsup_n \rho(x,y_n) \le \rho(x,y)$, leading to the contradiction $r \le 0$.

Consequently $\rho(x,y_n) < \rho(x,y) - r$, that is $y_n \in B_\rho(x, \rho(x,y) - r)$ for all $n \in \mathbb{N}$ excepting a finitely many. By hypothesis, $y \in B_\rho[x, \rho(x,y) - r]$, that is $\rho(x,y) \le \rho(x,y) - r$, leading again to the contradiction $r \le 0$. □

One can define other pairwise separation axioms. Call a bitopological space (T,τ,ν) *pairwise* T_0 if for any pair x,y of distinct points in T either there exists a τ-open set U such that $x \in U$ and $y \notin U$ or there exists $V \in \nu$ such that $y \in V$ and $x \notin V$. Again, by symmetry, it follows that the space (T,τ,ν) is pairwise T_0 if and only if

$$\begin{aligned} &[(\exists U \in \tau,\ x \in U \ \wedge\ y \notin U) \ \vee\ (\exists V \in \nu,\ y \in V \ \wedge\ x \notin V)] \\ \wedge\ &[(\exists U_1 \in \tau,\ y \in U_1 \ \wedge\ x \notin U_1) \ \vee\ (\exists V_1 \in \nu,\ x \in V_1 \ \wedge\ y \notin V_1)]\ . \end{aligned} \tag{1.1.14}$$

Taking into account the mutual distributivity of the operators $\wedge$ and $\vee$ it follows that this condition is equivalent to

$$\begin{aligned} &[(\exists U \in \tau,\ x \in U \ \wedge\ y \notin U) \ \wedge\ (\exists U_1 \in \tau,\ y \in U_1 \ \wedge\ x \notin U_1)] \\ \vee\ &[(\exists U \in \tau,\ x \in U \ \wedge\ y \notin U) \ \wedge\ (\exists V_1 \in \nu,\ x \in V_1 \ \wedge\ y \notin V_1)] \\ \vee\ &[(\exists V \in \tau,\ y \in V \ \wedge\ x \notin V) \ \wedge\ (\exists U_1 \in \tau,\ y \in U_1 \ \wedge\ x \notin U_1)] \\ \vee\ &[(\exists V \in \tau,\ y \in V \ \wedge\ x \notin V) \ \wedge\ (\exists V_1 \in \nu,\ x \in V_1 \ \wedge\ y \notin V_1)]\ . \end{aligned} \tag{1.1.15}$$

Similar conditions hold for the notion of pairwise T_1. A bitopological space (T, τ, ν) is called *pairwise* T_1 if for any pair x, y of distinct points in T the following condition holds:

$$\begin{aligned} &[(\exists U \in \tau,\ x \in U\ \wedge\ y \notin U)\ \wedge\ (\exists V \in \nu,\ y \in V\ \wedge\ x \notin V)] \\ \wedge\ &[(\exists U_1 \in \tau,\ y \in U_1\ \wedge\ x \notin U_1)\ \wedge\ (\exists V_1 \in \nu,\ x \in V_1\ \wedge\ y \notin V_1)]\ . \end{aligned} \tag{1.1.16}$$

In the case of a quasi-semimetric space the following proposition holds.

Proposition 1.1.9. *Let (X, ρ) be a quasi-semimetric space. If the associated bitopological space $(X, \tau_\rho, \tau_{\bar\rho})$ is pairwise T_0, then $\rho(x, y) > 0$ for any pair of distinct points $x, y \in X$.*

Proof. Suppose that there exists $x, y \in X$ with $x \neq y$ and $\rho(x, y) = 0$. Then $y \in B_\rho(x, r)$ and $x \in B_{\bar\rho}(x, r)$ for every $r > 0$, so that the condition (1.1.14) does not hold. □

Taking into account Proposition 1.1.8.3, one obtains the following corollary.

Corollary 1.1.10. *For a quasi-semimetric space (X, ρ) the following are equivalent.*

1. *The bitopological space $(X, \tau_\rho, \tau_{\bar\rho})$ is pairwise T_0.*
2. *The bitopological space $(X, \tau_\rho, \tau_{\bar\rho})$ is pairwise T_1.*
3. *The bitopological space $(X, \tau_\rho, \tau_{\bar\rho})$ is pairwise Hausdorff.*

Better continuity properties of the distance function holds in the class of the so-called balanced quasi-metric spaces. Following Doitchinov [63], a T_1 quasi-metric space (X, ρ) is called *balanced* if the following condition holds:

$$[\lim_{m,n} \rho(v_m, u_n) = 0\ \wedge\ \forall n,\ \rho(u, u_n) \le r\ \wedge\ \forall m,\ \rho(v_m, v) \le s]\ \Rightarrow \rho(u, v) \le r + s, \tag{1.1.17}$$

for all sequences (u_n), (v_m) in X and all $u, v \in X$.

A pair (X, ρ) where ρ is a balanced quasi-metric is called a *balanced quasi-metric space* or a *B-quasi-metric space.*

In the following proposition we collect some consequences of this definition.

Proposition 1.1.11 ([63])**.** *Let (X, ρ) be a B-quasi-metric space.*

1. *The following assertions hold for all sequences (x_m), (y_n) in X and all $x, y \in X$.*

$$\begin{aligned} &\text{(i)} \quad [\lim_n \rho(x, x_n) = 0\ \wedge\ \forall n,\ \rho(y, x_n) \le r]\ \Rightarrow \rho(y, x) \le r; \\ &\text{(ii)} \quad [\lim_n \rho(x_n, x) = 0\ \wedge\ \forall n,\ \rho(x_n, y) \le r]\ \Rightarrow \rho(x, y) \le r; \end{aligned} \tag{1.1.18}$$

$$\begin{aligned} &\text{(i)} \quad [\lim_n \rho(x, x_n) = 0\ \wedge\ \lim_n \rho(y, x_n) = 0]\ \Rightarrow x = y; \\ &\text{(ii)} \quad [\lim_n \rho(x_n, x) = 0\ \wedge\ \lim_n \rho(x_n, y) = 0]\ \Rightarrow x = y\ ; \end{aligned} \tag{1.1.19}$$

$$[\lim_n \rho(x_n,x)=0 \wedge \lim_n \rho(y,y_m)=0 \wedge \forall m,n,\, \rho(x_n,y_m)\le r] \Rightarrow \rho(x,y)\le r; \tag{1.1.20}$$

$$\begin{array}{ll} \text{(i)} & [\lim_n \rho(x_n,x)=0 \wedge \lim_m \rho(y,y_m)=0] \Rightarrow \lim_{m,n} \rho(x_m,y_n)=\rho(x,y); \\ \text{(ii)} & \lim_n \rho(x_n,x)=0 \;\Rightarrow\; \lim_n \rho(x_n,y)=\rho(x,y); \\ \text{(iii)} & \lim_n \rho(x,x_n)=0 \;\Rightarrow\; \lim_n \rho(y,x_n)=\rho(y,x)\;. \end{array} \tag{1.1.21}$$

2. *The topologies $\tau(\rho)$ and $\tau(\bar\rho)$ are T_2 (Hausdorff). For every fixed $y \in X$ the function $\rho(\cdot, y)$ is $\tau(\bar\rho)$ continuous on X and $\rho(y,\cdot)$ is $\tau(\rho)$-continuous on X.*

Proof. 1. To prove (1.1.18).(i) and (ii), take in (1.1.17) $u_n = x_n,\ v_m = x,\ u = y,\ v = x$, respectively, $u_m = x,\ v_n = x_n,\ u = x,\ v = y$ (with m,n having interchanged roles).

To prove (1.1.19).(i) let $\varepsilon > 0$. Since $\rho(y,x_n) \to 0$, there exists $n_\varepsilon \in \mathbb{N}$ such that $\rho(y,x_n) \le \varepsilon$ for all $n \ge n_\varepsilon$. Since $\rho(x,x_n) \to 0$, an application of (1.1.18).(i) yields $\rho(y,x) \le \varepsilon$. Since $\varepsilon > 0$ was arbitrarily chosen, this implies $\rho(y,x) = 0$ and so $x = y$ (in the definition of a balanced qm space we have required that the topology $\tau(\rho)$ is T_1). The assertion (ii) follows similarly.

The proof of (1.1.20). By (1.1.18).(ii) $\lim_n \rho(x_n,x) = 0$ and $\forall n,\ \rho(x_n,y_m) \le r$, imply $\rho(x,y_m) \le r$, for every $m \in \mathbb{N}$. Since $\lim_m \rho(y,y_m) = 0$, an application of (1.1.18).(i) yields $\rho(x,y) \le r$.

To prove (1.1.21).(i) let $\varepsilon > 0$. Observe first that the inequality

$$\rho(x_m,y_n) \le \rho(x_m,x) + \rho(x,y) + \rho(y,y_n)\;,$$

and the hypotheses imply the existence of $k_0 \in \mathbb{N}$ such that

$$\rho(x_m,y_n) \le \rho(x,y) + \varepsilon\;,$$

for all $m,n \ge k_0$. The proof will be complete if we prove the existence of $l_0 \in \mathbb{N}$ such that

$$\rho(x_m,y_n) > \rho(x,y) - \varepsilon\;,$$

for all $m,n \ge l_0$.

If contrary, then there exists the subsequences (x_{m_i}) of (x_m) and (y_{n_j}) of (y_n) such that

$$\rho(x_{m_i},y_{n_j}) \le \rho(x,y) - \varepsilon\;,$$

for all $i,j \in \mathbb{N}$. Since $\lim_i \rho(x_{m_i},x) = 0 = \lim_j \rho(y,y_{n_j})$, an application of (1.1.20) yields the contradiction $\rho(x,y) \le \rho(x,y) - \varepsilon$.

The assertions from (ii) and (iii) follow from (i).

2. The fact that the topologies $\tau(\rho)$ and $\tau(\bar\rho)$ are T_2 follows from (1.1.19).(i) and (ii), respectively.

The $\tau(\bar\rho)$-continuity of the function $\rho(\cdot,y)$ follows from (1.1.21).(ii) and the $\tau(\rho)$-continuity of the function $\rho(y,\cdot)$ follows from (1.1.21).(iii). □

As it is well known the distance function to a subset plays a key role in the study of metric spaces. As we shall see, the same is true in the asymmetric case where, due to the asymmetry, we have two kinds of distance functions.

Let (X,ρ) be a quasi-semimetric space. For a nonempty subset A of X and $x \in X$ put

$$\rho(x,A) = \inf\{\rho(x,y) : y \in A\} \quad \text{and} \quad \rho(A,x) = \inf\{\rho(y,x) : y \in A\} . \qquad (1.1.22)$$

It is obvious that $\rho(A,x) = \bar{\rho}(x,A)$. A sequence (y_n) in A such that $\lim_n \rho(x,y_n) = \rho(x,A)$ is called a *minimizing sequence* for $\rho(x,A)$, with a similar definition for minimizing sequences for $\rho(A,x)$. Since $A \neq \emptyset$, minimizing sequences always exist.

In the following proposition we collect the basic properties of the distance functions.

Proposition 1.1.12. *Let (X,ρ) be a quasi-semimetric space, A a nonempty subset of X and $x, x' \in X$. The following are true.*

1. $\rho(x,A) \leq \rho(x,x') + \rho(x',A)$ *and* $\rho(A,x) \leq \rho(A,x') + \rho(x',x)$.
2. $\rho(x,A) = 0 \iff x \in \tau_\rho\text{-cl}(A)$ *and* $\rho(A,x) = 0 \iff x \in \tau_{\bar{\rho}}\text{-cl}(A)$.
3. *The function $\rho(\cdot,A) : X \to \mathbb{R}$ is τ_ρ-lsc and $\tau_{\bar{\rho}}$-usc, and the function $\rho(A,\cdot) : X \to \mathbb{R}$ is τ_ρ-usc and $\tau_{\bar{\rho}}$-lsc.*

Proof. 1. For any $y \in A$,

$$\rho(x,A) \leq \rho(x,y) \leq \rho(x,x') + \rho(x',y) .$$

Taking the infimum with respect to $y \in A$ one obtains $\rho(x,A) \leq \rho(x,x') + \rho(x',A)$. The second inequality from 1 can be proved similarly.

2. The proof of the first assertion follows from the equivalences

$$\begin{aligned} \rho(x,A) = 0 &\iff \exists (y_n) \text{ in } A, \quad \lim_{n\to\infty} \rho(x,y_n) = 0 \\ &\iff \exists (y_n) \text{ in } A, \quad y_n \xrightarrow{\rho} x \iff x \in \tau\rho\text{- cl}(A) . \end{aligned}$$

The second equivalence from 2 can be proved similarly.

3. Suppose that for some $\alpha \in \mathbb{R}$, $\rho(x,A) > \alpha$. Taking $r := \rho(x,A) - \alpha > 0$, it follows that

$$\rho(x,A) \leq \rho(x,x') + \rho(x',A) < \rho(x,A) - \alpha + \rho(x',A) ,$$

for every $x' \in X$ with $\rho(x,x') < r$. Consequently, $\rho(x',A) > \alpha$ for every $x' \in B_\rho(x,r)$, proving the τ_ρ-lsc of the mapping $\rho(\cdot,A)$ at x.

Similarly, if $\rho(x,A) < \alpha$,

$$\rho(x',A) \leq \rho(x',x) + \rho(x,A) ,$$

implies $\rho(x', A) < \alpha$ for every $x' \in X$ with $\rho(x', x) < r$, where $r := \alpha - \rho(x, A) > 0$. So $\rho(x', A) < \alpha$ for every $x' \in B_{\bar\rho}(x, r)$, proving the $\tau_{\bar\rho}$-usc of the mapping $\rho(\cdot, A)$ at x.

The case of the distance function $\rho(A, \cdot)$ can be treated similarly. □

Based on the properties of the distance mapping we can prove further bitopological properties of quasi-semimetric spaces.

Proposition 1.1.13 (Kelly [111]). *Let (X, ρ) be a quasi-semimetric space.*

1. *The topology of a quasi-metric space is pairwise regular and pairwise normal.*
2. *If $\tau_\rho \subset \tau_{\bar\rho}$, then the topology $\tau_{\bar\rho}$ is semimetrizable.*
3. *If the mapping $\rho(x, \cdot) : X \to (\mathbb{R}, |\cdot|)$ is τ_ρ-continuous for every $x \in X$, then the topology τ_ρ is regular.*
   If $\rho(x, \cdot) : X \to (\mathbb{R}, |\cdot|)$ is $\tau_{\bar\rho}$-continuous for every $x \in X$, then the topology $\tau_{\bar\rho}$ is semi-metrizable.

Similar results hold for an asymmetric seminormed space (X, p).

Proof. 1. Since $\{B_\rho[x, r] : r > 0\}$ is a $\tau(\rho)$-neighborhood base of the point x formed of $\tau(\bar\rho)$-closed sets and $\{B_{\bar\rho}[x, r] : r > 0\}$ is a $\tau(\bar\rho)$-neighborhood base of the point x formed of $\tau(\rho)$-closed sets, it follows that the bitopological space $(X, \tau(\rho), \tau(\bar\rho))$ is pairwise regular.

To prove the normality of X, let $A, B \subset X$, , such that A is $\tau(\rho)$-closed, B is $\tau(\bar\rho)$-closed and $A \cap B = \emptyset$.

By the assertions 3 and 4 of Proposition 1.1.12,

$$A = \{x \in X : \rho(x, A) = 0\} \quad \text{and} \quad B = \{x \in X : \bar\rho(x, B) = 0\}\,.$$

Let

$$U = \{x \in X : \rho(x, A) < \bar\rho(x, B)\} \quad \text{and} \quad V = \{x \in X : \bar\rho(x, B) < \rho(x, A)\}\,.$$

It is obvious that $U \cap V = \emptyset$. By Proposition 1.1.12 the mapping $\rho(\cdot, A)$ is $\tau(\bar\rho)$-usc, so that the set U is $\tau(\bar\rho)$-open. By the same proposition, the mapping $\rho(A, \cdot)$ is $\tau(\rho)$-usc, so that the set V is $\tau(\rho)$-open.

If $x \in A$, then $x \notin B$, so that $\rho(x, A) = 0 < \bar\rho(x, B)$, showing that $A \subset U$. Similarly, $B \subset V$.

2. The semimetric topology τ_{ρ^s} is the smallest topology finer than τ_ρ and $\tau\bar\rho$, so that $\tau\bar\rho = \tau_{\rho^s}$.

3. If the mapping $\rho(x, \cdot) : X \to (\mathbb{R}, |\cdot|)$ is $\tau(\rho)$-continuous, then the balls $B_\rho[x, r] = \{y \in X : \rho(x, y) \le r\}$ are $\tau(\rho)$-closed for every $r > 0$. Since they form a $\tau(\rho)$-neighborhood base of the point x, it follows that the topology $\tau(\rho)$ is regular.

Suppose now that the mapping $\rho(x, \cdot) : X \to (\mathbb{R}, |\cdot|)$ is $\tau(\bar\rho)$-continuous for every $x \in X$. The $\tau(\bar\rho)$-continuity of $\rho(x, \cdot)$ implies that the spheres $B_\rho(x, r) =$

$\{y \in X : \rho(x,y) < r\}$ are $\tau(\bar{\rho})$-open for every $r > 0$, showing that the topology $\tau(\bar{\rho})$ is finer than $\tau(\rho)$.

The topology τ^s generated by the semimetric $\rho^s = \max\{\rho, \bar{\rho}\}$ is the smallest topology finer that both $\tau(\rho)$ and $\tau(\bar{\rho})$, that is $\tau^s = \tau(\rho) \vee \tau(\bar{\rho}) = \tau(\bar{\rho})$, implying the semimetrizability of $\tau(\bar{\rho})$. □

Remark 1.1.14. 1. The properties from the assertions 3, 4, 5 of Proposition 1.1.8 are taken from Kelly [111].

2. The lower and upper continuity properties from the assertion 4 of Proposition 1.1.8 are equivalent to the fact that the mapping $\rho(x,\cdot) : X \to \mathbb{R}$ is (τ_ρ, τ_u)-continuous, respectively $(\tau_{\bar{\rho}}, \tau_{\bar{u}})$-continuous. Similar equivalences hold for the semi-continuity properties of the mapping $\rho(\cdot, y)$.

3. The quasi-metric space $(\mathbb{R}, u)$ from Example 1.1.3 is not T_1 because, for instance, any neighborhood of 1 contains 0.

1.1.3 More on bitopological spaces

Kelly [111] proved several basic results on bitopological spaces including extensions of the classical theorems of Uryson and Tietze to semi-continuous functions defined on bitopological spaces. The Urysohn type theorem proved in [111] is the following.

Theorem 1.1.15. *Let (T, τ, ν) be a pairwise normal bitopological space, $A \subset T$ τ-closed and $B \subset T$ ν-closed with $A \cap B = \emptyset$. Then there exists a τ-lsc and ν-usc function $f : T \to [0;1]$ such that*

$$\forall t \in A, \quad f(t) = 0 \quad \textit{and} \quad \forall t \in B, \quad f(t) = 1\,. \tag{1.1.23}$$

Proof. The proof follows the line of the proof of the classical Urysohn theorem as given, for instance, in Pedersen [172]. Put $A_0 = A$ and $C_1 = T \setminus B$. Then A_0 is τ-closed, C_1 is ν-open and $A_0 \subset C_1$. By (1.1.12) there exists a τ-closed set $A_{1/2}$ and a ν-open set $C_{1/2}$ such that

$$A_0 \subset C_{1/2} \subset A_{1/2} \subset C_1\,.$$

Applying the same condition to the pairs $A_0 \subset C_{1/2}$ and $A_{1/2} \subset C_1$, we affirm the existence of two τ-closed sets $A_{1/4}, A_{3/4}$ and of two ν-open sets $C_{1/4}, G_{3/4}$ such that

$$A_0 \subset C_{1/4} \subset A_{1/4} \subset C_{1/2} \subset A_{1/2} \subset C_{3/4} \subset A_{3/4} \subset C_1\,.$$

Continuing in this manner, for any dyadic rational number $p \in \{i \cdot 2^{-k} : 1 = 1, 2, \ldots, 2^k - 1\}$, $k \in \mathbb{N}$, one finds a τ-closed set A_p and a ν-open set C_p. Putting, for convenience, $A_p = \emptyset$ for $p < 0$, $A_p = T$ for $p \geq 1$, $C_p = \emptyset$ for $p \leq 0$ and $C_p = T$ for $p > 1$, it follows that these sets satisfy the relations

$$C_p \subset C_q \subset A_q \subset A_r \text{ for all dyadic rational numbers } p \leq q \leq r, \text{ and}$$
$$A_p \subset C_q \text{ for all dyadic rational numbers } p < q\,.$$

Define the function $f : T \to \mathbb{R}$ by

$$f(t) = \inf\{p : t \in C_p\}, \quad t \in T .$$

It follows that

$$f(t) = \inf\{p : t \in A_p\}, \quad 0 \le f(t) \le 1, \quad \forall t \in T,$$
$$f(t) = 0, \quad t \in A, \quad \text{and} \quad f(t) = 1, \quad t \in B = T \setminus C_1 .$$

To prove that f is ν-usc we have to show that for every number α, $0 < \alpha \le 1$, the set $f^{-1}([0;\alpha))$ is ν-open. Let $0 < \alpha \le 1$ and $t \in T$ such that $f(t) < \alpha$. Using the definition of f in terms of the ν-open sets C_p, it follows that there exists a dyadic number p such that $p < \alpha$ and $t \in C_p$, that is

$$f(t) < \alpha \iff \exists p, \ p < \alpha, \ t \in C_p \iff t \in \cup_{p<\alpha} C_p .$$

Consequently, $f^{-1}([0;\alpha)) = \cup_{p<\alpha} C_p$ showing that the set $f^{-1}([0;\alpha))$ is ν-open.

To show that f is also τ-lsc we shall use the expression of f in terms of the τ-closed sets A_p, $f(t) = \inf\{p : t \in A_p\}$. We have to show that for every β, $0 \le \beta < 1$, the set $f^{-1}((\beta, 1])$ is τ-open. Let $t \in T$ such that $f(t) > \beta$. If $t \in A_p$ for every $p > \beta$, then, by the definition of f, $f(t) \le p$. Consequently, $f(t) \le p$ for every $p > \beta$, implying $f(t) \le \beta$. This shows that

$$f(t) > \beta \iff \exists p > \beta, \ t \in T \setminus A_p \iff t \in \cup_{p>\beta}(T \setminus A_p) .$$

It follows that the set $f^{-1}((\beta, 1]) = \cup_{p>\beta}(T \setminus A_p)$ is τ-open. □

The following theorem extends to bitopological spaces a result of Katětov [107, 108] asserting that a topological space T is normal if and only if for any pair of functions $f, g : T \to \mathbb{R}$ such that f is usc, g is lsc and $f \le g$, there exists a continuous function $h : T \to \mathbb{R}$ such that $f \le h \le g$. The proof given here follows the ideas suggested in Engelking [73, Exercise 2.7.2] for normal spaces.

Theorem 1.1.16 (Lane [144]). *A bitopological space (T, τ, ν) is pairwise normal if and only if for every pair $f, g : T \to \mathbb{R}$ of functions such that f is ν-usc, g is τ-lsc, and $f \le g$, there exists a τ-lsc and ν-usc function $h : T \to \mathbb{R}$ such that $g \le f \le h$.*

The following lemma extends to pairwise normal spaces a separation result valid in normal spaces.

Lemma 1.1.17. *Let (T, τ, ν) be a pairwise normal bitopological space. Let $A, B \subset T$ such that A is τ-F_σ, B is ν-F_σ and*

$$\overline{A}^{\tau} \cap B = \emptyset = A \cap \overline{B}^{\nu} . \tag{1.1.24}$$

Then there exist a ν-open set $U \supset A$ and a τ-open set $V \supset B$ such that $U \cap V = \emptyset$.

The sets U and V satisfy also the relations

$$A \subset U \subset \overline{U}^{\tau} \subset T \setminus B \quad \text{and} \quad B \subset V \subset \overline{V}^{\nu} \subset T \setminus A\,. \tag{1.1.25}$$

Proof. Let $A = \cup_{n=1}^{\infty} A_n$ and $B = \cup_{n=1}^{\infty} B_n$, with A_n τ-closed and B_n ν-closed. Applying (1.1.13) to the sets $A_1 \subset T \setminus \overline{B}^{\nu}$, there exists a ν-open set U_1 such that

$$A_1 \subset U_1 \subset \overline{U}_1^{\tau} \subset T \setminus \overline{B}^{\nu}\,.$$

By the same condition (with τ and ν interchanged), there exists a τ-open set V_1 such that

$$B_1 \subset V_1 \subset \overline{V}_1^{\nu} \subset T \setminus (\overline{A}^{\tau} \cup \overline{U}_1^{\tau})\,.$$

Continuing in this manner, we find inductively the ν-open sets U_n and the τ-open sets V_n such that

$$A_n \cup \overline{U}_1^{\tau} \cup \cdots \cup \overline{U}_{n-1}^{\tau} \subset U_n \subset \overline{U}_n^{\tau} \subset T \setminus (\overline{B}^{\nu} \cup \overline{V}_1^{\nu} \cup \cdots \cup \overline{V}_{n-1}^{\nu}),$$
$$B_n \cup \overline{V}_1^{\nu} \cup \cdots \cup \overline{v}_{n-1}^{\nu} \subset V_n \subset \overline{U}_n^{\nu} \subset T \setminus (\overline{A}^{\tau} \cup \overline{U}_1^{\tau} \cup \cdots \cup \overline{U}_n^{\tau})\,,$$

for all $n \in \mathbb{N}$, where $U_0 = V_0 = \emptyset$.

It follows that the set $U = \cup_{n=1}^{\infty} U_n$ is ν-open, $V = \cup_{n=1}^{\infty} V_n$ is τ-open, $A \subset U$, $B \subset V$ and $U \cap V = \emptyset$.

Since V is τ-open and $U \cap V = \emptyset$, it follows that $\overline{U}^{\tau} \cap V = \emptyset$, so that $\overline{U}^{\tau} \cap B = \emptyset$, i.e., $\overline{U}^{\tau} \subset T \setminus B$. The second set of inclusions in (1.1.25) follows similarly. □

Proof of Theorem 1.1.16. Suppose first that $f, g : T \to [0;1]$. For $r \in \mathbb{Q} \cap [0;1]$ let $A_r = g^{-1}([0;r))$ $(A_0 = \emptyset)$, $B_r = f^{-1}([0;r])$ and $C_r = T \setminus f^{-1}([0;r]) = f^{-1}((r;1])$. Since g is τ-lsc the set $g^{-1}([0;r'])$ is τ-closed, so that the set $A_r = \cup\{g^{-1}([0;r']) : r' \in \mathbb{Q},\ 0 < r' < r\}$ is τ-F_σ. Since f is ν-usc, the set $f^{-1}([r';1])$ is ν-closed, so that $C_r = \cup\{f^{-1}([r';1]) : r' \in \mathbb{Q},\ r < r' < 1\}$ is ν-F_σ and $B_r = T \setminus C_r$ is ν-G_δ. It is easy to check that

$$\overline{A}_r^{\tau} \cap C_r = \emptyset = A_r \cap \overline{C}_r^{\nu}\,. \tag{1.1.26}$$

Indeed, if $(t_i : i \in I)$ is a net in A_r that is τ-convergent to $t \in T$, then, by the τ-lsc of the function g, $g(t) \le \liminf_i g(t_i) \le r$, because $g(t_i) \le r$, $i \in I$. It follows that $t \notin C_r$. Similarly, using the ν-usc of the function f one obtains that $t \notin A$ for every $t \in \overline{C}_r^{\nu}$.

As in the proof of the bitopological Urysohn theorem (Theorem 1.1.15), we shall work with the set $\Lambda_2 = \{i \cdot 2^{-k} : i = 0, 1, 2, \ldots, 2^k,\ k \in \mathbb{N}\}$ of dyadic rational numbers in $[0;1]$ (including 0 and 1).

The equalities (1.1.26) show that the sets A_1 and C_1 satisfy the hypotheses of Lemma 1.1.17, so that, by (1.1.25) there exists a ν-open set U_1 such that

$$A_1 \subset U_1 \subset \overline{U}_1^{\tau} \subset B_1\,. \tag{1.1.27}$$

Similarly, there exists a ν-open set $U_{1/2}$ such that

$$A_{1/2} \subset U_{1/2} \subset \overline{U}^{\tau}_{1/2} \subset B_{1/2} \cap U_1 .$$

At the next step we find two ν-open sets $U_{1/4}, U_{3/4}$ such that

$$A_{1/4} \subset U_{1/4} \subset \overline{U}^{\tau}_{1/4} \subset B_{1/4} \cap U_{1/2}, \text{ and}$$
$$A_{3/4} \cup \overline{U}^{\tau}_{1/2} \subset U_{3/4} \subset \overline{U}^{\tau}_{3/4} \subset B_{3/4} \cap U_1 .$$

Continuing in this manner, one obtains by induction, the ν-open sets U_r satisfying (1.1.27) and

$$A_{(2i-1)\cdot 2^{-m}} \cup \overline{U}^{\tau}_{(i-1)\cdot 2^{-m+1}} \subset U_{(2i-1)\cdot 2^{-m}} \subset \overline{U}^{\tau}_{(2i-1)\cdot 2^{-m}} \subset B_{(2i-1)\cdot 2^{-m}} \cap U_{i\cdot 2^{-m+1}} ,$$

for $i = 1, 2, \ldots, 2^{m-1}$, $m \in \mathbb{N}$, where $U_0 = \emptyset$. It follows that the inclusions

$$\begin{aligned} &\text{(i)} \quad A_r \subset U_r \subset \overline{U}^{\tau}_r \subset B_r \\ &\text{(ii)} \quad U_r \subset \overline{U}^{\tau}_r \subset U_{r'} , \end{aligned} \tag{1.1.28}$$

hold for all $r, r' \in \Lambda_2$ with $r < r'$.

Put, for convenience, $U_r = T$ for $r > 1$ and define the function $h : T \to [0;1]$ by

$$h(t) = \inf\{r : r \in \Lambda_2 \cup (1, \infty) \text{ such that } t \in U_r\}, \quad t \in T .$$

Since the sets U_r are ν-open, reasoning as in the proof of the ν-usc of the function f in Theorem 1.1.15, one can show that the function h is ν-usc.

By (1.1.28).(ii),

$$h(t) = \inf\{r \in \Lambda_2 : t \in \overline{U}^{\tau}_r\} ,$$

and the sets $\overline{U}^{\tau}_r$ are τ-closed, so that, following again the ideas of the proof of the τ-lsc of the function f in Theorem 1.1.15, one can show that the function h is τ-lsc.

It remains to show that $f \le h \le g$. Let $t \in T$. If $r \in \Lambda_2$, $r < f(t)$, then, for every $r' \in \Lambda_2$, $0 \le r' < r$, $t \notin f^{-1}([o, r']) = B_{r'}$, and so $t \notin U_{r'}$, implying $h(t) \ge r$. That is $h(t) \ge r$ for every $r \in \Lambda_2$ with $r < f(t)$, and so $f(t) \le h(t)$.

If $r \in \Lambda_2$ is such that $g(t) < r$, then $t \in g^{-1}([0; r)) = A_r \subset U_r$, and so $h(t) \le r$. Since $h(t) \le r$ for every $r \in \Lambda_2$ with $g(t) < r$, it follows that $h(t) \le g(t)$.

The general case, when $f, g : T \to \mathbb{R}$, f is ν-usc, g is τ-lsc and $f \le g$, can be reduced to the preceding one by composing with the function$\psi : \mathbb{R} \to (0;1)$,

$$\psi(t) = \frac{|t| + t + 1}{2(|t| + 1)}, \quad , t \in \mathbb{R} .$$

The function ψ is a strictly increasing homeomorphism between $\mathbb{R}$ and $(0;1)$, and its inverse $\psi^{-1} : (0;1) \to \mathbb{R}$ is strictly increasing and continuous. Consequently, applying the proved result to the functions $\tilde{f} = \psi \circ f$ and $\tilde{g} = \psi \circ G$ we confirm the existence of a τ-lsc and ν-usc function $\tilde{h} : T \to (0;1)$ such that $\tilde{f} \le \tilde{h} \le \tilde{g}$. But then, the function $h = \psi^{-1} \circ \tilde{h} : T \to \mathbb{R}$ is τ-lsc, ν-usc and $f = \psi^{-1} \circ \tilde{f} \le h \le \psi^{-1} \circ \tilde{g} = g$.

To prove the converse, suppose that A, B are two disjoint subsets of T such that A is ν-closed and B is τ-closed. Then the characteristic function χ_A of the set A is ν-usc and $\chi_{T\setminus B}$ is τ-lsc. The inclusion $A \subset T \setminus B$ implies $\chi_A \le \chi_{T\setminus B}$, so that, by hypothesis, there exists a ν-usc and τ-lsc function $h : T \to [0;1]$ such that $\chi_A \le h \le \chi_{T\setminus B}$. By the τ-lsc of h, the set $U = h^{-1}((\frac{1}{2};1])$ is τ-open and $A \subset U$. By the ν-usc of the function h, the set $V = h^{-1}([\frac{1}{2};1])$ is ν-closed and $A \subset U \subset V \subset T \setminus B$. □

The proof of the following result follows that in topological spaces, see Engelking [73, Theorem 1.5.15].

Proposition 1.1.18 (Kelly [111]). *A pairwise regular bitopological space (T, τ, ν) satisfying the second countability axiom (i.e., both τ and ν satisfy this axiom) is pairwise normal.*

As it is well known, a relatively easy consequence of the Urysohn theorem is the Tietze extension theorem, but the proof given for topological spaces cannot be adapted to bitopological spaces. Kelly [111] proved a Tietze type theorem asserting that any real-valued τ-usc and ν-lsc function defined on τ-closed and ν-closed subset of a pairwise normal bitopological space T admits a τ-usc and ν-lsc extension to the whole space T. Lane [144] showed by a counterexample that the result is false in this form, and gave a proper formulation.

The example is the following.

Example 1.1.19. Let T be an uncountable set and $A = \{t_1, t_2, \dots\}$ be an infinite countable subset of T. Let τ be formed by the empty set and the complements of finite or countable subsets of T, and let ν be the discrete topology. Then the bitopological space (T, τ, ν) is pairwise normal, the function $f : A \to \mathbb{R}$, $f(t_i) = i$, $i \in \mathbb{N}$, is τ-lsc and ν-usc, but has no τ-lsc and ν-usc extension to T.

Indeed, if A, B are disjoint subsets of T such that A is τ-closed and B is ν-closed, then A is also ν-open and $T \setminus A$ is τ-open, so that (T, τ, ν) is pairwise normal. The set $A = \{t_1, t_2, \dots\}$ is τ- and ν-closed and the function $f : A \to \mathbb{R}$, $f(t_i) = i$, $i \in \mathbb{N}$, is τ-lsc and ν-usc. Suppose that $F : T \to \mathbb{R}$ is a τ-lsc and ν-usc extension of f. Then the sets $A_i = \{t \in T : F(t) \le i\}$, $i \in \mathbb{N}$, are τ-closed. Because $A_i \ne T$, A_i must be at most countable, implying that $T = \cup_{i\in\mathbb{N}} A_i$ is countable, in contradiction to the hypothesis.

A general Tietze type theorem holds only for bounded semi-continuous functions.

Theorem 1.1.20. *Let (T, τ, ν) be a pairwise normal bitopological space and A a τ-closed and ν-closed subset of T. Then every bounded, τ-lsc and ν-usc function*

$f : A \to \mathbb{R}$ admits a τ-lsc and ν-usc extension $F : T \to \mathbb{R}$ such that

$$\inf F(T) = \inf f(A) \quad \textit{and} \quad \sup F(T) = \sup f(A) . \tag{1.1.29}$$

Proof. Let

$$\alpha = \inf f(A) \quad \text{and} \quad \beta = \sup f(A) .$$

Define the functions $g, h : T \to \mathbb{R}$ by $g(t) = h(t) = f(t)$ for $t \in A$, $g(t) = \alpha$ and $h(t) = \beta$ for $t \in T \setminus A$. It is easily seen that g is ν-usc, h is τ-lsc and $g \le h$. By Theorem 1.1.16 there exists a τ-lsc and ν-usc function $F : T \to \mathbb{R}$ such that $g \le F \le h$. It follows that F is the desired extension of f. □

Lane [144] gave some sufficient conditions on the subset A of the bitopological space T in order for any τ-lsc and ν-usc real-valued function to have a τ-lsc and ν-usc extension to T. In particular this is true for quasi-metric spaces.

Proposition 1.1.21. *Let (X, ρ) be a quasi-semimetric space and A a nonempty τ_ρ- and $\tau_{\bar{\rho}}$-closed subset of X. Then any τ_ρ-lsc and $\tau_{\bar{\rho}}$-usc function $f : A \to (\mathbb{R}, |\cdot|)$ admits a τ_ρ-lsc and $\tau_{\bar{\rho}}$-usc extension to the whole space X.*

Proof. By Proposition 1.1.12, the function $h_1(x) = \rho(A, x)$, $x \in X$, is τ_ρ-usc and $\tau_{\bar{\rho}}$-lsc, while the function $h_2(x) = \rho(x, A)$, $x \in X$, is τ_ρ-lsc and $\tau_{\bar{\rho}}$-usc.

The function $g = f/(1 + |f|)$ is τ_ρ-lsc and $\tau_{\bar{\rho}}$-usc and $-1 < g(x) < 1$ for all $x \in X$. By Theorem 1.1.20 g has a τ_ρ-lsc and $\tau_{\bar{\rho}}$-usc extension $G : X \to [-1, 1]$. The function $G^+ = G \vee 0$ is τ_ρ-lsc, $\tau_{\bar{\rho}}$-usc and $0 \le G^+ \le 1$, while the function $G^- = -(G \wedge 0)$ is τ_ρ-usc, $\tau_{\bar{\rho}}$-lsc and $0 \le G^- \le 1$.

The function $G_1 = G^+/(1 + h_1)$ is τ_ρ-lsc, $\tau_{\bar{\rho}}$-usc and $0 \le G_1(x) < 1$, $x \in X$, while the function $G_2 = G^+/(1 + h_2)$ is τ_ρ-usc, $\tau_{\bar{\rho}}$-lsc and $0 \le G_2(x) < 1$, $x \in X$. Also, for every $x \in A$,

$$G_1(x) = G^+(x) \quad \text{and} \quad G_2(x) = G^-(x) .$$

It follows that the function $\overline{G} = G_1 - G_2$ is τ_ρ-lsc, $\tau_{\bar{\rho}}$-usc and $-1 < \overline{G}(x) < 1$, $x \in X$. Also, for every $x \in A$,

$$\overline{G}(x) = G_1(x) - G_2(x) = G^+(x) - G^-(x) = G(x) = g(x) ,$$

that is $\overline{G}$ is a τ_ρ-lsc and $\tau_{\bar{\rho}}$-usc extension of g. But then the function $F = \overline{G}/(1-|\overline{G}|)$ is τ_ρ-lsc and $\tau_{\bar{\rho}}$-usc extension of the function f. □

Remark 1.1.22. The proof of Theorem 1.1.20, based on Theorem 1.1.16, is taken from [144]. Another proof, based on the Urysohn theorem for bitopological spaces (Theorem 1.1.15) is given in [78]. In [78] some conditions on the set $A \subset T$ ensuring the existence of a τ-usc and ν-lsc extension F of an arbitrary τ-usc and ν-lsc F on A are given too.

A bitopological space (T, τ, ν) is called *pairwise perfectly normal* if it is pairwise normal and every τ-closed (ν-closed) subset of T is ν-G_δ (τ-G_δ).

The following characterization of pairwise perfectly normal bitopological spaces extends a well-known result in topology, see, e.g., Engelking [73, Theorem 1.5.19].

Theorem 1.1.23 ([78], [144]). *A bitopological space (T, τ, ν) is pairwise perfectly normal if and only if for any pair A, B of subsets of T such that A is τ-closed, B is ν-closed and $A \cap B = \emptyset$, there exists a τ-lsc and ν-usc function $f : T \to [0; 1]$ such that $A = f^{-1}(0)$ and $B = f^{-1}(1)$.*

Bîrsan studied in the paper [24] pairwise completely regular bitopological spaces and extended to this setting Hausdorff's embedding theorem. A bitopological space (X, τ, ν) is called *τ-completely regular with respect to ν* if for every $x \in X$ and every τ-open neighborhood U of x there exits a τ-lsc and ν-usc function $f : X \to [0; 1]$ such that $fx) = 1$ and $f(X \setminus U) = \{0\}$. Equivalently, (X, τ, ν) is τ-completely regular with respect to ν if and only if for every τ-closed set Z and every point $x \in X \setminus Z$ there exists a τ-lsc and ν-usc function $f : X \to [0; 1]$ such that $f(x) = 1$ and $f(Z) = \{0\}$.

Consider the bitopological space $(\mathbb{R}, \tau_u, \tau_{\bar{u}})$ from Example 1.1.3 and let $I = [0; 1]$ with the induced induced topologies.

By definition a *bitopological cube* is a product I^A of A copies of the bitopological space I, where A an arbitrary nonempty set. Any bitopological cube is pairwise completely regular and pairwise compact.

A bitopological space (X, τ, ν) is called *weakly pairwise Hausdorff* if for every pair x, y of distinct points from X there exist a τ-open neighborhood U of one of these points and a ν-open neighborhood V of the other one with $U \cap V = \emptyset$. The space $(\mathbb{R}, \tau_u, \tau_{\bar{u}})$ is weakly pairwise Hausdorff but not pairwise Hausdorff.

Theorem 1.1.24 ([24]). *A weakly pairwise Hausdorff bitopological space is pairwise completely regular if and only if it is bi-homeomorphic to a subspace of a bitopological cube.*

A mapping $f : (X, \tau_1, \tau_2) \to (Y, \nu_1, \nu_2)$ between two bitopological spaces is called *bi-continuous* if it is both (τ_1, ν_1)-continuous and (τ_2, ν_2)-continuous. If f is a bijection such that both f and f^{-1} are bi-continuous, then f is called a *bi-homeomorphism.*

The paper [25] is concerned with bitopological groups and [26] with some properties relating quasi-uniform and bitopological spaces.

An important problem in topology is that of metrizability of topological spaces. As we have seen, quasi-semimetric spaces are bitopological spaces, and so the corresponding problem in this setting would be that of quasi-metrizability of bitopological spaces. The following result is the analog of Urysohn metrizability theorem and improves Proposition 1.1.18.

Theorem 1.1.25 (Kelly [111]). *If (T,τ,ν) is a pairwise regular bitopological space such that both τ and ν satisfy the second axiom of countability, then it is quasi-semimetrizable. If further, T is pairwise Hausdorff, then it is quasi-metrizable.*

Lane [144] proved a bitopological version of the Nagata-Smirnov metrizability theorem.

1.1.4 Compactness in bitopological spaces

We shall briefly present some results on compactness in bitopological spaces obtained by Bîrsan [23].

If $\mathcal{A},\mathcal{B}$ are two covers of a set X one says that $\mathcal{B}$ *refines* $\mathcal{A}$ (or that $\mathcal{B}$ is a *refinement* of $\mathcal{A}$) if

$$\forall B\in\mathcal{B},\ \exists A\in\mathcal{A},\qquad B\subset A\,. \tag{1.1.30}$$

Following [23], a bitopological space (X,τ,ν) is called:

- *τ-compact with respect to ν* if every τ-open cover of X admits a finite ν-open refinement that covers X;
- *pairwise compact* if it is τ-compact with respect to ν and ν-compact with respect to τ.

It is obvious that a topological space (X,τ) is compact if and only if for every open cover of X there exists a finite refinement of X which is an open cover of X, so that the above definitions are natural extensions of the usual notion of compactness. The bitopological compactness can be characterized in terms of closed sets.

Proposition 1.1.26 ([23]). *Let (X,τ,ν) be a bitopological space. The following are equivalent.*

1. *The space X is τ-compact with respect to ν.*
2. *If a family $F_i,\ i\in I,$ of τ-closed sets has empty intersection, then there exists a finite family $H_j,\ j\in J,$ of ν-closed sets having empty intersection and such that for every $j\in J$ there exists $i\in I$ with $H_j\supset F_i$.*

Proof. $1\Rightarrow 2$. Let $F_i,\ i\in I$, be a family of τ-closed sets with empty intersection. Then their complements $G_i=\complement(F_i),\ i\in I$, are τ-open and cover X, so there exists a finite refinement $Z_j,\ j\in J$, of $\{G_i\}$ with ν-open sets that covers X. Putting $H_j=\complement(Z_j),\ j\in J$, it follows that each H_j is ν-closed, $\cap_{j\in J}H_j=\complement\left(\cup_{j\in J}Z_j\right)=\emptyset$. Also, for every $j\in J$ there exists $i\in I$ such that $Z_j\subset G_i\iff H_j\supset F_i$.

The implication $2\Rightarrow 1$ is proved similarly. □

Proposition 1.1.27 ([23]). *If the bitopological space (X,τ,ν) is pairwise Hausdorff and (X,τ) is compact, then $\tau\subset\nu$.*

Proof. It is sufficient to show that every τ-closed subset of X is ν-closed. Let $Z \subset X$ be τ-closed and $y \notin Z$. Then for every $z \in Z$ there exists a τ-open set U_z containing z and a ν-open set V_z containing y such that $U_z \cap V_z = \emptyset$. By the τ-compactness of X, the τ-open cover U_z, $z \in Z$, of Z contains a finite subcover $U_{z_1}, \ldots, U_{z_n}$. It follows that $V = \cap_{k=1}^n V_{z_k}$ is a ν-open neighborhood of y and $V \cap Z \subset V \cap (\cup_{k=1}^n U_{z_k}) = \emptyset$, so that $V \subset X \setminus Z$. Consequently $X \setminus Z$ is ν-open and Z is ν-closed. □

This proposition has the following corollary.

Corollary 1.1.28. *Let (X, τ, ν) be a pairwise Hausdorff bitopological space.*

1. *If both the topologies τ and ν are compact, then $\tau = \nu$.*
2. *If the space X is τ-compact with respect to ν, then $\tau \subset \nu$.*
3. *If (X, τ, ν) is pairwise compact, then $\tau = \nu$.*

It is known that a subset of a topological space is compact if and only if it is compact with respect to the induced topology, a result that does not hold in bitopological spaces.

Call a subset Y of a bitopological space (X, τ, ν) *τ-compact with respect to ν* if it is $\tau|_Y$-compact with respect to $\nu|_Y$ as a bitopological subspace of X.

Proposition 1.1.29 ([23]). *Let (X, τ, ν) be a bitopological space and $Y \subset X$.*

1. *If every τ-open cover of Y admits a finite refinement that is a ν-open cover of Y, then Y is τ-compact with respect to ν.*
2. *If the set Y is ν-open, then the converse is also true.*

It is shown by an example [23, Example 8] that the second assertion from the above proposition is not true for arbitrary subsets.

The preservation of compactness by continuous mappings takes the following form.

Proposition 1.1.30 ([23]). *Let (X, τ_1, τ_2) and (X, ν_1, ν_2) be bitopological spaces. Suppose that X is τ_1-compact with respect to τ_2. If the mapping $f : X \to Y$ is (τ_1, ν_1)-continuous and (τ_2, ν_2)-open, then $f(X)$ is ν_1-compact with respect to ν_2.*

Proof. One shows that the hypotheses from Proposition 1.1.29.1 are fulfilled. □

In the same paper [23], Bîrsan extended to this setting Alexander's subbase theorem and Tikhonov's theorem on the compactness of the product of compact topological spaces.

Theorem 1.1.31 (Alexander's subbase theorem, [23]). *Let (X, τ, ν) be a bitopological space, $\mathcal{A}$ a subbase of the topology τ and $\mathcal{B}$ a subbase of the topology ν.*

1. *If every cover of X with sets in $\mathcal{A}$ admits a finite refinement with elements from ν that covers X, then X is τ-compact with respect to ν.*

2. *If every cover of X with sets in $\mathcal{A}$ admits a finite refinement with elements from $\mathcal{B}$ that covers X, then the bitopological space X is pairwise compact.*

The *bitopological product* of a family (X_i, τ_i, ν_i), $i \in I$, of bitopological spaces is the bitopological space (X, τ, ν), where $X = \prod_{i\in I} X_i$, $\tau = \prod_{i\in I} \tau_i$ and $\nu = \prod_{i\in I} \nu_i$. The so-defined bitopological product has properties similar to that of the usual topological product.

Proposition 1.1.32. *Let (X_i, τ_i, ν_i), $i \in I$, be a family of bitopological spaces, (X, τ, ν) their bitopological product.*

1. *For each $i \in I$ the projection $p_i : X \to X_i$ is bi-continuous and bi-open,* i.e., *(τ, τ_i)-continuous, (ν, ν_i)-continuous and (τ, τ_i)-open and (ν, ν_i)-open.*
2. *If (X, τ, ν) is τ-compact with respect to ν (pairwise compact), then for each $i \in I$ the bitopological space (X_i, τ_i, ν_i) is τ_i-compact with respect to ν_i (respectively, pairwise compact).*

Theorem 1.1.33 (Tikhonov compactness theorem, [23]). *Let (X_i, τ_i, ν_i), $i \in I$, be a family of bitopological spaces. If each bitopological space X_i is τ_i-compact with respect to ν_i (pairwise compact), then their bitopological product (X, τ, ν) is τ-compact with respect to ν (respectively, pairwise compact).*

In spite of the good results holding for this notion of compactness, it has a serious drawback – a finite set need not be compact – as it is shown by the following example.

Example 1.1.34 ([23]). Consider a three point set $X = \{a, b, c\}$ with the topologies $\tau = \{\emptyset, \{a\}, \{b, c\}, X\}$ and $\nu = \{\emptyset, \{c\}, \{a, b\}, X\}$. Then the topologies τ and ν are compact, but τ is not compact with respect to ν, nor ν is compact with respect to τ.

The paper [23] contains also examples of

- a pairwise compact, not pairwise normal, bitopological space (X, τ, ν) with $\tau \neq \nu$;
- a pairwise Hausdorff bitopological space (X, τ, ν) which is τ-compact with respect to ν, but not pairwise normal;
- a pairwise normal, pairwise compact bitopological space (X, τ, ν) which is not pairwise regular.

There are also other notions of compactness in bitopological spaces, a good analysis of various relations holding between them is done in the paper [52]. Note that our terminology differs from that in [52]. For convenience we shall call a bitopological space *pairwise B-compact* if it is pairwise compact in the sense considered by Bîrsan.

Let (X, τ, ν) be a bitopological space. A cover $\mathcal{A} \subset \tau \cup \nu$ of X is called *(τ, ν)-open*. If, in addition, $\mathcal{A}$ contains a nonempty set from τ and a nonempty set from ν, then $\mathcal{A}$ is called *pairwise open*.

The bitopological space (X, τ, ν) is called:

- *pairwise S-compact* provided that every pairwise open cover of X contains a finite subcover (Swart [231]);
- *semi-compact* provided that every (τ, ν)-open cover of X contains a finite subcover (Datta [53, 54]);
- *pseudo-compact* provided that every bi-continuous function $f : (X, \tau, \nu) \to (\mathbb{R}, \tau_u, \tau_{\bar{u}})$ is bounded (Saegrove [213]);
- *pairwise real-compact* provided it is bi-homeomorphic to the intersection of a $\prod_{i \in I} \tau_u$-closed subset and a $\prod_{i \in I} \tau_{\bar{u}}$-closed subset of a product of I copies of $(\mathbb{R}, \tau_u, \tau_{\bar{u}})$ (Saegrove [213]);
- *bicompact* provided it is both pseudo-compact and pairwise real-compact (Saegrove [213]).

In the following theorem we collect some results from [52].

Theorem 1.1.35. *Let (X, τ, nu) be a bitopological space.*

1. *The bitopological space (X, τ, ν) is semi-compact if and only if it is compact with respect to the topology $\tau \vee \nu$.*
2. *The bitopological space (X, τ, ν) is semi-compact if and only if it is pairwise compact, τ-compact and ν-compact.*
3. *The bitopological space (X, τ, ν) is pairwise S-compact if and only if each τ-closed proper subset of X is ν-compact and each ν-closed proper subset of X is τ-compact.*
4. *If the bitopological space (X, τ, ν) is semi-compact or pairwise B-compact, then it is pseudo-compact.*
5. *If the bitopological space (X, τ, ν) is pairwise Hausdorff and either semi-compact, bicompact or pairwise B-compact, then $\tau = \nu$.*
5. *If the bitopological space (X, τ, ν) is either semi-compact, bicompact or pairwise B-compact, then both of the topologies τ and ν are compact.*

Remark 1.1.36.

1. Any finite bitopological space is semi-compact and pairwise S-compact. Consequently, Example 1.1.34 furnishes also an example of a semi-compact and pairwise S-compact bitopological space which is not pairwise B-compact.
2. Example 3 in [52] gives a pairwise B-compact bitopological space which is neither pairwise S-compact nor semi-compact.
3. Example 1 in [52] gives a pairwise S-compact bitopological space (X, τ, ν) such that (X, ν) is not compact.

Paracompactness is even a more delicate matter to be treated within the framework of bitopological spaces. For some attempts and discussions see the papers [29], [55], [124], [178] and [179].

We mention also the following metrizability conditions.

Theorem 1.1.37. *Let (T, τ, ν) be a bitopological space and (X, q) a quasi-metric space.*

1. *If the bitopological space T is quasi-metrizable and τ is locally (countably) compact with respect to ν, then (T, ν) is metrizable. The same conclusion holds if locally compact is replaced by paracompact.*
2. *If the topology $\tau(\bar{\rho})$ is sequentially compact, then the topology $\tau(\rho)$ is metrizable. The same is true if $\tau(\bar{\rho})$ is compact.*
3. *If (X, ρ) is sequentially compact and Hausdorff, then the topology $\tau(\rho)$ is metrizable.*

Remark 1.1.38. Assertions 1 and 2 are taken from [178] while the third one is from [227].

1.1.5 Topological properties of asymmetric seminormed spaces

In general the topology generated by an asymmetric norm is T_0 but not T_1. Indeed, it is easy to check that the space $(\mathbb{R}, u)$ from Example 1.1.3 is T_0 but -1 belongs to every neighborhood of 1. A condition that the topology of an asymmetric normed space be Hausdorff was found by García-Raffi, Romaguera and Sánchez-Pérez [91], in terms of a functional $p^\diamond : X \to [0; \infty)$ associated to an asymmetric seminorm p defined on a real vector space X, a result that was extended to asymmetric locally convex spaces in [41], see Subsection 1.1.7. The separation properties of an asymmetric seminormed space will be presented in Proposition 1.1.40.

The functional $p^\diamond$ is defined by the formula

$$p^\diamond(x) = \inf\{p(x') + p(x' - x) : x' \in X\},\ x \in X. \tag{1.1.31}$$

In the following proposition we present the properties of $p^\diamond$.

Proposition 1.1.39. *The functional $p^\diamond$ is a (symmetric) seminorm on X, $p^\diamond \le p$, and $p^\diamond$ is the greatest seminorm on X majorized by p.*

Proof. First observe that, replacing x' by $x' - x$ in (1.1.31), we get

$$\begin{aligned} p^\diamond(-x) &= \inf\{p(x') + p(x' + x) : x' \in X\} \\ &= \inf\{p(x' - x) + p((x' - x) + x) : x' \in X\} = p^\diamond(x)\ , \end{aligned}$$

so that $p^\diamond$ is symmetric. The positive homogeneity of $p^\diamond$, $p^\diamond(\alpha x) = \alpha p^\diamond(x),\ x \in X, \alpha \ge 0,$ is obvious. For $x, y \in X$ and arbitrary $x', y' \in X$ we have

$$p^\diamond(x + y) \le p(x' + y') + p(x' + y' - x - y) \le p(x') + p(x' - x) + p(y') + p(y' - y)\ ,$$

so that, passing to infimum with respect to $x', y' \in X,$ we obtain the subadditivity of $p^\diamond$,

$$p^\diamond(x + y) \le p^\diamond(x) + p^\diamond(y)\ .$$

Suppose now that there exists a seminorm q on X such that $q \leq p$, i.e., $\forall z \in X,\ q(z) \leq p(z)$, and $p^\diamond(x) < q(x) \leq p(x)$, for some $x \in X$. Then, by the definition of $p^\diamond$, there exists $x' \in X$ such that $p^\diamond(x) < p(x') + p(x' - x) < q(x)$, leading to the contradiction

$$q(x) \leq q(x') + q(x - x') = q(x') + q(x' - x) \leq p(x') + p(x' - x) < q(x) . \qquad \square$$

In the following proposition we collect the separation properties of an asymmetric seminormed space.

Proposition 1.1.40 ([91]). *Let (X, p) be an asymmetric seminormed space.*

1. *The topology τ_p is T_0 if and only if for every $x \in X$, $x \neq 0$, $p(x) > 0$ or $p(-x) > 0$, in other words, if and only if p is an asymmetric norm.*
2. *The topology τ_p is T_1 if and only if $p(x) > 0$ for every $x \in X$, $x \neq 0$.*
3. *The topology τ_p is Hausdorff if and only if $p^\diamond(x) > 0$ for every $x \neq 0$.*

Proof. The assertion from 1 follows from Proposition 1.1.8.3.

2. If $p^\diamond(x) > 0$ whenever $x \neq 0$, then $p^\diamond$ is a norm on X, so that the topology $\tau_{p^\diamond}$ generated by $p^\diamond$ is Hausdorff. The inequality $p^\diamond \leq p$ implies that the topology τ_p is finer than $\tau_{p^\diamond}$, so it is Hausdorff, too.

Suppose $p^\diamond(x) = 0$ for some $x \neq 0$. By the definition (1.1.31) of $p^\diamond$, there exists a sequence (x_n) in X such that $\lim_n [p(x_n) + p(x_n - x)] = p^\diamond(x) = 0$. This implies $\lim_n p(x_n) = 0$ and $\lim_n p(x_n - x) = 0$, showing that the sequence (x_n) has two limits with respect to τ_p. Consequently, the topology τ_p is not Hausdorff. $\square$

Remark 1.1.41. It is known that a T_0 topological vector space (TVS) is Hausdorff and completely regular (see [149, Theorem 2.2.14]), a result that is no longer true in asymmetric normed spaces. An easy example illustrating this situation is that of the space $(\mathbb{R}, u)$ from Example 1.1.3, which is T_0 but not T_1.

The following proposition shows that an asymmetric normed space is not necessarily a topological vector space.

Proposition 1.1.42. *If (X, p) is an asymmetric normed space, then the topology τ_p is translation invariant, so that the addition $+ : X \times X \to X$ is continuous. Also any additive mapping between two asymmetric normed spaces $(X, p), (Y, q)$ is continuous if and only if it is continuous at $0 \in X$ (or at an arbitrary point $x_0 \in X$).*

The scalar multiplication is not always continuous, so that an asymmetric normed space need not be a topological vector space.

Proof. The fact that the topology τ_p is translation invariant follows from the formulae (1.1.5). Example 1.1.3 shows that the multiplication by scalars need not be continuous. $\square$

Another example was given by Borodin [28].

Example 1.1.43. In the space $X = C_0[0;1]$, where

$$C_0[0;1] = \left\{ f \in C[0;1] : \int_0^1 f(t)dt = 0 \right\} ,$$

with the asymmetric seminorm $p(f) = \max f([0;1])$, the multiplication by scalars is not continuous at the point $(-1,0) \in \mathbb{R} \times X$.

To prove this, we show that $(-1)B_p[0,r]$ is not contained in $B_p[0,1]$ for any $r > 0$. Indeed, let $t_n = 1/n$ and

$$f_n(t) = \begin{cases} r(n-1)(nt-1), & 0 \le t \le t_n, \\ r\frac{n}{n-1}\left(t - \frac{1}{n}\right), & t_n < t \le 1 , \end{cases}$$

for $n \in \mathbb{N}$. Then $f_n \in C_0[0;1]$, $p(f_n) = r$, $-f_n \in C_0[0;1]$ and $p(-f_n) = (n-1)r > 1$ for sufficiently large n.

Remark 1.1.44. It is known (and easy to check) that if a topology is separable, then any coarser topology is separable. In particular, if the semimetric space (X, ρ^s) is separable, then the quasi-metric space (X, ρ) is also separable (with respect to the topology τ_ρ).

The following example shows that the converse is not true, in general.

Example 1.1.45 (Borodin [28]). There exists an asymmetric normed space (X,p) which is τ_p-separable but not τ_{p^s}-separable.

Take

$$X = \left\{ x \in \ell_\infty : x = (x_k),\ \sum_{k=1}^{\infty} \frac{x_k}{2^k} = 0 \right\} , \tag{1.1.32}$$

with the asymmetric norm $p(x) = \sup_k x_k$. It is clear that p is an asymmetric norm on X satisfying $p(x) > 0$ whenever $x \neq 0$ and $p^s(x) = \|x\|_\infty = \sup_k |x_k|$ is the usual sup-norm on ℓ_∞. Because $\varphi(x) = \sum_{k=1}^{\infty} 2^{-k} x_k$, $x = (x_k) \in \ell_\infty$, is a continuous linear functional on ℓ_∞, it follows that $X = \ker \varphi$ is a codimension one closed subspace of ℓ_∞. Since ℓ_∞ is nonseparable with respect to p^s, X is also nonseparable with respect to p^s.

Let us show that X is p-separable. Consider the set Y formed of all $y = (y_k)$ such that $y_k \in \mathbb{Q}$ for all k and there exists $n = n(y)$ such that $y_k = y_{n+1}$ for all $k > n$. It is clear that Y is contained in X and that Y is countable. To show that Y is p-dense in X, let $x \in X$ and $\varepsilon > 0$.

Choose $n \in \mathbb{N}$ such that

$$\sum_{i=1}^{n} 2^{-i}\varepsilon - \sum_{i=1}^{n} 2^{-i} x_i > \|x\|_\infty \sum_{j=n+1}^{\infty} 2^{-j} . \tag{1.1.33}$$

This is possible because the left-hand side of (1.1.33) tends to ε for $n \to \infty$, while the right-hand side tends to 0. Choose $y_k \in \mathbb{Q} \cap (x_k - 2\varepsilon; x_k - \varepsilon)$ for $k = 1, \ldots, n$, and let

$$y_k = \alpha := -\left(\sum_{i=1}^{n} 2^{-i} y_i\right) : \left(\sum_{j=n+1}^{\infty} 2^{-j}\right),$$

for $k > n$. Then $y = (y_k) \in Y$, $\varepsilon < y_k - x_k < 2\varepsilon$, for $k = 1, \ldots, n$ and, by (1.1.33),

$$y_k = \alpha > \left(\sum_{i=1}^{n} 2^{-i}(\varepsilon - x_i)\right) : \left(\sum_{j=n+1}^{\infty} 2^{-j}\right) > \|x\|_\infty ,$$

for $k > n$. It follows that $x_k - y_k \le \|x\|_\infty - y_k < 0$ for $k > n$, so that $p(x - y) = \max\{x_k - y_k : 1 \le k \le n\} < 2\varepsilon$.

As it is well known, by the classical Banach-Mazur theorem, any separable real Banach space can be linearly and isometrically embedded in the Banach space $C[0; 1]$ of all continuous real-valued functions on $[0; 1]$ with the sup-norm, see, for instance, [74, Theorem 5.17]. In other words, $C[0; 1]$ is a universal space in the category of separable real Banach spaces. The validity of this result in the case of asymmetric normed spaces was discussed by Alimov [11] and Borodin [28]. The above example shows that some attention must be paid to the notion of separability that we are using.

Denote by $(C[0; 1], 1, 0)$ the space of all real-valued continuous functions on $[0; 1]$ functions with the asymmetric norm

$$\|f| = \max\{f_+(t) : t \in [0; 1]\}, \tag{1.1.34}$$

for $f \in C[0; 1]$, where $f_+(t) = \max\{f(t), 0\}$, $t \in [0; 1]$.

Theorem 1.1.46 ([28]). *Any T_1 asymmetric normed space (X, p) such that the associated normed space (X, p^s) is separable is isometrically isomorphic to a subspace of the asymmetric normed space $(C[0; 1], 1, 0)$.*

Since any linear isometry from (X, p) to $(C[0; 1], 1, 0)$ induces a linear isometry from (X, p^s) to the usual Banach space $C[0; 1]$, the separability condition with respect to the symmetric norm p^s is necessary for the validity of the Banach-Mazur theorem.

Some complements to this result were given by Alimov [11].

Theorem 1.1.47 ([11]). *A T_1 asymmetric normed space (X, p) is isometrically isomorphic to an affine variety Z of the usual Banach space $C[0; 1]$ if and only if the topology τ_p is metrizable and separable.*

Proof. Note that the topology τ_p of a T_1 asymmetric normed space (X, p) is metrizable if and only if $\tau_p = \tau_{\bar{p}} = \tau_{p^s}$, so that the topology τ_p is generated by the associated norm p^s. Also, in this case, it does not matter with respect to which of the topologies $\tau_p, \tau_{\bar{p}}$, or τ_{p^s}, is considered the separability.

Suppose that the topology τ_p is metrizable and separable. Then the topology generated by the norm p^s agrees with that generated by the asymmetric norm p, implying the existence of two numbers $0 < R < R'$ such that $RB_{p^s} \subset B_p \subset R'B_{p^s}$, where B_{p^s}, B_p denote the corresponding closed unit balls. Since $\alpha\delta B_{\delta p^s} = \delta\, B_{p^s}$, we can suppose, replacing, if necessary, p^s by $q = \alpha p^s$ for some $0 < \alpha < 1$, that there exist a norm q on X and the numbers $0 < r < r' < 1$ such that

$$rB_q \subset B_p \subset r'B_q \subset B_q.$$

In the space $X \times \mathbb{R}$ consider the sets $U = \mathrm{co}\,[(B_p \times \{1\}) \cup (B_q \times \{0\})]$, $V = U \cup \mathrm{co}(\{v\} \cup B_p \times \{1\})$, where $v = (0, 1 + r)$, and $W = V \cup (-V)$. It follows that the sets U, V are convex and W is a bounded absolutely convex body, so that it generates a norm $\|\cdot\|$ on $X \times \mathbb{R}$. Since $W \cap (X \times \{0\}) = B_q \times \{0\}$, it follows that $\|(x, 0)\| = q(x)$ for every $x \in X$. If Y is a countable dense subset of (X, q), then $Y \times \mathbb{Q}$ is a countable dense subset of $(X \times \mathbb{R}, \|\cdot\|)$. By the classical Banach-Mazur theorem there exists an isometric linear mapping $\varphi : (X \times \mathbb{R}, \|\cdot\|) \to (C[0;1], \|\cdot\|_\infty)$. The image $Z = \varphi(X \times \{1\})$ of the hyperplane $X \times \{1\}$ by φ is an affine variety in $C[0;1]$. Since $W \cap (X \times \{1\}) = B_p \times \{1\}$ and φ is an isometry, it follows that the set $K = Z \cap B_{C[0;1]}$ is isometric to B. In particular, the Minkowski functional $P_{K,\xi}$ of the set K with respect to the point $\xi = \varphi(0, 1)$ agrees with the Minkowski functional of the set B, that is with the asymmetric norm p. □

Remark 1.1.48. Let K be a convex subset of a linear space X. The set K is called absorbing with respect to a point $\xi \in K$ if for every $x \in X$ there exists $t > 0$ such that $x \in \xi + t(K - \xi)$. The Minkowski functional $p_{K,\xi}$ of the set K with respect to ξ is defined by

$$p_{K,\xi}(x) = p_{K-\xi}(x - \xi) = \inf\{t > 0 : x \in \xi + t(K - \xi)\}. \tag{1.1.35}$$

The isometry mentioned at the end of the proof of Theorem 1.1.46 is given by the formula

$$p_{K,\xi}(\varphi(x)) = p_{K-\xi}(\varphi(x - v)) = p(x), \quad x \in X.$$

A corollary of the Banach-Mazur theorem asserts that any separable metric space can be isometrically embedded in $C[0;1]$ (see [74, Corollary 5.18]). Kleiber and Pervin [117] extended this result to metric spaces of arbitrary density character α, where α is an uncountable cardinal number, by proving that such a space can be isometrically embedded in the space $C([0;1]^\alpha)$. The density character of a topological space T is the smallest cardinal number α such that T contains a dense subset of cardinality α. As remarked by Alimov [11], these results hold also for a quasi-metric space (X, ρ), the density character being that of the associated metric space (X, ρ^s).

1.1.6 Quasi-uniform spaces

Quasi-semimetric spaces are particular cases of quasi-uniform spaces. A quasi-uniformity on a set X is a filter $\mathcal{U}$ on $X \times X$ such that

$$\text{(QU1)} \qquad \Delta(X) \subset U, \ \forall U \in \mathcal{U};$$
$$\text{(QU1)} \qquad \forall U \in \mathcal{U}, \ \exists V \in \mathcal{U}, \text{ such that } \ V \circ V \subset U \,,$$

where $\Delta(X) = \{(x,x) : x \in X\}$ denotes the diagonal of X and, for $M, N \subset X \times X$,

$$M \circ N = \{(x,z) \in X \times X : \exists y \in X, \ (x,y) \in M \text{ and } (y,z) \in N\} \,.$$

The composition $V \circ V$ is denoted sometimes simply by V^2. Since every entourage contains the diagonal $\Delta(X)$, the inclusion $V^2 \subset U$ implies $V \subset U$.

If the filter $\mathcal{U}$ satisfies also the condition

$$\text{(U3)} \qquad \forall U, \ U \in \mathcal{U} \ \Rightarrow \ U^{-1} \in \mathcal{U} \,,$$

where

$$U^{-1} = \{(y,x) \in X \times X : (x,y) \in U\} \,,$$

then $\mathcal{U}$ is called a *uniformity* on X. The sets in $\mathcal{U}$ are called *entourages*. The pair $(X,\mathcal{U})$ is called a *quasi-uniform space*.

For $U \in \mathcal{U}$, $x \in X$ and $Z \subset X$ put

$$U(x) = \{y \in X : (x,y) \in U\} \quad \text{and} \quad U[Z] = \bigcup\{U(z) : z \in Z\} \,.$$

A quasi-uniformity $\mathcal{U}$ generates a topology $\tau(\mathcal{U})$ on X for which the family of sets

$$\{U(x) : U \in \mathcal{U}\}$$

is a base of neighborhoods of the point $x \in X$. A mapping f between two quasi-uniform spaces $(X,\mathcal{U})$, $(Y,\mathcal{W})$ is called *quasi-uniformly continuous* if for every $W \in \mathcal{W}$ there exists $U \in \mathcal{U}$ such that $(f(x), f(y)) \in W$ for all $(x,y) \in U$. By the definition of the topology generated by a quasi-uniformity, it is clear that a quasi-uniformly continuous mapping is continuous with respect to the topologies $\tau(\mathcal{U})$, $\tau(\mathcal{W})$.

The family of sets

$$\mathcal{U}^{-1} = \{U^{-1} : U \in \mathcal{U}\} \tag{1.1.36}$$

is another quasi-uniformity on X called the *conjugate quasi-uniformity*. With respect to the topologies $\tau(\mathcal{U})$ and $\tau(\mathcal{U}^{-1})$, X is a bitopological space. The equivalences of the separation axioms from Corollary 1.1.10 holds in this case too.

Proposition 1.1.49. *For a quasi-uniform space $(X,\mathcal{U})$ the following are equivalent.*

1. *The bitopological space $(X, \tau(\mathcal{U}), \tau(\mathcal{U}^{-1}))$ is pairwise T_0.*
2. *The bitopological space $(X, \tau(\mathcal{U}), \tau(\mathcal{U}^{-1}))$ is pairwise T_1.*
3. *The bitopological space $(X, \tau(\mathcal{U}), \tau(\mathcal{U}^{-1}))$ is pairwise Hausdorff.*

Proof. Obviously, it is sufficient to prove the implication $1 \Rightarrow 3$. Suppose, for instance, that for $x \neq y$ in X, there exists $U \in \mathcal{U}$ such that $y \notin U(x)$. Taking $V \in \mathcal{U}$ such that $V^2 \subset U$, it follows that $V(x) \cap V^{-1}(y) = \emptyset$. Indeed

$$z \in V(x) \cap V^{-1}(y) \iff (x,z) \in V \wedge (z,y) \in V \Rightarrow (x,y) \in V^2 \subset U \Rightarrow y \in U(x) ,$$

a contradiction. As the other cases (see (1.1.15)) can be treated similarly, it follows that the bitopological space $(X, \tau(\mathcal{U}), \tau(\mathcal{U}^{-1}))$ is pairwise Hausdorff. □

If (X, ρ) is a quasi-semimetric space, then

$$V_\varepsilon = \{(x,y) \in X \times X : \rho(x,y) < \varepsilon\},\ \varepsilon > 0 ,$$

is a basis for a quasi-uniformity $\mathcal{U}_\rho$ on X. The family

$$V_\varepsilon^- = \{(x,y) \in X \times X : \rho(x,y) \leq \varepsilon\},\ \varepsilon > 0 ,$$

generates the same quasi-uniformity. Since $V_\varepsilon(x) = B_\rho(x,\varepsilon)$ and $V_\varepsilon^-(x) = B_\rho[x,\varepsilon]$, it follows that the topologies generated by the quasi-semimetric ρ and by the quasi-uniformity $\mathcal{U}_\rho$ agree, i.e., $\tau_\rho = \tau(\mathcal{U}_\rho)$.

If P is a family of quasi-semimetrics, then the family of sets $V_{d,\varepsilon},\ d \in P,\ \varepsilon > 0$, generates a quasi-uniformity on X, the sets

$$V_{d,\varepsilon} = \{(x,y) \in X \times X : d(x,y) < \varepsilon\}, \quad \varepsilon > 0,\ d \in P ,$$

being a base for this quasi-uniformity.

The following proposition is an adaptation of [110, Ch. 6, Thm. 11].

Proposition 1.1.50. *Let $(X,\mathcal{U})$ be a quasi-uniform space and d a quasi-semimetric on X. Then the function d is quasi-uniformly continuous on X if and only if $V_{d,\varepsilon} \in \mathcal{U}$ for every $\varepsilon > 0$.*

Proof. The space $X \times X$ is considered equipped with the quasi-uniformity generated by the base $\{U \times U : U \in \mathcal{U}\}$. Consequently, due to the symmetry in the definition of product quasi-uniformity, the function $d : (X \times X, \mathcal{U} \times \mathcal{U}) \to (\mathbb{R}, \mathcal{U}_u)$ is quasi-uniformly continuous if and only if for every $\varepsilon > 0$ there exists $U \in \mathcal{U}$ such that for every $(x,y),(u,v) \in U$, $|d(x,y) - d(u,v)| < \varepsilon$.

Now, if d is supposed quasi-uniformly continuous, then for every $\varepsilon > 0$ there exists $U \in \mathcal{U}$ such that the above condition is satisfied. Taking $(u,v) = (y,y) \in U$ it follows that $(x,y) \in V_{d,\varepsilon}$ for every $(x,y) \in U$, that is $U \subset V_{d,\varepsilon}$, and so $V_{d,\varepsilon} \in \mathcal{U}$.

Conversely, suppose that $V_{d,\varepsilon} \in \mathcal{U}$ for every $\varepsilon > 0$ and prove that the quasi-semimetric d is quasi-uniformly continuous on X. Given $\varepsilon > 0$, $(x,y),(u,v) \in V_{d,\varepsilon}$ imply $d(x,y) < \varepsilon$ and $d(u,v) < \varepsilon$, so that $d(x,y) - d(u,v) < \varepsilon$ and $d(u,v) - d(x,y) < \varepsilon$, that is $|d(x,y) - d(u,v)| < \varepsilon$, proving the quasi-uniformity of the function d on $X \times X$. □

To prove that every quasi-uniformity is generated by a family of quasi-seminorms in the way described above, we need the following result in general topology.

Proposition 1.1.51 (Kelley [110]). *Let X be a nonempty set and $\{U_n : n \in \mathbb{N}\}$ a family of nonempty subsets of $X \times X$ such that $U_0 = X \times X$, each U_n contains the diagonal $\Delta(X)$ and*

$$U_{n+1} \circ U_{n+1} \circ U_{n+1} \subset U_n \,, \tag{1.1.37}$$

for all $n \in \mathbb{N} \cup \{0\}$. Then there exists a quasi-semimetric d on X such that

$$U_{n+1} \subset \{(x,y) \in X \times X : d(x,y) < 2^{-n}\} \subset U_n \,, \; n \in \mathbb{N} \cup \{0\} \,. \tag{1.1.38}$$

If, in addition, each U_n is symmetric, then there exists a semimetric d satisfying (1.1.38).

Proof. For the convenience of the reader, we include a proof of this result. Observe that the fact that each U_n contains the diagonal and (1.1.37) implies that $U_{n+1} \subset U_n$ for all $n \in \mathbb{N} \cup \{0\}$.

Define a function $f : X \times X \to [0;\infty)$ by

$$\begin{aligned} &f(x,y) = 2^{-n} \;\text{ if }\; (x,y) \in U_n \setminus U_{n+1}, \text{ for some } n \in \mathbb{N} \cup \{0\}, \text{ and} \\ &f(x,y) = 0 \;\text{ if }\; (x,y) \in \cap_n U_n \,. \end{aligned} \tag{1.1.39}$$

Observe that there exists at most one $n \in \mathbb{N} \cup \{0\}$ such that $(x,y) \in U_n \setminus U_{n+1}$, so that the function f is well defined.

From the definition of the function f it is clear that

$$f(x,y) \le 2^{-n} \iff (x,y) \in U_n \,. \tag{1.1.40}$$

A sequence $x_0, x_1, \ldots, x_{n+1}$ of points in X with $x_0 = x$ and $x_{n+1} = y$ is called a chain connecting x and y and n is called the length of the chain. Define $d : X \times X \to [0;\infty)$ by

$$d(x,y) = \inf \sum_{i=0}^{n} f(x_i, x_{i+1}) \,, \tag{1.1.41}$$

where the infimum is taken over all chains connecting x and y. From the definition it is clear that $d(x,y) \le f(x,y)$, so that, taking into account (1.1.40),

$$U_{n+1} \subset \{(x,y) \in X \times X : d(x,y) < 2^{-n}\} \,.$$

Indeed, $(x,y) \in U_{n+1}$ implies $f(x,y) \le 2^{-n-1}$, so that $d(x,y) \le 2^{-n-1} < 2^{-n}$.

Let us prove that

$$f(x_0, x_{n+1}) \le 2 \sum_{i=0}^{n} f(x_i, x_{i+1}) \,, \tag{1.1.42}$$

for all chains $x_0, x_1, \ldots, x_{n+1}$ in X.

The proof will be done by induction over the length n of the chain. For $n = 0$ the inequality (1.1.42) is trivial. Suppose that it holds for every m, $0 \le m < n$, and prove it for n. Let $a = \sum_{i=0}^{n} f(x_i, x_{i+1})$. The case $a = 0$ is trivial, as well as the case when $f(x_i, x_{i+1}) = a$ for some $i \in \{0, \ldots, n\}$.

Excluding these situations, let k be the greatest number between 0 and n such that $\sum_{i=0}^{k} f(x_i, x_{i+1}) \le a/2$.

If $0 \le k < n - 1$, then $\sum_{i=k+2}^{n} f(x_i, x_{i+1}) \le a/2$ so that, by the induction hypothesis,

$$f(x_0, x_{k+1}) \le a, \quad f(x_{k+2}, x_{n+1}) \le a \, ,$$

and, obviously, $f(x_{k+1}, x_{k+2}) \le a$.

Let $m \ge 0$ be the smallest integer such that $2^{-m} \le a$. Then

$$f(x_0, x_{k+1}) \le 2^{-m}, \quad f(x_{k+1}, x_{k+2}) \le 2^{-m} \quad \text{and} \quad f(x_{k+2}, x_{n+1}) \le 2^{-m} \, ,$$

so that, by (1.1.40), $(x_0, x_{k+1}), (x_{k+1}, x_{k+2}), (x_{k+2}, x_{n+1}) \in U_m$, implying

$$(x_0, x_{n+1}) \in U_m \circ U_m \circ U_m \subset U_{m-1},$$

which, by the same equivalence, yields

$$f(x_0, x_{n+1}) \le 2^{-m+1} = 2 \cdot 2^{-m} \le 2a \, .$$

If $k = n - 1$, then, reasoning as above,

$$f(x_0, x_n) \le 2^{-m} \qquad \text{and} \quad f(x_n, x_{n+1}) \le 2^{-m},$$

implying $(x_0, x_n), (x_n, x_{n+1}) \in U_m$, so that

$$(x_0, x_{n+1}) \in U_m \circ U_m \subset U_m \circ U_m \circ U_m \subset U_{m-1}.$$

(The inclusion $U_m \circ U_m \subset U_m \circ U_m \circ U_m$ follows from the fact that $\Delta(X) \subset U_m$.)

Now, (1.1.42) and the definition (1.1.41) of d imply $f(x, y) \le 2d(x, y)$.

Consequently, if $d(x, y) < 2^{-n}$, then

$$f(x, y) < 2^{-n+1} \iff f(x, y) \le 2^{-n} \iff (x, y) \in U_n \, ,$$

so that the second inclusion in (1.1.38) holds too.

If, in addition, each U_n is symmetric, then, obviously, the function f and also d, are symmetric. □

Theorem 1.1.52. *Any quasi-uniform space is generated by a family of quasi-uniformly continuous quasi-semimetrics in the way described above.*

Proof. It is clear that a family P of quasi-semimetrics generates a quasi-uniformity $\mathcal{U}_P$ and that each quasi-semimetric in P is quasi-uniformly continuous with respect to $\mathcal{U}_P$.

Given a quasi-uniform space $(X,\mathcal{U})$, let P denote the family of all quasi-uniformly continuous quasi-semimetrics on X. By Proposition 1.1.50, $V_{d,\varepsilon} \in \mathcal{U}$ for every $d \in P$ and every $\varepsilon > 0$, implying $\mathcal{U}_P \subset \mathcal{U}$.

Conversely, for $U \in \mathcal{U}$, choose a family $U_n \in \mathcal{U},\ n \in \mathbb{N},$ of entourages such that $U_1 = U$ and $U_{n+1} \circ U_{n+1} \circ U_{n+1} \subset U_n$ for all $n \in \mathbb{N}$. By Proposition 1.1.51 there exists a quasi-semimetric d satisfying (1.1.38). The first inclusions in (1.1.38) show that d is quasi-uniformly continuous, while $\{(x,y) \in X \times X : d(x,y) < 2^{-2}\} \subset U_1 = U$ shows that U belongs to $\mathcal{U}_P$, and so $\mathcal{U} \subset \mathcal{U}_P$. □

The following quasi-metrizability criterium for a quasi-uniform space is the analog of a well-known one for uniform spaces.

Corollary 1.1.53. *A quasi-uniform space $(X,\mathcal{U})$ is quasi-semimetrizable if and only if the quasi-uniformity $\mathcal{U}$ has a countable basis.*

Proof. The necessity is clear. Suppose now that the quasi-uniformity $\mathcal{U}$ has a countable basis $B_k,\ k \in \mathbb{N}$. Apply Proposition 1.1.51 for $U_1^k = B_k$ to find a family d_k of quasi-uniformly continuous semimetrics $d_k,\ k \in \mathbb{N}$, satisfying (1.1.38). It follows that the family $\{d_k : k \in \mathbb{N}\}$ generates the quasi-uniformity $\mathcal{U}$. It is a standard procedure to check that $d = \sum_k 2^{-k} d_k/(1+d_k)$ is a quasi-semimetric on X which generates $\mathcal{U}$. □

Remark 1.1.54. For other quasi-metrizability and metrizability results for quasi-uniform spaces see Künzi [137] and the references given therein.

As it was shown by Pervin [175], every topological space is quasi-uniformizable.

Theorem 1.1.55 ([175]). *Let (X,τ) be a topological space. Then the family of subsets*

$$\mathcal{B} = \{(G \times C) \cup ((X \setminus G) \times X) : G \in \tau\} \tag{1.1.43}$$

is a subbase of a quasi-uniformity on X that generates the topology X.

As Pervin remarked in [175] the quasi-uniformity generating the topology is not unique.

Example 1.1.56. Let $\mathbb{R}$ with the natural quasi-uniformity $\mathcal{U}_d$ be generated by the quasi-metric $d(x,y) = \max\{y - x, 0\}$. Then the quasi-uniformity $\mathcal{U}$ generated by the subbase (1.1.43) is not even comparable with the quasi-uniformity $\mathcal{U}_d$, although both generate the same topology $\tau_d = \tau_u$ on $\mathbb{R}$ (see Example 1.1.3).

An account of the theory of quasi-uniform and quasi-metric spaces up to 1982 is given in the book by Fletcher and Lindgren [80]. The survey papers by Künzi [132, 133, 134, 135, 136, 137] are good guides for subsequent developments. Another book on quasi-uniform spaces is [153]. The properties of bitopologies and quasi-uniformities on function spaces were studied in [197].

A function $f : (X,\mathcal{U}) \to (Y,\mathcal{V})$ between two quasi-uniform spaces is called *quasi-uniformly continuous* on a subset A of X if

$$\forall V \in \mathcal{V},\ \ \exists U \in \mathcal{U},\ \ \forall (x,y),\ \ (x,y) \in (A\times A)\cap U \ \Rightarrow (f(x),f(y)) \in V\ . \quad (1.1.44)$$

A *quasi-uniform isomorphism* is a bijective quasi-uniformly continuous function f such that the inverse function f^{-1} is also quasi-uniformly continuous.

As it is well known, if $(X,\mathcal{U})$, $(Y,\mathcal{V})$ are uniform spaces, then any continuous functions $f : X \to Y$ is uniformly continuous on every compact subset of X. This result is no longer true in the case of quasi-uniform spaces. Lambrinos [141] gives an example of a continuous function from a compact quasi-uniform space $(X,\mathcal{U})$ to a a quasi-uniform space $(X,\mathcal{V})$ that is not quasi-uniformly continuous and gives a proper formulation to this result.

Theorem 1.1.57. *A continuous function from a compact quasi-uniform space to a uniform space is uniformly continuous.*

This result corrects an example given without proof in [153, p. 5] of a continuous function from a compact quasi-uniform space to a uniform space which is not quasi-uniformly continuous.

In fact Lambrinos proves a slightly more general result concerning continuous functions on topologically bounded sets, a notion considered by him in the paper [140]. A subset A of a topological space (X,τ) is called *τ-bounded* (or *topologically bounded*) if any family of open sets covering X contains a finite subfamily covering A. Obviously, a compact set is topologically bounded and the space X is topologically bounded if and only if it is compact. Lambrinos [140] gives an example of a topologically bounded set that is not relatively compact.

Theorem 1.1.57 will be a consequence of the following more general result.

Theorem 1.1.58. *A continuous function from a quasi-uniform space $(X,\mathcal{U})$ to a uniform space $(X,\mathcal{V})$ is uniformly continuous on every $\tau(\mathcal{U})$-bounded subset of X.*

Proof. Suppose that A is a $\tau(\mathcal{U})$-bounded subset of X. For an arbitrary $V \in \mathcal{V}$ let W be a symmetric entourage in $\mathcal{V}$ such that $W^2 \subset V$.

By the continuity of f, for every $x \in X$ there exists $U_x \in \mathcal{U}$ such that $f(U_x(x)) \subset W(f(x))$. Let $S_x \in \mathcal{U}$ such that $S_x^2 \subset U_x$, $x \in X$. Since the family $\{\tau(\mathcal{U})\text{-}\operatorname{int} S_x(x) : x \in X\}$ is an open cover of X, there exists $x_1, \dots, x_n \in X$ such that $A \subset \cup_{k=1}^n S_{x_k}(x_k)$.

Put $U := \cap_{k=1}^n S_{x_k}$ and show that the condition (1.1.44) is satisfied by U and V.

For $(x,y) \in U \cap (A\times A)$ there exists $j \in \{1,\dots,n\}$ such that $x \in S_{x_j}(x_j) \iff (x_j,x) \in S_{x_j}$. Since $S_{x_j}^2 \subset U_{x_j}$, it follows that $S_{x_j} \subset U_{x_j}$ and $x \in U_{x_j}(x_j)$.

Since $(x,y) \in U = \cap_{k=1}^n S_{x_k}$, it follows that $(x,y) \in S_{x_j}$, so that $(x_j,y) \in$

$S_{x_j}^2 \subset U_{x_j}$. Consequently, $x, y \in U_{x_j}(x_j)$, implying

$$\begin{aligned} f(x) \in W(f(x_j)) &\iff (f(x_j, f(x)) \in W \\ &\iff (f(x), f(x_j)) \in W \quad (W \text{ is symmetric}), \\ f(y) \in W(f(x_j)) &\iff (f(x_j), f(y)) \in W \ , \end{aligned}$$

so that $(f(x), f(y)) \in W^2 \subset U$. □

1.1.7 Asymmetric locally convex spaces

In this subsection we shall present some properties of asymmetric locally convex spaces. Note that they were considered as early as 1997 in the paper [5]. In our presentation we shall follow the paper [41]. A more general approach, based on the theory of ordered locally convex cones, is considered in [109] and [212].

Let P be a family of asymmetric seminorms on a real vector space X. Denote by $\mathcal{F}(P)$ the family of all nonempty finite subsets of P and, for $F \in \mathcal{F}(P)$, $x \in X$, and $r > 0$ let

$$B_F[x, r] = \{y \in X : p(y - x) \le r, \ p \in F\} = \cap\{B_p[x, r] : p \in F\}$$

and

$$B_F(x, r) = \{y \in X : p(y - x) < r, \ p \in F\} = \cap\{B_p(x, r) : p \in F\}$$

denote the closed, respectively open *multiball* of center x and radius r. It is immediate that these multiballs are convex absorbing subsets of X.

The *asymmetric locally convex* topology associated to the family P of asymmetric seminorms on a real vector space X is the topology τ_P having as basis of neighborhoods of any point $x \in X$ the family $\mathcal{N}(x) = \{B_F(x, r) : r > 0, \ F \in \mathcal{F}(P)\}$ of open multiballs. The family $\{B_F[x, r] : r > 0, \ F \in \mathcal{F}(P)\}$ of closed multiballs is also a neighborhood basis at x for τ_P.

We shall abbreviate locally convex space as LCS.

It is easy to check that for every $x \in X$ the family $\mathcal{N}(x)$ fulfills the requirements of a neighborhood basis at x so that it defines a topology τ_P on X (or $\tau(P)$).

Obviously, for $P = \{p\}$ we obtain the topology τ_p of an asymmetric seminormed space (X,p) considered above, i.e., $\tau_{\{p\}} = \tau_p$.

The sets

$$U_{F,\varepsilon} = \{(x, y) \in X \times X : p(y - x) < \varepsilon, \ p \in F\} \ , \quad F \in \mathcal{F}(P), \ \varepsilon > 0 \ , \qquad (1.1.45)$$

form a subbasis of a quasi-uniformity $\mathcal{U}_P$ on X generating the topology τ_P.

We say that the family P is *directed* if for any $p_1, p_2 \in P$ there exists $p \in P$ such that $p \ge p_i$, $i = 1, 2$, where $p \ge q$ stands for the pointwise ordering: $p(x) \ge q(x)$ for all $x \in X$. If the family P is directed then for any τ_P-neighborhood of a point $x \in X$ there exist $p \in P$ and $r > 0$ such that $B_p(x,r) \subset V$ (respectively $B_p[x,r] \subset V$). Indeed, if $B_F(x,r) \subset V$ then there exists $p \in P$ such that $p \geq q$ for all $q \in F$ so that $B_p(x,r) \subset B_F(x,r) \subset V$. Also $W \in \mathcal{U}_P$ if and only if there exist $p \in P$ and $\varepsilon > 0$ such that $U_{p,\varepsilon} \subset W$, that is the family of sets (1.1.45) is a basis for the quasi-uniformity $\mathcal{U}_P$.

For $F \in \mathcal{F}(P)$ let

$$p_F(x) = \max\{p(x) : p \in F\}, \ x \in X. \tag{1.1.46}$$

Then p_F is an asymmetric seminorm on X and

$$B_F[x,r] = B_{p_F}[x,r] \quad \text{and} \quad B_F(x,r) = B_{p_F}(x,r). \tag{1.1.47}$$

The family

$$P_d = \{p_F : F \in \mathcal{F}(P)\}, \tag{1.1.48}$$

where p_F is defined by (1.1.46), is a directed family of asymmetric seminorms generating the same topology as P, i.e., $\tau_{P_d} = \tau_P$. Therefore, without restricting the generality, we can always suppose that the family P of asymmetric seminorms is directed.

Because $B_F[x,r] = x + B_F[0,r]$ and $B_F(x,r) = x + B_F(0,r)$, the topology τ_P is translation invariant

$$\mathcal{V}(x) = \{x + V : V \in \mathcal{V}(0)\} \ ,$$

where by $\mathcal{V}(x)$ we have denoted the family of all neighborhoods with respect to τ_P of a point $x \in X$.

The addition $+ : X \times X \to X$ is continuous. Indeed, for $x, y \in X$ and the neighborhood $B_F(x+y,r)$ of $x+y$ we have $B_F(x,r/2)+B_F(y,r/2) \subset B_F(x+y,r)$.

As we have seen in Proposition 1.1.42, the multiplication by scalars need not be continuous, even in asymmetric seminormed spaces.

For an asymmetric seminorm p on a vector space put

$$\bar{p}(x) = p(-x), \quad \text{and} \quad p^s(x) = \max\{p(x), \bar{p}(x)\} \ ,$$

for all $x \in X$.

If P is a family of asymmetric seminorms on X, then

$$\bar{P} = \{\bar{p} : p \in P\} \quad \text{and} \quad P^s = \{p^s : p \in P\} \ .$$

The following proposition contains some simple properties of asymmetric LCS.

Proposition 1.1.59. *Let P be a directed family of asymmetric seminorms on a real vector space X. Then*

1. *For every $p \in P$, $x \in X$ and $r > 0$,*

$$B_{\bar{p}}(0,r) = -B_p(0,r), \qquad B_{\bar{p}}[0,r] = -B_p[0,r],$$
$$B_{p^s}(x,r) = B_p(x,r) \cap B_{\bar{p}}(x,r) \quad \text{and} \quad B_{p^s}[x,r] = B_p[x,r] \cap B_{\bar{p}}[x,r] \, .$$

2. *Any $\tau(P)$-open set is $\tau(P^s)$-open and any $\tau(\bar{P})$-open set is $\tau(P^s)$-open, that is $\tau(P) \subset \tau(P^s)$ and $\tau(\bar{P}) \subset \tau(P^s)$. The same inclusions hold for the corresponding closed sets.*
3. *Any $\tau(P)$-continuous (or $\tau(\bar{P})$-continuous) mapping f from X to a topological space T is $\tau(P^s)$-continuous.*
4. *A ball $B_p(x,r)$ is $\tau(P)$-open. A ball $B_p[x,r]$ is $\tau(\bar{P})$-closed and it need not be $\tau(P)$-closed.*

Proof. The assertions 1 and 2 are immediate and 3 is a consequence of 2, so we only need to prove 4.

For $y \in B_p(x,r)$ let $r' := r - p(y-x) > 0$. Since $p(z-y) < r'$ implies $p(z-x) \leq p(z-y)+p(y-x) < r'+p(y-x) = r$ it follows that $B_p(y,r') \subset B_p(x,r)$.

To prove that $B_p[x,r]$ is $\tau(\bar{P})$-closed let $y \in X \setminus B_p[x,r]$. Then $r' := p(y-x) - r > 0$ and $B_{\bar{p}}(y,r') \subset X \setminus B_p[x,r]$. Indeed, if there exists an element $z \in B_p[x,r] \cap B_{\bar{p}}(y,r')$, then one obtains the contradiction

$$p(y-x) \leq p(y-z) + p(z-x) = \bar{p}(y-z) + p(z-x) < r' + r = p(y-x) \, .$$

Consequently, $X \setminus B_p[x,r]$ is $\tau(\bar{P})$-open and so $B_p[x,r]$ is $\tau(\bar{P})$-closed. □

Example 1.1.60. In $\mathbb{R}$ with the upper topology τ_u, where $u(x) = \max\{x,0\}$, $x \in \mathbb{R}$, we have $B_u[0,1] = (-\infty;1]$ and $\mathbb{R} \setminus B_u[0,1] = (1;+\infty)$ is $\tau_{\bar{u}}$-open, but not τ_u-open.

The following property is easy to check.

Proposition 1.1.61. *Let (X,P) be an asymmetric LCS. A net $(x_i : i \in I)$ is τ_P-convergent to x if and only if*

$$\forall p \in P, \qquad \lim_i p(x_i - x) = 0 \, . \tag{1.1.49}$$

The topology τ_p generated by an asymmetric norm is not always Hausdorff. A necessary and sufficient condition in order that τ_p be Hausdorff is given in Proposition 1.1.40.

The following characterization of the Hausdorff separation property for locally convex spaces is well known, see, e.g., [238, Lemma VIII.1.4].

Proposition 1.1.62. *Let (X,Q) be a locally convex space, where Q is a family of seminorms generating the topology τ_Q of X.*

The topology τ_Q is Hausdorff separated if and only if for every $x \in X$, $x \neq 0$, there exists $q \in Q$ such that $q(x) > 0$.

In the case of asymmetric locally convex spaces we have the following separation properties.

Proposition 1.1.63. *Let P be a family of asymmetric seminorms on a real vector space X.*

1. *The topology τ_P is T_0 if and only if for every $x \neq 0$ in X there exists $p \in P$ such that $p(x) > 0$ or $p(-x) > 0$.*
2. *The topology τ_P is T_1 if and only if for every $x \neq 0$ in X there exists $p \in P$ such that $p(x) > 0$.*
3. *The topology τ_P is T_2 if and only if for every $x \neq 0$ in X there exists $p \in P$ such that $p^\diamond(x) > 0$, where $p^\diamond$ is given by* (1.1.31).

Proof. 1. Let $x \neq y$ in X. Then $x - y \neq 0$ and $y - x \neq 0$, so that, by hypothesis, there exists $p \in P$ such that $p(x-y) > 0$ or $p(y-x) > 0$. If, say $r := p(y-x) > 0$, then $y \notin B_p(x, r)$. Similarly, $r' := p(x-y) > 0$ implies $x \notin B_p(y, r')$.

Conversely, if τ_P is T_0, then for $x \neq 0$ there exists $U \in \mathcal{V}(0)$ such that $x \notin U$, or there exists $U' \in \mathcal{V}(x)$ such that $0 \notin U'$. Let $F \in \mathcal{F}(P)$ and $r > 0$ such that $B_F(o, r) \subset U$. Since $x \notin U$, there exists $p \in F$ such that $p(x) \geq r > 0$. In the second case, let $F' \in \mathcal{F}(P)$ and $r' > 0$ such that $B_{F'}(x, r) \subset U'$. Since $x \notin U'$, there exists $q \in F'$ such that $q(-x) = q(0 - x) \geq r' > 0$.

2. If $x \neq y$, then there exists $p_1, p_2 \in P$ such that $r_1 := p_1(y - x) > 0$ and $r_2 := p_2(x - y) > 0$, implying $y \notin B_{p_1}(x, r_1)$ and $x \notin B_{p_2}(x, r_2)$, that is the topology τ_P is T_1.

Conversely, if τ_P is T_1, then for every $x \neq 0$ in X there exists $U \in \mathcal{V}(0)$ such that $x \notin U$. If $F \in \mathcal{F}(P)$ and $r > 0$ are such that $B_F(0, r) \subset U$, then $x \notin B_F(0, r)$, so there exists $p \in F$ with $p(x) = p(x - 0) \geq r > 0$.

3. Suppose that P is directed and let

$$P^\diamond = \{p^\diamond : p \in P\},$$

where for $p \in P$, $p^\diamond$ is defined by (1.1.31).

By Proposition 1.1.39, $p^\diamond$ is a seminorm on X. Denote by $\tau_{P^\diamond}$ the locally convex topology on X generated by the family $P^\diamond$ of seminorms. The topology τ_P is finer than $\tau_{P^\diamond}$. Indeed, $G \in \tau_{P^\diamond}$ is equivalent to the fact that for every $x \in G$ there exist $p \in P$ and $r > 0$ such that $B_{p^\diamond}(x, r) \subset G$. Because, by Proposition 1.1.39, $p^\diamond(y - x) \leq p(y - x) < r$ it follows that $B_p(x, r) \subset B_{p^\diamond}(x, r) \subset G$, so that $G \in \tau_P$. If for every $x \in X$, $x \neq 0$, there exists $p \in P$ such that $p^\diamond(x) > 0$, then, by Proposition 1.1.62, the locally convex topology $\tau_{P^\diamond}$ is separated Hausdorff, and so will be the finer topology τ_P.

Conversely, suppose that the topology τ_P is Hausdorff and show that $p^\diamond(x) = 0$ for all $p \in P$ implies $x = 0$.

Let $x \in P$ be such that $p^\diamond(x) = 0$ for all $p \in P$. By the definition (1.1.31) of the seminorm $p^\diamond$, for every $p \in P$ and $n \in \mathbb{N}$ there exists an element $x_{(p,n)} \in X$

such that

$$p(x_{(p,n)}) + p(x_{(p,n)} - x) < \frac{1}{n}. \tag{1.1.50}$$

Define the order on $P \times \mathbb{N}$ by $(p,n) \leq (q,m)$ if and only if $p \leq q$ and $n \leq m$. Since the family P of asymmetric seminorms is directed, the set $P \times \mathbb{N}$ is also directed with respect to the order we just defined. Therefore, $\{x_{(p,n)} : (p,n) \in P \times \mathbb{N}\}$ is a net in X and by (1.1.50) we have

$$p(x_{(p,n)}) < \frac{1}{n} \quad \text{and} \quad p(x_{(p,n)} - x) < \frac{1}{n}, \tag{1.1.51}$$

for all $(p,n) \in P \times \mathbb{N}$.

We shall prove that the net $\{x_{(p,n)}\}$ converges to both 0 and x. Since the topology τ_P is Hausdorff this will imply $x = 0$.

To prove that the net $\{x_{(p,n)}\}$ converges to 0 we have to show that for every $p \in P$ the net $\{p(x_{(p,n)})\}$ tends to 0, i.e.,

$$\begin{gathered} \forall p \in P, \forall \varepsilon > 0, \ \exists (p_0, n_0) \in P \times \mathbb{N}, \ \forall (q,n) \in P \times \mathbb{N}, \\ (q,n) \geq (p_0, n_0) \ \Rightarrow \ p(x_{(q,n)}) < \varepsilon \,. \end{gathered}$$

Let $p \in P$ and $\varepsilon > 0$. Put $p_0 = p$ and let $n_0 \in \mathbb{N}$ be such that $1/n_0 < \varepsilon$. Then for every $(q,n) \in P \times \mathbb{N}$ such that $q \geq p$ and $n \geq n_0$ we have

$$p(x_{(q,n)}) \leq q(x_{(q,n)}) < \frac{1}{n} \leq \frac{1}{n_0} < \varepsilon \,.$$

The convergence of $\{p(x_{(p,n)} - x)\}$ to 0, which is equivalent to the τ_P-convergence of $\{x_{(p,n)}\}$ to x, can be proved similarly, using the second inequality in (1.1.51). □

The following proposition emphasizes some continuity properties of the algebraic operations in an asymmetric LCS.

Proposition 1.1.64. *Let (X, P) be a symmetric LCS, where P is a directed family of asymmetric seminorms generating the topology of X.*

1. *The addition $+ : X \times X \to X$ is continuous, so that for every $a \in X$ the mapping $\psi_a : X \to X$ given by $\psi_a(x) = x + a$, $x \in X$, is a homeomorphism of X.*
2. *For each fixed x the mapping $f : \mathbb{R} \to X$, $f(t) = tx$, is continuous.*
3. *The mapping $g : (0;\infty) \times X \to X$, $g(t,x) = tx$, is continuous.*

Proof. 1. The continuity of $+$ follows from the fact that the topology generated by P is translation invariant. Since both the mapping ψ_a and its inverse ψ_{-a} are continuous, it follows that ψ_a is a homeomorphism.

2. Let $x \in X$, $\alpha \in \mathbb{R}$, $p \in P$ and $\varepsilon > 0$. Then for every $|t-\alpha| < \delta$,

$$p(tx - \alpha x) \le p^s(tx - \alpha x) = |t-\alpha| p^s(x) \le \delta p^s(x) < \varepsilon ,$$

provided that $\delta > 0$ is chosen so that $\delta p^s(x) < \varepsilon$.

3. Let $\alpha > 0$, $x \in X$, $p \in P$ and $\varepsilon > 0$ be given. For $0 < \delta < \alpha$ and $r > 0$ let $|t-\alpha| < \delta$ and $p(y-x) < r$. Then

$$p(ty - \alpha x) \le tp(y-x) + |t-\alpha| p^s(x) < (\alpha+\delta)r + \delta p^s(x) .$$

If, in addition, we choose δ, r such that $\delta p^s(x) < \varepsilon/2$ and $r < \varepsilon/2(\alpha+\delta)$, then $p(ty - \alpha x) < \varepsilon$, proving the continuity of g at (α, x). □

Example 1.1.65. The multiplication by scalars need not be continuous from $\mathbb{R} \times X$ to X, as the space $(\mathbb{R}, u)$ from Example 1.1.3 shows.

Indeed, $(-1) \cdot 1 = -1$. If $0 < \varepsilon < 1$, then $W = (-\infty; -1+\varepsilon)$ is a u-neighborhood of $-1 = (-1) \cdot 1$, $(-1,-1) \in U \times V$ for any neighborhood U of -1 and V of 1, but $(-1)(-1) = 1 > -1+\varepsilon$, that is $(-1)(-1) \notin W$.

Remark 1.1.66. As a consequence of the assertion 1 from Proposition 1.1.64, if Y is a closed subset of an asymmetric LCS (X, P), then $Y + Z$ is closed for every finite subset Z of X.

In a symmetric Hausdorff topological vector space X, if $Y \subset X$ is closed, then $Y + Z$ is closed for every compact subset Z of X. I do not know if this assertion is true in an asymmetric LCS.

Indeed, for each $z \in Z$ the mapping $\psi_z(x) = x + z$, $x \in X$, is a homeomorphism, so that $Y + z = \psi_z(Y)$ is closed and so will be the finite union $Y + Z = \cup\{Y + z : z \in Z\}$.

The difficulty when Z is compact arises from the following fact. If $(y_i + z_i : i \in I)$ is a net in $Y + Z$ converging to some $x \in X$, then the net (z_i) admits a subnet $(z_{i_j} : j \in J)$ converging to some $z \in Z$. But this does not imply that the net $y_{i_j} = (y_{i_j} + z_{i_j}) - z_{i_j}$, $j \in J$, converges to $y := x - z \in Y$, as in the symmetric case. We do not know a concrete example illustrating this situation.

Convex sets in asymmetric locally convex spaces have some properties similar to those valid in locally convex spaces.

Proposition 1.1.67. *Let (X, P) be an asymmetric LCS, where P is a directed family of asymmetric seminorms generating the topology of X and $Y \subset X$.*

1. *If $\mathcal{B}(0)$ is a base of 0-neighborhoods, then*

$$\overline{Y} = \cap\{Y - B : B \in \mathcal{B}(0)\} . \tag{1.1.52}$$

2. *If the subset Y is convex, then $\overline{Y}$ is convex too.*

3. *If the subset Y is convex and $\overset{\circ}{Y} \neq \emptyset$, then*
$$(1-\alpha)\overset{\circ}{Y} + \alpha Y \subset \overset{\circ}{Y}\,, \tag{1.1.53}$$
for every $0 < \alpha < 1$. Consequently, if Y is convex with nonempty interior, then the interior of Y is a convex set too.

4. *If the subset Y is convex, absorbing and $\overset{\circ}{Y} \neq \emptyset$, then $0 \in \overset{\circ}{Y}$.*

Proof. 1. If $x \in \overline{Y}$, then for every $B \in \mathcal{B}(0)$, $(x+B)\cap Y \neq \emptyset$. If $u \in B$ and $y \in Y$ are such that $x + u = y$, then $x = y - u \in Y - B$.

Conversely, suppose that x belongs to the intersection from the right-hand side of (1.1.52) and let $V \in \mathcal{V}(x)$. Then there exists $B \in \mathcal{B}(0)$ such that $x+B \subset V$. Since $x \in Y - B$, there exists $y \in Y$ and $u \in B$ such that $x = y - u$ implying, $y = x + u \in (x + B) \cap Y \subset V \cap Y$. It follows that $x \in \overline{Y}$.

2. Take $\mathcal{B}(0)$ to be a base formed of convex neighborhoods of 0 (for instance, p-balls for $p \in P$). Then $Y - B$ is convex for every $B \in \mathcal{B}(0)$ and so will be the intersection (1.1.52).

3. Let $x \in \overset{\circ}{Y}$, $y \in Y$ and $0 < \alpha < 1$. If $U \in \mathcal{V}(0)$ is such that $x + U \subset Y$, then $(1-\alpha)U \in \mathcal{V}(0)$ and, by the convexity of Y,
$$(1-\alpha)x + \alpha y + (1-\alpha)U = (1-\alpha)(x+U) + \alpha y \subset Y\,,$$
showing that $(1-\alpha)x + \alpha y \in \overset{\circ}{Y}$.

4. Let $x_0 \in \overset{\circ}{Y}$. Since Y is absorbing and convex, there exists $\alpha > 0$ such that $-\alpha x_0 = \alpha(-x_0) \in Y$. By 3,
$$0 = \frac{1}{\alpha+1}(-\alpha x_0) + \frac{\alpha}{\alpha+1}x_0 \in \overset{\circ}{Y}\,. \qquad \square$$

It is well known that any finite-dimensional Hausdorff topological vector space X is topologically isomorphic to the Euclidean $\mathbb{R}^m$, where $m = \dim X$. García-Raffi [86] proved that the result still holds for finite-dimensional T_1 asymmetric normed spaces, a result that was subsequently extended to T_1 asymmetric LCS in [44]. The isomorphism result is the following.

Proposition 1.1.68. *Let (X,P) be an asymmetric LCS whose topology $\tau(P)$ is T_1. If X is finite dimensional with* $\dim X = m$, *then it is topologically isomorphic to the Euclidean space $\mathbb{R}^m$.*

As we shall work with nets we shall present some properties related to nets in LCS. A net is an application of a directed set $(I,\leq)$ to a set X, $\varphi : I \to X$,, also denoted by $(x_i : i \in I)$, where $x_i = \varphi(i)$, $i \in I$. A subset J of the directed set I is called *cofinal* in I provided that for every $i \in I$ there exists $j \in J$ with $i \leq j$. One says that a net $\psi : J \to X$ is a *subnet* of the net $\varphi : I \to X$ if there exists a monotone mapping $\lambda : J \to I$ (i.e., $j_1 \leq_J j_2$ implies $\lambda(j_1) \leq_I \lambda(j_2)$) such that $\psi = \varphi \circ \lambda$ and $\lambda(J)$ is cofinal in I. A subnet of the net $(x_i : i \in I)$ is denoted also by $(x_{\lambda(j)} : j \in J)$. If J is a cofinal subset of I, then $(x_j : j \in J)$ is a subnet of $(x_i : i \in I)$. It is clear that, if J_1 is a cofinal subset of the directed set I and J_2 is a cofinal subset of J_1, then J_2 is also a cofinal subset of I.

The following result is probably well known, but for the sake of reader convenience we include a proof.

Lemma 1.1.69. *Let $(I, \leq)$ be a directed set. If $I = J_1 \cup \cdots \cup J_m$, where J_k are nonempty subsets of I, $k = 1, \ldots, m$, then at least one of the sets J_k is cofinal in I.*

Proof. If J_1 is a cofinal subsets of I, then we are done. If J_1 is not cofinal in I, then there exists $i_1 \in I$ such that there is no $j \in J_1$ with $i_1 \leq j$. Putting

$$I_1 = \{i \in I : i \neq i_1 \text{ and } i_1 \leq i\} ,$$

it follows that $I_1 \subset J_2 \cup \cdots \cup J_m$. We pose two distinct cases.

I. $I_1 = \emptyset$. In this case i_1 is the greatest element of I. Indeed for $i \in I$ there exists $i_2 \in I$ such that $i_1 \leq i_2$ and $i \leq i_2$. By hypothesis $i_2 = i_1$, so that $i \leq i_1$. If $k \in \{1, \ldots, m\}$ is such that $i_1 \in J_k$, then J_k is a cofinal subset of I.

II. $I_1 \neq \emptyset$. In this case the set I_1 is cofinal in I. Indeed, since I_1 is nonempty there exists an element $j_1 \in I_1$. If $i \in I$ is arbitrary, then there exists $j \in I$ such that $i \leq j$ and $j_1 \leq j$, implying $j \in I_1$ and $i \leq j$. Since I_1 is contained in $J_2 \cup \cdots \cup J_m$, it follows that $J_2 \cup \cdots \cup J_m$ is also a cofinal subset of I.

Repeating the argument with $J_2 \cup \cdots \cup J_m$ instead of I and continuing in this manner, we get that some J_k is a cofinal subset of I. □

Proposition 1.1.68 will be an immediate consequence of the following lemma.

Lemma 1.1.70. *Let (X, P) be an asymmetric LCS of finite dimension $m \geq 1$ with basis $e_1, \ldots, e_m$ and let $x_i = \alpha_{1,i}\, e_1 + \cdots + \alpha_{m,i}\, e_m$, $i \in I$, be a net in X.*

1. *If for every $k \in \{1, \ldots, m\}$ the net $(\alpha_{k,i})$ converges in $\mathbb{R}$ to some $\alpha_k \in \mathbb{R}$, then the net (x_i) converges to $x = \alpha_1 e_1 + \cdots + \alpha_m e_m$ with respect to the topology $\tau(P)$.*
2. *If the topology $\tau(P)$ is T_1 and the net (x_i) converges with respect to $\tau(P)$ to $x = \alpha_1 e_1 + \cdots + \alpha_m e_m$, then the net $(\alpha_{k,i})$ converges in $\mathbb{R}$ to α_k for every $k \in \{1, \ldots, m\}$.*

Proof. 1. For any $p \in P$,

$$\begin{aligned} p(x_i - x) &= p\left(\sum_{k=1}^m (\alpha_{k,i} - \alpha_k) e_k\right) \leq p^s\left(\sum_{k=1}^m (\alpha_{k,i} - \alpha_k) e_k\right) \\ &\leq \sum_{k=1}^m |\alpha_{k,i} - \alpha_k| p^s(e_k) \to 0, \quad \text{for} \quad i \in I \, . \end{aligned}$$

Here $p^s(x) = \max\{p(x), p(-x)\}$ denotes the symmetric norm associated to p.

2. Suppose, by contradiction, that $p(x_i) \to 0$ for every $p \in P$, but at least one of the nets $(\alpha_{k,i})$, say $(\alpha_{1,i})$, does not converge to 0 in $\mathbb{R}$. Then there exists $\varepsilon > 0$ such that for every $i \in I$ there exists $j \in I$, $j \geq i$, such that $|\alpha_{1,j}| \geq \varepsilon$. This implies that the set $J = \{j \in I : |\alpha_{1,j}| \geq \varepsilon\}$ is cofinal in I, and, consequently, $(x_j : j \in J)$ is a subnet of $(x_i : i \in I)$, so it also converges to 0 with respect to $\tau(P)$. It follows also that $M_j := \max\{|\alpha_{1,j}| : 1 \leq k \leq m\} \geq \varepsilon$ for all $j \in J$.

If $y_j := M_j^{-1} x_j$, $j \in J$, then

$$p(y_j) = \frac{1}{M_j} p(x_j) \leq \frac{1}{\varepsilon} p(x_j) \to 0,\ j \in J\,.$$

Writing $y_j = \beta_{1,j}\, e_1 + \cdots + \beta_{m,j}\, e_m$, $j \in J$, it follows that for every $j \in J$, $|\beta_{k,j}| \leq 1$, $k = 1, \ldots, m$, and at least one of the numbers $\beta_{k,j}$ has modulus 1.

If $J_k := \{j \in J : |\beta_{k,j}| = 1\}$, $k = 1, \ldots, m$, then, by Lemma 1.1.69, at least one of the sets J_k, say J_1, is cofinal in J. By the same lemma, one of the sets $J_1^s = \{j \in J_1 : \beta_{1,j} = (-1)^s\}$, $s = 1, 2$, is cofinal in J_1. Denote it by A_1. Since the net $(\beta_{2,j} : j \in A_1)$ is bounded, there exists a subnet $(\beta_{2,\alpha} : \alpha \in A_2)$ of it converging to some $\beta_2 \in \mathbb{R}$. Similarly, the bounded net $(\beta_{3,\alpha} : \alpha \in A_2)$ contains a subnet $(\beta_{3,\alpha} : \alpha \in A_3)$ converging to some $\beta_3 \in \mathbb{R}$. Continuing in this way we obtain the subnets $(\beta_{k,\alpha} : \alpha \in A_m)$ converging to $\beta_k \in \mathbb{R}$ for every $k = 1, \ldots, m$, with $|\beta_1| = 1$. Let $z_\alpha = \beta_{1,\alpha} e_1 + \cdots + \beta_{m,\alpha} e_m$, $\alpha \in A_m$, and $z := \beta_1 e_1 + \cdots + \beta_m e_m \neq 0$. By the first part of the lemma, the net $(-z_\alpha)$ is $\tau(P)$-convergent to $-z$, which is equivalent to $p(-z_\alpha + z) \to 0$ for every $p \in P$. Since $\tau(P)$ is T_1, there exists $p_0 \in P$ such that $p_0(z) > 0$. It follows that

$$0 < p_0(z) \leq p_0(-z_\alpha + z) + p_0(z_\alpha)\,,$$

in contradiction to the fact that $p_0(z_\alpha) \to 0$. □

Remark 1.1.71. There are infinite-dimensional asymmetric normed spaces which are Hausdorff but not normable. An example is given in [91]. On the space X of all sequences of real numbers with finite support,

$$p(x) = \|x^+\|_1 + \|x^+\|_2,\ x \in X,$$

is an asymmetric norm on X, the induced topology τ_p is Hausdorff, but (X, p) is not isomorphic to any normed space, in other words, the topology τ_p is not normable. Here $\|\cdot\|_1$ and $\|\cdot\|_2$ are the ℓ^p-norms for $p = 1, 2$.

1.2 Completeness and compactness in quasi-metric and quasi-uniform spaces

Completeness, total boundedness and compactness look very different in quasi-metric and quasi-uniform spaces with respect to metric or uniform spaces. The lack of symmetry in the definition of quasi-metric and quasi-uniform spaces causes a lot of troubles, mainly concerning completeness, compactness and total boundedness in such spaces. There are several notions of completeness in quasi-metric and quasi-uniform spaces, all agreeing with the usual notion of completeness in the case of metric or uniform spaces, each of them having its advantages and weaknesses. Also, countable compactness, sequential compactness and compactness do not agree in quasi-metric spaces, in contrast to the metric case.

The aim of this section is to present these notions of completeness as well as the relations existing between total boundedness, completeness and compactness in the setting of quasi-metric and of quasi-uniform spaces.

1.2.1 Various notions of completeness for quasi-metric spaces

We shall describe briefly some of these notions of completeness along with some of their properties.

In the case of a quasi-metric space (X, ρ) there are several completeness notions, which we present following [185], starting with the definitions of Cauchy sequences.

A sequence (x_n) in (X, ρ) is called

(a) *left (right) ρ-Cauchy* if for every $\varepsilon > 0$ there exist $x \in X$ and $n_0 \in \mathbb{N}$ such that

$$\forall n \geq n_0, \quad \rho(x, x_n) < \varepsilon$$

(respectively $\rho(x_n, x) < \varepsilon$);

(b) *ρ^s-Cauchy* if it is a Cauchy sequence is the semimetric space (X, ρ^s), that is for every $\varepsilon > 0$ there exists $n_0 \in \mathbb{N}$ such that

$$\forall n, k \geq n_0, \quad \rho^s(x_n, x_k) < \varepsilon,$$

or, equivalently, $\forall n, k \geq n_0, \quad \rho(x_n, x_k) < \varepsilon$;

(c) *left (right) K-Cauchy* if for every $\varepsilon > 0$ there exists $n_0 \in \mathbb{N}$ such that

$$\forall n, k, \quad n_0 \leq k \leq n \Rightarrow \rho(x_k, x_n) < \varepsilon$$

(respectively $\rho(x_n, x_k) < \varepsilon$);

(d) *weakly left (right) K-Cauchy* if for every $\varepsilon > 0$ there exists $n_0 \in \mathbb{N}$ such that

$$\forall n \geq n_0, \quad \rho(x_{n_0}, x_n) < \varepsilon$$

(respectively $\rho(x_n, x_{n_0}) < \varepsilon$).

Sometimes, to emphasize the quasi-metric ρ, we shall say that a sequence is left K-ρ-Cauchy, etc.

It seems that K in the definition of a left K-Cauchy sequence comes from Kelly [111] who was the first to consider this notion (see also [50]).

Some remarks are in order.

Proposition 1.2.1 ([185]). *Let (X, ρ) be a quasi-semimetric space.*

1. *These notions are related in the following way:*

$$\begin{aligned}\rho^s\text{-Cauchy} &\Rightarrow \text{left } K\text{-Cauchy} \\ &\Rightarrow \text{weakly left } K\text{-Cauchy} \Rightarrow \text{left } \rho\text{-Cauchy},\end{aligned}$$

   The same implications hold for the corresponding right notions.
   No one of the above implications is reversible.

2. *A sequence is left Cauchy (in some sense) with respect to ρ if and only if it is right Cauchy (in the same sense) with respect to $\bar{\rho}$.*
3. *A sequence is ρ^s-Cauchy if and only if it is both left and right K-Cauchy.*
4. *A ρ-convergent sequence is left ρ-Cauchy and a $\bar{\rho}$-convergent sequence is right ρ-Cauchy.*
5. *If each convergent sequence in a regular quasi-metric space (X, ρ) admits a left K-Cauchy subsequence, then X is metrizable* ([138]).

A quasi-semimetric space (X, ρ) is called:

- *ρ-sequentially complete* if every ρ^s-Cauchy sequence is τ_ρ-convergent ([9]);
- *left ρ-sequentially complete* if every left ρ-Cauchy sequence is τ_ρ-convergent ([185]);
- *bicomplete* if the associated semimetric space (X, ρ^s) is complete, i.e., every ρ^s-Cauchy sequence is τ_{ρ^s}-convergent;
- *left (right) Smyth sequentially complete* if every left (right) K-Cauchy sequence is ρ^s-convergent.

A bicomplete asymmetric normed space (X, p) is called a *biBanach space*.

To each of the other notions of Cauchy sequence corresponds a notion of sequential completeness, by asking that each corresponding Cauchy sequence be convergent in (X, τ_ρ).

By the assertion 4 from above, each ρ-convergent sequence is left ρ-Cauchy, but for each of the other notions there are examples of ρ-convergent sequences that are not Cauchy, which is a major inconvenience. Another one is that a complete (in some sense) subspace of a quasi-metric space need not be closed. The assertion 5 from Proposition 1.2.1 shows that putting too many conditions on a quasi-metric, or on a quasi-uniform space, in order to obtain results similar to those in the symmetric case, there is the danger of forcing the quasi-uniformity a uniformity. In fact, this is a general problem when dealing with generalizations.

It follows that the implications between these completeness notions are obtained by reversing the implications between the corresponding notions of Cauchy sequence from Proposition 1.2.1.1.

Proposition 1.2.2. *These notions of completeness are related in the following way:*

ρ-sequentially complete $\Rightarrow$ weakly left K-sequentially complete $\Rightarrow$ left K-sequentially complete $\Rightarrow$ left ρ-complete.

The same implications hold for the corresponding notions of right completeness.

In spite of the obvious fact that left ρ-Cauchy is equivalent to right $\bar{\rho}$-Cauchy, left ρ- and right $\bar{\rho}$-completeness do not agree, due to the fact that right $\bar{\rho}$-completeness means that every left ρ-Cauchy sequence converges in $(X, \bar{\rho})$, while left ρ-completeness means the convergence of such sequences in the space (X, ρ). For concrete examples and counterexamples, see [185].

In fact, as remarked Mennucci [151, §3.ii.2], starting from these seven notions of Cauchy sequence, one can obtain (taking into account the symmetry between ρ and $\bar{\rho}$) 14 different notions of completeness, by asking that every sequence which is Cauchy in some sense for ρ converges with respect to one of the topologies $\tau(\rho)$, $\tau(\bar{\rho})$ or $\tau(\rho^s)$. Mennucci [151] works with the notion of left Smyth completeness considered by Smyth [225] (see also [216], [229], [230]) in connection with some questions in theoretical computer science.

We mention the following example from [185].

Example 1.2.3. On the set $X = \mathbb{N}$ define the quasi-metric ρ by $\rho(m,n) = 0$ if $m = n$, $\rho(m,n) = n^{-1}$ if $m > n$, m even, n odd, and $\rho(m,n) = 1$ otherwise. Since there are no non-trivial right K-Cauchy sequences, X is right K-sequentially complete. The quasi-metric space X is not right ρ-sequentially complete because the sequence $\{2, 4, 6, \dots\}$ is right ρ-Cauchy but not convergent. This sequence is left $\bar{\rho}$-Cauchy but not weakly left K-Cauchy in $(X, \bar{\rho})$. Also, the space $(X, \bar{\rho})$ is left K-sequentially complete but not left $\bar{\rho}$-sequentially complete.

The following simple remarks concerning Cauchy sequences in quasi-semimetric spaces are true.

Proposition 1.2.4 ([27, 186]). *Let (x_n) be a left K-Cauchy sequence in a quasi-semimetric space (X, ρ).*

1. *If (x_n) has a subsequence which is τ_ρ-convergent to to x, then (x_n) is τ_ρ-convergent to x.*
2. *If (x_n) has a subsequence which is $\tau_{\bar{\rho}}$-convergent to x, then (x_n) is $\tau_{\bar{\rho}}$-convergent to x.*
3. *If (x_n) has a subsequence which is ρ^s-convergent to x, then (x_n) is ρ^s-convergent to x.*

Proof. 1. Suppose that (x_n) is left K-Cauchy and (x_{n_k}) is a subsequence of (x_n) such that $\lim_k \rho(x, x_{n_k}) = 0$. For $\varepsilon > 0$ choose n_0 such that $n_0 \le m < n$ implies

$\rho(x_m, x_n) < \varepsilon$, and let $k_0 \in \mathbb{N}$ be such that $n_{k_0} \geq n_0$ and $\rho(x, x_{n_k}) < \varepsilon$ for all $k \geq k_0$. Then, for $n \geq n_{k_0}$, $\rho(x, x_n) \leq \rho(x, x_{n_{k_0}}) + \rho(x_{n_{k_0}}, x_n) < 2\varepsilon$.

2. Suppose that (x_n) is left K-Cauchy such that there exists a subsequence (x_{n_k}) which is $\bar{\rho}$-convergent to some $x \in X$. For $\varepsilon > 0$ let $k_0 \in \mathbb{N}$ be such that

$$\forall k \geq k_0, \quad \rho(x_{n_k}, x) < \varepsilon, \tag{1.2.1}$$

and let $n_0 \in \mathbb{N}$ be such that

$$\forall n, m \in \mathbb{N}, \;\; n_0 \leq m < n \;\Rightarrow\; \rho(x_m, x_n) < \varepsilon\,. \tag{1.2.2}$$

For $n \geq n_0$ let $k \geq k_0$ be such that $n_k > n$. Then by (1.2.1) and (1.2.2)

$$\rho(x_n, x) \leq \rho(x_n, x_{n_k}) + \rho(x_{n_k}, x) < 2\varepsilon\,,$$

proving that the sequence (x_n) is $\bar{\rho}$-convergent to x.

3. Suppose there is a subsequence $(x_{n_k})_{k\in\mathbb{N}}$ of (x_n) such that $x_{n_k} \xrightarrow{\rho^s} x$. Then, by Proposition 1.1.8.2, $x_{n_k} \xrightarrow{\rho} x$ and $x_{n_k} \xrightarrow{\bar{\rho}} x$, so that, taking into account the assertions 1 and 2, $x_n \xrightarrow{\rho} x$ and $x_{n_k} \xrightarrow{\bar{\rho}} x$. Appealing again to Proposition 1.1.8.2, it follows that $x_n \xrightarrow{\rho^s} x$. □

Remark 1.2.5. Property 3 from Proposition 1.2.4 was kindly communicated to us by M.D. Mabula.

A series $\sum_n x_n$ in an asymmetric seminormed space (X, p) is called *convergent* if there exists $x \in X$ such that $x = \lim_{n\to\infty} \sum_{k=1}^{n} x_k$. The series $\sum_n x_n$ is called *absolutely convergent* if $\sum_{n=1}^{\infty} p(x_n) < \infty$. It is well known that a normed space is complete if and only if every absolutely convergent series is convergent. A similar result holds in the asymmetric case too.

Proposition 1.2.6.

1. *If a sequence (x_n) in a quasi-semimetric space (X, ρ) satisfies*

$$\textstyle\sum_{n=1}^{\infty} \rho(x_n, x_{n+1}) < \infty,$$

then it is left K-Cauchy.

2. *An asymmetric seminormed space (X, p) is left K-sequentially complete if and only if every absolutely convergent series is convergent.*

Proof. 1. For $\varepsilon > 0$ let $n_0 \in \mathbb{N}$ be such that $\sum_{i=0}^{\infty} \rho(x_{n_0+i}, x_{n_0+i+1}) < \varepsilon$. Then for $n \geq n_0$ and $k \in \mathbb{N}$, $\rho(x_n, x_{n+k}) \leq \sum_{i=0}^{k-1} \rho(x_{n+i}, x_{n+i+1}) < \varepsilon$.

2. Suppose that (X, p) is left K-sequentially complete and let (x_n) be a sequence in X such that $\sum_{n=1}^{\infty} p(x_n) < \infty$ and let $X_n = x_1 + \cdots + x_n$, $n \in \mathbb{N}$. For $\varepsilon > 0$ let $n_0 \in \mathbb{N}$ be such that $\sum_{i=0}^{\infty} p(x_{n_0+i}) < \varepsilon$. Then for $n \geq n_0$ and $k \in \mathbb{N}$, $p(X_{n+k} - X_n) = p(x_{n+1} + \cdots + x_{n+k}) \leq p(x_{n+1}) + \cdots + p(x_{n+k}) < \varepsilon$,

showing that the sequence (X_n) is left K-Cauchy. By the left K-completeness of the space X there exists $x \in X$ such that $x = \lim_{n\to\infty} X_n = \sum_{k=1}^{\infty} x_k$.

Conversely, suppose that every absolutely convergent series in (X, p) is convergent and let (x_n) be a left K-Cauchy sequence in X. Let $n_1 \in \mathbb{N}$ be such that $p(x_m - x_n) < 2^{-1}$ for all $n, m \in \mathbb{N}$ with $n_1 \le n < m$. Now let $n_2 > n_1$ be such that $p(x_m - x_n) < 2^{-1}$ for all $n, m \in \mathbb{N}$ with $n_2 \le n < m$. Continuing in this manner we find a sequence of indices $n_1 < n_2 < \cdots$ such that $p(x_{n_{k+1}} - x_{n_k}) < 2^{-k}$ for all $k \in \mathbb{N}$. It follows that $\sum_{k=1}^{\infty} p(x_{n_{k+1}} - x_{n_k}) = 1$, so that, by hypothesis, there exists $x \in X$ such that $y = \sum_{k=1}^{\infty}(x_{n_k} - x_{n_{k-1}})$. But $Y_k = (x_{n_2} - x_{n_1}) + \cdots + (x_{n_{k+1}} - x_{n_k}) = x_{n_{k+1}} - x_{n_1}$, for every $k \in \mathbb{N}$. It follows that $x_{n_k} \xrightarrow{p} y + x_{n_1}$ as $k \to \infty$. By Proposition 1.2.4, the sequence (x_n) converges to $y + x_{n_1}$. □

By Proposition 1.2.2 a weakly left K-sequentially complete quasi-semimetric space is left K-sequentially complete. As remarked by Romaguera [190] these notions are in fact equivalent.

Proposition 1.2.7 ([190], Proposition 1). *A quasi-semimetric space is weakly left K-sequentially complete if and only if it is left K-sequentially complete.*

Proof. It remains to show that a left K-complete quasi-semimetric space is weakly left K-complete. Suppose that the space (X, ρ) is left K-complete and let (x_n) be a weakly left K-Cauchy sequence in X. We have to show that (x_n) is ρ-convergent to some $x \in X$.

Let $n(1)$ be the smallest natural number such that

$$\forall n \ge n(1), \quad \rho(x_{n(1)}, x_n) < 1 . \tag{1.2.3}$$

If $\rho(x_{n(1)}, x_n) = 0$ for all $n \ge n(1)$, then $x_n \xrightarrow{\rho} x_{n(1)}$. Supposing that $\rho(x_{n(1)}, x_{m(1)}) > 0$ for some $m(1) > n(1)$, let $k_2 \in \mathbb{N}$ be such that

$$\frac{1}{k_2} \le \rho(x_{n(1)}, x_{m(1)}) < 1 , \tag{1.2.4}$$

and let $n(2)$ be the smallest natural number such that

$$\forall n \ge n(2), \quad \rho(x_{n(2)}, x_n) < 1/k_2 . \tag{1.2.5}$$

By the choice of $n(1)$, $n(2) \ge n(1)$, and by (1.2.4), $n(2) \ne n(1)$ so that $n(2) > n(1)$.

Again, if $\rho(x_{n(2)}, x_n) = 0$ for all $n \ge n(2)$, then $x_n \xrightarrow{\rho} x_{n(2)}$. If not, choose $m(2) > n(2)$ such that $\rho(x_{n(2)}, x_{m(2)}) > 0$, then pick $k_3 \in \mathbb{N}$ such that

$$\frac{1}{k_3} \le \rho(x_{n(1)}, x_{m(1)} < \frac{1}{k_2} ,$$

and take $n(3)$ to be the smallest natural number such that

$$\forall n \geq n(3), \quad \rho(x_{n(2)}, x_n) < 1/k_3 .$$

Continuing in this manner we can get at some step i an element $x_{n(i)}$ such that $\rho(x_{n(i)}, x_n) = 0$ for all $n \geq n(i)$, implying $x_n \xrightarrow{\rho} x_{n(i)}$.

If such an i does not exist, we find the sequences of natural numbers

$$1 = k_1 < k_2 < \cdots \qquad \text{and} \qquad n(1) < n(2) < \cdots$$

such that

$$\forall n \geq n(i), \quad \rho(x_{n(i)}, x_n) < 1/k_i , \tag{1.2.6}$$

for all $i \in \mathbb{N}$.

It is easy to check that the condition (1.2.6) implies that the sequence $(x_{n(i)})_{i\in\mathbb{N}}$ is left K-Cauchy, so that, by the left K-completeness of the space (X, ρ), it is ρ-convergent to some $x \in X$.

Let us show that the sequence (x_n) is ρ-convergent to x. For $\varepsilon > 0$ let $i_0 \in \mathbb{N}$ be such that $k_{i_0}^{-1} < \varepsilon$ and $\rho(x, x_{n(i)}) < \varepsilon$, for all $i \geq i_0$. Then for every $n \geq n(i_0)$,

$$\rho(x, x_n) \leq \rho(x, x_{n(i_0)}) + \rho(x_{n(i_0)}, x_n) < \varepsilon + \frac{1}{k_{i_0}} < 2\varepsilon ,$$

proving the ρ-convergence of (x_n) to x. □

Concerning Baire's characterization of completeness in terms of descending sequences of closed sets we mention the following result, [185, Th. 10]. The *diameter* of a subset A of X is defined by

$$\operatorname{diam}(A) = \sup\{\rho(x, y) : x, y \in A\} . \tag{1.2.7}$$

It is clear that the diameter, as defined, is in fact the diameter with respect to the associated semimetric ρ^s.

Theorem 1.2.8 ([185], Theorem 10). *A quasi-semimetric space (X, ρ) is ρ-sequentially complete if and only if each decreasing sequence $F_1 \supset F_2 \ldots$ of nonempty closed sets with* $\operatorname{diam} F_n \to 0$ *as $n \to \infty$ has nonempty intersection, which is a singleton if ρ is a quasi-metric.*

Proof. Let $F_1 \supset F_2 \supset \cdots$ be a sequence of nonempty τ_ρ-closed sets with $\operatorname{diam}(F_n) \to 0$ for $n \to \infty$. Choosing $x_n \in F_n$, $n \in \mathbb{N}$, it follows that the sequence (x_n) is ρ^s-Cauchy, so that it is τ_ρ-convergent to some $x \in X$. Since for every $n \in \mathbb{N}$, $x_{n+k} \in F_{n+k} \subset F_n$ for all $k \in \mathbb{N}$, letting $k \to \infty$ and taking into account the closedness of the set F_n it follows that $x \in F_n$ for all $n \in \mathbb{N}$, that is $x \in \cap_n F_n$.

Suppose that ρ is a quasi-metric. Since ρ is a quasi-metric if and only if ρ^s is a metric, the hypothesis $\operatorname{diam}(F_n) \to 0$ implies that $\cap_n F_n$ can contain at most one element.

We shall prove the converse by contradiction. Suppose that there exists a ρ^s-Cauchy sequence (x_n) in X which is not ρ-convergent. Then (x_n) is left K-ρ-Cauchy, so that, by Proposition 1.2.4 it is ρ-convergent provided it contains a ρ-convergent subsequence. Consequently (x_n) does not contain ρ-convergent subsequences, so that the set $F_n = \{x_k : k \geq n\}$ is τ_ρ-closed for every $n \in \mathbb{N}$. Since (x_n) is ρ^s-Cauchy it follows that $\operatorname{diam}(F_n) \to 0$ for $n \to \infty$. If $x \in \cap_{n=1}^\infty F_n$, then the inequalities $\rho(x, x_n) \leq \operatorname{diam}(F_n)$, $n \in \mathbb{N}$, imply $\lim_n \rho(x, x_n) = 0$, that is $x_n \xrightarrow{\rho} x$, in contradiction to the hypothesis. Consequently $\cap_{n=1}^\infty F_n = \emptyset$, which ends the proof of the reverse implication. □

The following characterization of right K-completeness was obtained in [35] using a different terminology.

Proposition 1.2.9. *A quasi-semimetric space (X, ρ) is right K-sequentially complete if and only if any decreasing sequence of closed $\bar{\rho}$-balls*

$$B_{\bar{\rho}}[x_1, r_1] \supset B_{\bar{\rho}}[x_1, r_1] \supset \cdots \quad \textit{with} \quad \lim_{n\to\infty} r_n = 0$$

has nonempty intersection.

If the topology τ_ρ is Hausdorff, then $\cap_{n=1}^\infty B_{\bar{\rho}}[x_n, r_n]$ contains exactly one element.

Proof. Suppose that (X, p) is right K-complete and let $B_{\bar{\rho}}[x_n, r_n]$, $n \in \mathbb{N}$, be a sequence of closed $\bar{\rho}$-balls satisfying the requirements of the proposition.

We show first that the sequence (x_n) formed with their centers is right K-Cauchy. For $\varepsilon > 0$ there exists $n_0 \in \mathbb{N}$ such that $r_n < \varepsilon$ for all $n \geq n_0$. If $n_0 \leq n < m$, then $x_m \in B_{\bar{\rho}}[x_n, r_n]$ so that

$$\bar{\rho}(x_n, x_m) \leq r_n < \varepsilon \iff \rho(x_m, x_n) \leq r_n < \varepsilon ,$$

showing that (x_n) is right K-Cauchy. It follows that there exists $x \in X$ such that $x_n \xrightarrow{\rho} x$. For every $k \in \mathbb{N}$, $x_n \in B_{\bar{\rho}}[x_k, r_k]$ for all $n \geq k$. Since, by Proposition 1.1.8, the ball $B_{\bar{\rho}}[x_k, r_k]$ is τ_ρ-closed, it follows that $x = \rho\text{-}\lim_{n\to\infty} x_n \in B_{\bar{\rho}}[x_k, r_k]$, showing that $x \in \cap_{k=1}^\infty B_{\bar{\rho}}[x_k, r_k]$.

If $y \in \cap_{n=1}^\infty B_{\bar{\rho}}[x_n, r_n]$, then for all $n \in \mathbb{N}$, $\bar{\rho}(x_n, y) \leq r_n \iff \rho(y, x_n) \leq r_n$, so that $x_n \xrightarrow{\rho} y$. If the topology τ_ρ is Hausdorff, then $y = x$.

To prove the converse, let (x_n) be a right K-Cauchy sequence in (X, ρ). Then there exists n_1 such that $n_1 \leq n < m$ implies $\rho(x_m, x_n) \leq 1/2 \iff \bar{\rho}(x_n, x_m) \leq 1/2$. In particular $\bar{\rho}(x_{n_1}, x_n) \leq 1/2$ for all $n \geq n_1$. Consider the ball $B_{\bar{\rho}}[x_{n_1}, 1]$.

Let now $n_2 > n_1$ be such that $n_2 \leq n < m$ implies $\rho(x_m, x_n) \leq 1/2^2 \iff \bar{\rho}(x_n, x_m) \leq 1/2^2$, so that $\bar{\rho}(x_{n_2}, x_n) \leq 1/2^2$ for all $n \geq n_2$.

It follows that $B_{\bar\rho}[x_{n_2}, 1/2] \subset B_{\bar\rho}[x_{n_1}, 1]$. Indeed, $\rho(x_{n_2}, y) \le 1/2$ implies

$$\bar\rho(x_{n_1}, y) \le \bar\rho(x_{n_1}, x_{n_2}) + \bar\rho(x_{n_2}, y) \le \frac{1}{2} + \frac{1}{2} = 1 .$$

Continuing in this manner we obtain a sequence $n_1 < n_2 < \cdots$ such that

$$\forall n \ge n_k, \quad \bar\rho(x_{n_k}, x_n) \le \frac{1}{2^k} .$$

It follows that $\bar\rho(x_{n_k}, x_{n_{k+1}}) \le 2^{-k}$ and the balls $B_{\bar\rho}[x_{n_k}, 2^{-(k-1)}]$ satisfy $B_{\bar\rho}[x_{n_{k+1}}, 2^{-k}] \subset B_{\bar\rho}[x_{n_k}, 2^{-(k-1)}], \ k \in \mathbb{N}$.

By hypothesis there exists $x \in \cap_{k=1}^{\infty} B_{\bar\rho}[x_{n_k}, 2^{-(k-1)}]$.

The relations $\rho(x, x_{n_k}) = \bar\rho(x_{n_k}, x) \le 2^{-(k-1)} \to 0$ as $k \to \infty$, show that ρ-$\lim_{k\to\infty} x_{n_k} = x$. By Proposition 1.2.4.2, the right K-Cauchy sequence (x_n) is ρ-convergent to x. □

Remark 1.2.10. It is obvious that the conclusion of Proposition 1.2.9 remains true for every family $F_1 \supset F_2 \supset \cdots$ of nonempty $\tau_{\bar\rho}$-closed sets in a right K-sequentially complete quasi-metric space (X, ρ) for which there exists the balls $B_\rho[x_n, r_n] \supset F_n, \ n \in \mathbb{N}$, satisfying the hypotheses of the proposition.

As it is well known in the case of semimetric spaces, the sequential completeness is equivalent to the completeness defined in terms of filters or of nets. Romaguera [190] proved a similar result for left K-complete quasi-semimetric spaces.

A filter $\mathcal{F}$ in a quasi-semimetric space (X, ρ) is called *left K-Cauchy* if for every $\varepsilon > 0$ there exists $F_\varepsilon \in \mathcal{F}$ such that

$$\forall x \in F_\varepsilon, \quad B_\rho(x, \varepsilon) \in \mathcal{F} . \tag{1.2.8}$$

Also a net $(x_i : i \in I)$ is called *left K-Cauchy* if for every $\varepsilon > 0$ there exists $i_0 \in I$ such that

$$\forall i, j \in I, \ \ i_0 \le i \le j \ \Rightarrow \ \rho(x_i, x_j) < \varepsilon . \tag{1.2.9}$$

Proposition 1.2.11. *For a quasi-semimetric space (X, ρ) the following are equivalent.*

1. *The space (X, ρ) is left K-sequentially complete.*
2. *Every left K-Cauchy filter in X is τ_ρ-convergent.*
3. *Every left K-Cauchy net in X is τ_ρ-convergent.*

Proof. 1 $\Rightarrow$ 2. If $\mathcal{F}$ is a left K-Cauchy filter in (X, ρ), then for every $n \in \mathbb{N}$ there exists $F_n \in \mathcal{F}$ such that

$$\forall x \in F_n, \quad B_\rho(x, 2^{-n}) \in \mathcal{F} .$$

Pick $x_1 \in F_1$ and $x_n \in F_n \cap \left(\cap_{k=1}^{n-1} B_\rho(x_k, 2^{-k})\right)$ for $n > 1$.

The sequence (x_n) is left K-Cauchy. Indeed, for $\varepsilon > 0$ let $k \in \mathbb{N}$ be such that $2^{-k} < \varepsilon$. Then, by the choice of x_n, for $k \le m < n$, $x_n \in B_\rho(x_m, 2^{-m})$, implying $\rho(x_m, x_n) < 2^{-m} \le 2^{-k} < \varepsilon$.

By hypothesis, there exists $x \in X$ such that

$$x_n \xrightarrow{\rho} x \iff \rho(x, x_n) \to 0 . \tag{1.2.10}$$

We want to show that $x = \lim \mathcal{F}$ which is equivalent to the condition

$$\forall k \in \mathbb{N}, \quad B_\rho(x, 2^{-k}) \in \mathcal{F} .$$

Let $k \in \mathbb{N}$. By (1.2.10) there exists $n > k$ such that $\rho(x, x_n) < 2^{-k-1}$. If $\rho(x_n, y) < 2^{-n}$, then

$$\rho(x, y) \le \rho(x, x_n) + \rho(x_n, y) < \frac{1}{2^{k+1}} + \frac{1}{2^n} \le \frac{1}{2^k} ,$$

showing that $B_\rho(x_n, 2^{-n}) \subset B_\rho(x, 2^{-k})$. Since $B_\rho(x_n, 2^{-n}) \in \mathcal{F}$, it follows that $B_\rho(x, 2^{-k}) \in \mathcal{F}$.

2 $\Rightarrow$ 3. If $(x_i : i \in I)$ is a left K-Cauchy net in X, then the sets $F_i = \{x_j : j \ge i\}$, $i \in I$, form the base of a filter $\mathcal{F}$ on X. For $\varepsilon > 0$ let i_0 be such that $i_0 \le i \le j$ implies $\rho(x_i, x_j) < \varepsilon$. If $i \ge i_0$, then $\rho(x_i, x_j) < \varepsilon$ for every $j \ge i$, showing that $F_i \subset B_\rho(x_i, \varepsilon)$. Consequently, $B_\rho(x_i, \varepsilon) \in \mathcal{F}$ for every $x_i \in F_{i_0}$.

By hypothesis there exists $x \in X$ such that $x = \lim \mathcal{F}$. Using the definition of the filter $\mathcal{F}$, it is easy to check that $x = \lim_i x_i$. Indeed, for $\varepsilon > 0$, $B_\rho(x, \varepsilon) \in \mathcal{F}$, so there exists $i_0 \in I$ such that $F_{i_0} \subset B_\rho(x, \varepsilon)$, implying $\rho(x, x_i) < \varepsilon$ for every $i \ge i_0$.

The implication 3 $\Rightarrow$ 1 is obvious. □

A filter $\mathcal{F}$ in a quasi-uniform space $(X, \mathcal{U})$ is called *left $\mathcal{U}$-Cauchy* if for every $U \in \mathcal{U}$ there exists $x \in X$ such that $U(x) \in \mathcal{F}$. The quasi-uniform space $(X, \mathcal{U})$ is called *left $\mathcal{U}$-complete* if every left $\mathcal{U}$-Cauchy filter is convergent with respect to the induced topology $\tau(\mathcal{U})$. In quasi-semimetric spaces these notions correspond to those of left ρ-Cauchy sequence and sequential left ρ-completeness. Call a quasi-semimetric space *left ρ-complete* if the associated quasi-uniform space $(X, \mathcal{U}_\rho)$ is left $\mathcal{U}_\rho$-complete. We have seen in Proposition 1.2.11 that a quasi-semimetric is left K-complete if and only if the associated quasi-uniform space $(X, \mathcal{U}_\rho)$ is left K-complete.

In the case of left ρ-completeness this equivalence does not hold in general.

Proposition 1.2.12 (Künzi [131]). *A Hausdorff quasi-metric space (X, ρ) is left ρ-sequentially complete if and only if the associated quasi-uniform space $(X, \mathcal{U}_\rho)$ is left $\mathcal{U}_\rho$-complete.*

Proof. Suppose that (X,ρ) is left ρ-sequentially complete and let $\mathcal{F}$ be a left $\mathcal{U}_\rho$-Cauchy filter. Then, for every $k \in \mathbb{N}$, there exists $x_k \in X$ such that $B_k := B_\rho(x_k, 2^{-k}) \in \mathcal{F}$. Let

$$K = \cap_{n=1}^{\infty} \tau_\rho\text{-}\operatorname{cl}\left(\cap_{k=1}^{n} B_k\right) . \tag{1.2.11}$$

Claim I. Any sequence $y_n \in \cap_{k=1}^{n} B_k,\ n \in \mathbb{N}$, is left ρ-Cauchy.

For $\varepsilon > 0$ let $k \in \mathbb{N}$ be such that $2^{-k} < \varepsilon$. If $n \geq k$, then $y_n \in B_k$, so that $\rho(x_k, y_n) < 2^{-k} < \varepsilon$.

Claim II. The set K is a singleton.

Let $y_n \in \cap_{k=1}^{n} B_k,\ n \in \mathbb{N}$. Then the sequence (y_n) is left ρ-Cauchy, so that, by hypothesis, there exists $x \in X$ such that $y_n \xrightarrow{\rho} x$. Since K is τ_ρ-closed it follows that $x \in K$.

Let $y \in K$. By the definition (1.2.11) of the set K, for every $n \in \mathbb{N}$ there exists the elements

$$x_n \in B_\rho(x, 2^{-n}) \cap B_1 \cap \cdots \cap B_n \quad \text{and} \quad z_n \in B_\rho(y, 2^{-n}) \cap B_1 \cap \cdots \cap B_n .$$

It follows that $x_n \xrightarrow{\rho} x$ and $z_n \xrightarrow{\rho} y$. Define a sequence (w_n) by $w_{2k-1} = x_k$ and $w_{2k} = z_k$. By Claim I the sequence (w_n) is left ρ-Cauchy, because $w_n \in B_1 \cap \cdots \cap B_n,\ n \in \mathbb{N}$, so that it is ρ-convergent to some $w \in X$. Since τ_ρ is Hausdorff this limit is unique, implying $x = \lim_k w_{2k-1} = w$ and $y = \lim_k w_{2k} = w$, that is $y = x$.

Claim III. $\lim \mathcal{F} = x$.

Supposing the contrary, there exists an open neighborhood V of x such that $V \notin \mathcal{F}$. It follows that $B_1 \cap \cdots \cap B_n \not\subseteq V$, that is $B_1 \cap \cdots \cap B_n \setminus V \neq \emptyset$ for every $n \in \mathbb{N}$. Let $y_n \in B_1 \cap \cdots \cap B_n \setminus V,\ n \in \mathbb{N}$. By Claim I the sequence (y_n) is left ρ-Cauchy, so that there exists $y \in X$ such that $y_n \xrightarrow{\rho} y$. It follows that $y \in K = \{x\}$, so that $y_n \xrightarrow{\rho} x$, in contradiction to the fact that $y_n \notin V \in \mathcal{V}_\rho(x)$ for all $n \in \mathbb{N}$.

The converse is easy to prove and holds in general: any left $\mathcal{U}_\rho$ complete quasi-semimetric space (X,ρ) is ρ-sequentially complete.

If (x_n) is a left ρ-Cauchy sequence in (X,ρ), then the sets $F_n = \{x_k : k \geq n\},\ n \in \mathbb{N}$, form the base of a filter $\mathcal{F}$ on X. It is easy to check that $\mathcal{F}$ is $\mathcal{U}_\rho$-Cauchy, so it is τ_ρ-convergent to some $x \in X$, implying $x_n \xrightarrow{\rho} x$. Indeed, for $\varepsilon > 0$, $B_\rho(x,\varepsilon) \in \mathcal{F}$, so there exists $n_0 \in \mathbb{N}$ such that $F_{n_0} \subset B_\rho(x,\varepsilon)$, implying $x_n \in B_\rho(x,\varepsilon)$ for all $n \geq n_0$. □

Künzi [131] considers a more general notion of completeness, namely that of *cluster completeness* meaning that every Cauchy (of some kind) filter or net, has a cluster point. If one considers sequences, then one says *cluster sequentially complete.* (Our terminology differs from that in [131] where a quasi-uniform space such that any Cauchy filter is convergent is called convergence complete and a quasi-uniform space such that any Cauchy filter has a cluster point is called complete.)

Using this terminology, by Proposition 1.2.4 a left K-cluster sequentially complete quasi-semimetric space is left K-sequentially complete.

Künzi [131] gives examples of

- a T_1 quasi-metric space (X, ρ) that is $\mathcal{U}_\rho$-cluster complete, but not left ρ-sequentially complete (Example 1);
- a T_1 quasi-metric space (X, ρ) that is left ρ-sequentially complete, but not $\mathcal{U}_\rho$-cluster complete (Example 2);
- a T_1 quasi-metric space (X, ρ) that is left ρ-sequentially complete and $\mathcal{U}_\rho$-cluster complete, but not $\mathcal{U}_\rho$-complete (Example 3).

Call a quasi-semimetric space (X, ρ)

- *point-symmetric* if $\tau(\rho) \subset \tau(\bar{\rho})$, and
- *locally symmetric* if for every $x \in X$ and every $\varepsilon > 0$ there exists $\delta > 0$ such that
$$\cup\{B_{\bar{\rho}}(y, \delta) : y \in B_\rho(x, \delta)\} \subset B_\rho(x, \varepsilon), \tag{1.2.12}$$
or, equivalently, if
$$\forall z \in X, \; (\exists y \in X, \;\; \rho(x, y) < \delta \;\wedge\; \rho(z, y) < \delta \;\Rightarrow\; \rho(x, z) < \varepsilon) \; . \tag{1.2.13}$$

A point-symmetric quasi-semimetric is called also a *strong* quasi-semimetric. Observe that if ρ is point-symmetric, then the topology $\tau(\bar{\rho})$ is semimetrizable (metrizable if ρ is a quasi-metric), see Proposition 1.1.13.

We mention the following result.

Proposition 1.2.13.

1. *A locally symmetric quasi-semimetric space is point-symmetric.*
2. *A quasi-semimetric space (X, ρ) is point-symmetric if and only if*
$$x_n \xrightarrow{\bar{\rho}} x \;\Rightarrow\; x_n \xrightarrow{\rho} x \;, \tag{1.2.14}$$
for every sequence (x_n) in X.
3. ([10]) *A weakly right K-sequentially complete T_1 quasi-metric space is point-symmetric*
4. ([10]) *A countably compact T_1 quasi-metric space is point-symmetric.*

Proof. 1. If $Y \subset X$ is $\tau(\rho)$-open, then for every $x \in Y$ there exists $\varepsilon > 0$ such that $B_\rho(x, \varepsilon) \subset Y$. Choosing $\delta > 0$ according to (1.2.12), it follows that the $\tau(\bar{\rho})$-open set $\cup\{B_{\bar{\rho}}(y, \delta) : y \in B_\rho(x, \delta)\}$ contains x and is contained in Y, so that Y is $\tau(\bar{\rho})$-open.

The equivalence from 2 follows from the fact that in a quasi-semimetric space all the topological properties can be expressed in terms of sequences, a property valid in any first countable topological space.

3. The proof is based on 2. The space (X,ρ) is T_1 if and only if $\rho(x,y) = 0 \iff x = y$ for all $x,y \in X$. Let (x_n) be a sequence in X that is $\tau(\bar{\rho})$-convergent to $x \in X$. Define the sequence (y_n) by $y_{2k-1} = x$ and $y_{2k} = x_k$ for $k \in \mathbb{N}$. It is obvious that (y_n) is weakly right K-Cauchy, so there exists $y \in X$ such that $y_n \xrightarrow{\rho} y$. Since $\rho(y,x) = \rho(y, y_{2k-1}) \to 0$ as $k \to \infty$, it follows that $\rho(y,x) = 0$, so that $y = x$. The relations $\rho(x, x_k) = \rho(y, y_{2k}) \to 0$ for $k \to \infty$ show that $x_n \xrightarrow{\rho} x$.

4. Suppose, by contradiction, that there exists $G \in \tau(\rho) \setminus \tau(\bar{\rho})$. It follows that

$$\begin{aligned} &\text{(i)} \quad \forall x \in G,\ \exists n_x \in \mathbb{N}, \quad B_\rho(x, 2^{-n_x}) \subset G, \\ &\text{(ii)} \quad \exists x_0 \in G,\ \forall n \in \mathbb{N}, \quad B_{\bar{\rho}}(x, 2^{-n}) \cap \complement(G) \neq \emptyset, \end{aligned} \tag{1.2.15}$$

where for $Z \subset X$, $\complement(Z) = X \setminus Z$.

Since a topological space is countably compact if and only if every filter with a countable base has a cluster point, it follows that the filter $\mathcal{F}$ generated by the countable filter base $\{B_{\bar{\rho}}(x, 2^{-n}) \cap \complement(G) : n \in \mathbb{N}\}$ has a cluster point $y \in X$, implying

$$y \in B_\rho(y, 2^{-n}) \cap B_{\bar{\rho}}(x_0, 2^{-n}) \cap \complement(G)\ , \tag{1.2.16}$$

for every $n \in \mathbb{N}$.

From $y \in \complement(G)$, it follows that $y \neq x_0$, so that there exists $n \in \mathbb{N}$ with

$$x_0 \notin B_\rho(y, 2^{-n}) \iff \rho(y, x_0) \geq 2^{-n}\ .$$

By (1.2.16) there exists $z \in B_\rho(y, 2^{-n-1}) \cap B_{\bar{\rho}}(x_0, 2^{-n-1})$, leading to the contradiction

$$\begin{aligned} 2^{-n} &\leq \rho(y, x_0) \leq \rho(y, z) + \rho(z, x_0) \\ &= \rho(y, z) + \bar{\rho}(x_0, z) < 2^{-n-1} + 2^{-n-1} = 2^{-n}\ . \end{aligned}$$

□

We have seen in Proposition 1.2.7 that sequential weak left K-completeness and sequential left K-completeness are equivalent notions in any quasi-semimetric space. The following example shows that this result is not true for right completeness.

Example 1.2.14 (Császár, see [10]). Let $X = \{0\} \cup \{1/n : n \in \mathbb{N}\}$ and let the quasi-semimetric d be defined by $d(0, (2k)^{-1}) = (2k)^{-1}$, $d(n^{-1}, (2k)^{-1}) = (2k)^{-1} - n^{-1}$ if $n > 2k$, $d(x,x) = 0$, for all $x \in X$ and $d(x,y) = 0$ otherwise. Then $\tau(\bar{d})$ is the discrete topology, so that $\tau(d) \subset \tau(\bar{d})$, that is the space (X,d) is point-symmetric. A right K-Cauchy sequence in (X,d) is either eventually constant or a subsequence of $(1/2k)$, say $x_i = 1/2k_i$, $i \in \mathbb{N}$. In this last case $d(0, x_i) = (2k_i)^{-1} \to 0$ as $i \to \infty$. Consequently (X,d) is right K-sequentially complete. However the sequence $(1/n)$ is weakly right K-Cauchy without being $\tau(d)$-convergent.

Although, an equivalence result can be obtained under some supplementary conditions on the space X.

Proposition 1.2.15 ([10]). *A right K-sequentially complete locally symmetric quasi-semimetric space is weakly right K-sequentially complete,*

Proof. Suppose that (x_n) is a weakly right K-Cauchy sequence in a locally symmetric quasi-semimetric space (X, ρ). Let n_1 be the first natural number such that

$$\forall n > n_1, \quad \rho(x_n, x_{n_1}) < 1/2 . \tag{1.2.17}$$

If $\rho(x_n, x_{n_1}) = 0$ for all $n > n_1$, then $x_n \xrightarrow{\bar{\rho}} x_{n_1}$, so that, by Proposition 1.2.13.2, $x_n \xrightarrow{\rho} x_{n_1}$. Else, let m_1 be the first natural number greater than n_1 such that $\rho(x_{m_1}, x_{n_1}) > 0$, and let $k_2 \in \mathbb{N}$ be such that

$$\frac{1}{2^{k_2}} \leq \rho(x_{m_1}, x_{n_1}) < \frac{1}{2^{k_2-1}} . \tag{1.2.18}$$

Since $\rho(x_{m_1}, x_{n_1}) < 1/2$, it follows that $k_2 > 1$.

Let n_2 be the first natural number such that

$$\forall n > n_2, \quad \rho(x_n, x_{n_2}) < \frac{1}{2^{k_2}} .$$

If $n_2 < n_1$, then, by the definition of the number n_1, there exists m, $n_2 < m < n_1$, such that $\rho(x_m, x_{n_2}) \geq 1/2$, leading to the contradiction $1/2 \leq \rho(x_m, x_{n_2}) < 1/2^{k_2}$.

If $n_1 \leq n_2 < m_1$, then $\rho(x_{n_2}, x_{n_1}) = 0$, implying

$$\frac{1}{2^{k_2}} \leq \rho(x_{m_1}, x_{n_1}) \leq \rho(x_{m_1}, x_{n_2}) + \rho(x_{n_2}, x_{n_1}) = \rho(x_{m_1}, x_{n_2}) < \frac{1}{2^{k_2}} ,$$

a contradiction, again.

Consequently $n_1 \geq m_1 > n_1$, and $\rho(x_{n_2}, x_{n_1}) < 1/2$.

Continuing in this manner, it is possible that at some step i, $\rho(x_n, x_{n_i}) = 0$ for all $n \geq n_i$, implying $x_n \xrightarrow{\rho} x_{n_i}$. If such an i does not exist, then we find the increasing sequences of natural numbers

$$n_1 < n_2 < \cdots \qquad \text{and} \qquad k_1 < k_2 < \cdots$$

such that

$$\forall i \in \mathbb{N}, \ \forall n > n_i, \quad \rho(x_n, x_{n_i}) < \frac{1}{2^{k_i}} .$$

It follows that

$$\forall i \in \mathbb{N}, \quad \rho(x_{n_{i+1}}, x_{n_i}) < \frac{1}{2^{k_i}} . \tag{1.2.19}$$

It is easy to check that (1.2.19) implies that the sequence $(x_{n_i})_{i \in \mathbb{N}}$ is right K-Cauchy, so there exists $x \in X$ such that $x_{n_i} \xrightarrow{\rho} x \iff \lim_{i \to \infty} \rho(x, x_{n_i}) = 0$.

For $\varepsilon > 0$ choose $\delta > 0$ according to (1.2.13).

Let $i_0 \in \mathbb{N}$ be such that

$$\forall i \geq i_0, \quad \rho(x, x_{n_i}) < \delta ,$$

and let $j \geq i_0$ be such that $2^{-k_j} < \delta$. Then for $n > n_j$,

$$\rho(x, x_{n_j}) < \delta \quad \text{and} \quad \rho(x_n, x_{n_j}) < \frac{1}{2^{k_j}} < \delta .$$

Applying (1.2.13) with $y = x_{n_j}$ and $z = x_n$, it follows that $\rho(x, x_n) < \varepsilon$ for all $n > n_j$, that is $x_n \xrightarrow{\rho} x$. □

The analog of Proposition 1.2.11 can also be obtained only under some supplementary hypotheses on the quasi-semimetric space X.

A filter $\mathcal{F}$ in a quasi-semimetric space (X, ρ) is called a *right K-Cauchy filter* if for every $\varepsilon > 0$ there exists $F_\varepsilon \in \mathcal{F}$ such that $B_{\bar{\rho}}(x, \varepsilon) \in \mathcal{F}$ for every $x \in F_\varepsilon$. A net $(x_i : i \in I)$ is called a *right K-Cauchy net* if for every $\varepsilon > 0$ there exists $i_0 \in I$ such that $\rho(x_j, x_i) < \varepsilon$ for all $i, j \in I$ with $i_0 \leq i \leq j$.

Call the space (X, ρ) *right K-complete* if every right K-Cauchy filter in X is τ_ρ-convergent to some $x \in X$.

The quasi-semimetric space (X, ρ) is called R_1 if for all $x, y \in X$, $\tau_\rho\text{-cl}\{x\} \neq \tau_\rho\text{-cl}\{y\}$ implies the existence of two disjoint τ_ρ-open sets U, V such that $x \in U$ and $y \in V$.

Proposition 1.2.16 ([10]). *Let (X, ρ) be a quasi-semimetric space. The following are true.*

1. *If X is right K-complete, then every right K-Cauchy net in X is convergent. In particular, every right K-complete quasi-semimetric space is right K-sequentially complete.*
2. *If the quasi-semimetric space (X, ρ) is R_1 then X is right K-complete if and only if it is right K-sequentially complete.*

Proof. 1. If $(x_i : i \in I)$ is a right K-Cauchy net in X, then $F_i = \{x_j : j \in I,\ i \leq j\}$, $i \in I$, is the base of a filter $\mathcal{F}$ in X. For $\varepsilon > 0$ let $i_0 \in I$ be such that $\rho(x_j, x_i) < \varepsilon \iff \bar{\rho}(x_i, x_j) < \varepsilon$ for all $i, j \in I$ with $i_0 \leq i \leq j$. Then for every $n \geq n_0$, $F_n \subset B_{\bar{\rho}}(x_n, \varepsilon)$, implying $B_{\bar{\rho}}(x_i, \varepsilon) \in \mathcal{F}$ for every $i \geq i_0$, that is the filter $\mathcal{F}$ is right K-Cauchy. By hypothesis, $\mathcal{F}$ is convergent to some $x \in X$. It is easy to check that the net $(x_i : i \in I)$ is ρ-convergent to x.

2. We have to prove only that a right K-sequentially complete quasi-semimetric space (X, ρ) is right K-complete. Let $\mathcal{F}$ be a right K-Cauchy filter on X. Then for every $n \in \mathbb{N}$ there exists $F_n \in \mathcal{F}$ such that $B_{\bar{\rho}}(x, 2^{-n}) \in \mathcal{F}$ for all $x \in F_n$. Let $x_1 \in F_1$ and $x_n \in F_n \cap \bigcap_{k=1}^{n-1} B_{\bar{\rho}}(x_k, 2^{-k})$, $n > 1$. It follows that the so-constructed sequence (x_n) is right K-Cauchy, so it is ρ-convergent to some $x \in X$. We shall show that the filter $\mathcal{F}$ converges with respect to τ_ρ to x. Supposing

the contrary, there exists $m \in \mathbb{N}$ such that $B_\rho(x, 2^{-m}) \notin \mathcal{F}$. We construct a new sequence (y_n) in the following way: take $y_1 \in F_1 \cap B_{\bar{\rho}}(x_1, 2^{-1}) \setminus B_\rho(x, 2^{-m})$ and

$$y_n \in F_n \cap B_{\bar{\rho}}(x_n, 2^{-n}) \cap \bigcap_{k=1}^{n-1} B_{\bar{\rho}}(y_k, 2^{-k}) \setminus B_\rho(x, 2^{-m}) ,$$

for $n > 1$. The relations $y_n \in \bigcap_{k=1}^{n-1} B_{\bar{\rho}}(y_k, 2^{-k}),\ n > 1$, imply that (y_n) is also a right K-Cauchy sequence, so it is ρ-convergent to some $y \in X$.

If $\tau_\rho\text{-cl}\{x\} = \tau_\rho\text{-cl}\{y\}$, then $x \in \tau_\rho\text{-cl}\{y\}$, implying $\rho(x, y) = 0$. The relations $\rho(x, y_n) \le \rho(x, y) + \rho(y, y_n) = \rho(y, y_n) \to 0,\ n \to \infty$, imply that $y_n \xrightarrow{\rho} x$, in contradiction with the fact that $y_n \notin B_\rho(x, 2^{-m})$ for all $n \in \mathbb{N}$.

If $\tau_\rho\text{-cl}\{x\} \ne \tau_\rho\text{-cl}\{y\}$, then there exists $\varepsilon > 0$ such that $B_\rho(x, \varepsilon) \cap B_\rho(y, \varepsilon) = \emptyset$. Since $y_n \in B_{\bar{\rho}}(x_n, 2^{-n})$ it follows that $\rho(y_n, x_n) < 2^{-n} \to 0,\ n \to \infty$. But then

$$\rho(y, x_n) \le \rho(y, y_n) + \rho(y_n, x_n) \to 0, \quad n \to \infty .$$

Consequently, $x_n \xrightarrow{\rho} x$ and $x_n \xrightarrow{\rho} y$, implying that, for sufficiently large $n,\ x_n \in B_\rho(x, \varepsilon) \cap B_\rho(y, \varepsilon) = \emptyset$, a contradiction. □

Stoltenberg [226] considered a more general notion of a right Cauchy net. A net $(x_i : i \in I)$ in a quasi-semimetric space (X, ρ) is called *right S_t-Cauchy* provided that for every $\varepsilon > 0$ there exists $i_\varepsilon \in I$ such that

$$\forall i, j \ge i_\varepsilon,\ \ i \le j \ \vee\ i \not\sim j \ \Rightarrow\ \rho(x_j, x_i) < \varepsilon , \tag{1.2.20}$$

where $i \not\sim j$ means that i, j are incomparable (that is no one of the relations $i \le j$ and $j \le i$ holds). Observe that

$$\rho(x_j, x_i) < \varepsilon \quad \text{and} \quad \rho(x_j, x_i) < \varepsilon , \tag{1.2.21}$$

for all $i, j \ge i_\varepsilon,\ i \not\sim j$.

It is obvious that, particularized to sequences, the notions of right K-completeness and right S_t-completeness agree.

Proposition 1.2.17 ([226]). *A T_1 quasi-metric space (X, ρ) is right K-sequentially complete if and only if every right S_t-Cauchy net in X is τ_ρ-convergent.*

Proof. We have only to prove that the sequential right K-completeness implies that every right S_t-Cauchy net in X is τ_ρ-convergent.

Let $(x_i : i \in I)$ be a right S_t-Cauchy net in X. Let $i_k \ge i_{k-1},\ k \ge 2$, such that (1.2.20) holds for $\varepsilon = 1/2^k,\ k \in \mathbb{N}$.

The inequalities $\rho(x_{i_{k+1}}, x_{i_k}) < 2^{-k},\ k \in \mathbb{N}$, imply that the sequence $(x_{i_k})_{k\in\mathbb{N}}$ is right K-Cauchy, so it converges to some $x \in X$.

We distinguish two cases.

Case I. $\exists i_0 \in I,\ \exists k_0 \in \mathbb{N}, \quad \forall k \geq k_0,\ x_{i_k} \leq i_0.$

Then for every $k \geq k_0$ and $i_0 \leq i$ we have $i_k \leq i_0 \leq i$, so that $\rho(x_i, x_{i_0}) < 2^{-k}$, implying $\rho(x_i, x_{i_0}) = 0$. Since the quasi-metric space (X, ρ) is T_1, it follows that $x_i = x_{i_0}$ for all $i \geq i_0$ (see Proposition 1.1.8), so that the net $(x_i : i \in I)$ is τ_ρ-convergent to x_{i_0}.

Case II. $\forall i \in I,\ \forall k \in \mathbb{N},\ \exists k' \geq k, \quad i_{k'} \not\leq i.$

For $\varepsilon > 0$ let $k_0 \in \mathbb{N}$ be such that $2^{-k_0} < \varepsilon$ and $\rho(x, x_{i_k}) < \varepsilon$ for all $k \geq k_0$.

For $i \geq i_{k_0}$ let $k \geq k_0$ be such that $i_k \not\leq i$. Again there are two situations. If $i_k \nsim i$, then, since $i_{k_0} \leq i_k$ and $i_{k_0} \leq i$, it follows that $\rho(x_{i_k}, x_i) < 2^{-k_0} < \varepsilon$. If i_k, i are comparable, then we must have $i \leq i_k$, so that $\rho(x_{i_k}, x_i) < 2^{-k_0} < \varepsilon$.

In both cases

$$\rho(x, x_i) \leq \rho(x, x_{i_k}) + \rho(x_{i_k}, x_i) < 2\varepsilon\ ,$$

for all $i \geq i_{k_0}$, showing that the net $(x_i : i \in I)$ is τ_ρ-convergent to x. □

Stoltenberg gives in [226] examples of

- a right K-sequentially complete T_1 quasi-metric space (X, ρ) containing a right K-Cauchy net that is not τ_ρ-convergent (Example 2.4);
- a right K-sequentially complete T_1 quasi-semimetric space (X, ρ) containing a right S_t-Cauchy net that is not τ_ρ-convergent, showing that Proposition 1.2.17 does not hold for quasi-semimetric spaces (Example 2.6).

1.2.2 Compactness, total boundedness and precompactness

A subset Y of a quasi-metric space (X, ρ) is called *precompact* if for every $\varepsilon > 0$ there exists a finite subset Z of Y such that

$$Y \subset \cup\{B_\rho(z, \varepsilon) : z \in Z\}\ . \tag{1.2.22}$$

If for every $\varepsilon > 0$ there exists a finite subset Z of X such that (1.2.22) holds, then the set Y is called *outside precompact.*

One obtains the same notions if one works with closed balls $B_\rho[z, \varepsilon]$, $z \in Z$.

Proposition 1.2.18. *Let (X, ρ) be a quasi-semimetric space.*

1. *A subset Y of X is (outside) precompact if and only if for every $\varepsilon > 0$ there exists a finite subset Z of Y (resp. of X) such that $Y \subset \cup\{B_\rho[z, \varepsilon] : z \in Z\}$.*
2. *A subset Y of X is (outside) precompact if and only if the set $\tau_{\bar{\rho}}$-cl(Y) is (outside) precompact.*
3. *A subset Y of X is precompact if and only if for every $\varepsilon > 0$ there exists a finite subset $x_1, \ldots, x_n$ of X such that $Y \subset \cup_{i=1}^n B_\rho(x_i, \varepsilon)$ and $Y \cap B_{\bar{\rho}}(x_i, \varepsilon) \neq \emptyset$, for all $i = 1, \ldots, n$.*

Proof. The nontrivial part of the assertion 1 follows from the inclusion $B[x,\varepsilon] \subset B(x,2\varepsilon)$.

2. Let $\varepsilon > 0$. Since Y is precompact there exists a finite set $Z \subset Y$ such that $Y \subset \cup\{B_\rho[z,\varepsilon] : z \in Z\}$. Since every ball $B_\rho[z,\varepsilon]$ is $\tau_{\bar{\rho}}$-closed (see Proposition 1.1.8, it follows that $\tau_{\bar{\rho}}$-cl$(Y) \subset \cup\{B_\rho[z,\varepsilon] : z \in Z\}$.

Conversely, if $\tau_{\bar{\rho}}$-cl(Y) is precompact, then for given $\varepsilon > 0$ there exists a finite subset $\{z_1,\ldots,z_k\}$ of $\tau_{\bar{\rho}}$-cl(Y) such that $\tau_{\bar{\rho}}$-cl$(Y) \subset \cup_{i=1}^k B_\rho(z_i,\varepsilon)$. For every $i \in \{1,2,\ldots,k\}$ there exists $y_i \in Y$ such that $\rho(y_i,z_i) = \bar{\rho}(z_i,y_i) < \varepsilon$. For $y \in Y$ there exists $i \in \{1,2,\ldots,k\}$ such that $y \in B(z_i,\varepsilon)$, implying

$$\rho(y_i,y) \leq \rho(y_i,z_i) + \rho(z_i,y) < 2\varepsilon\ ,$$

showing that $Y \subset \cup_{i=1}^k B_\rho(y_i,2\varepsilon)$, that is Y is precompact.

3. For $\varepsilon > 0$ let $\{x_1,\ldots,x_n\} \subset X$ such that the conditions hold for $\varepsilon/2$. If $y_i \in Y \cap B_{\bar{\rho}}(x_i,\varepsilon/2),\ i = 1,\ldots,n$, then $Y \subset \cup_{i=1}^n B_\rho(x_i,\varepsilon)$.

Indeed, for any $y \in Y$ there exists $k \in \{1,\ldots,n\}$ such that $\rho(x_k,y) < \varepsilon/2$, implying

$$\rho(y_k,y) \leq \rho(y_k,x_k) + \rho(x_k,y) = \bar{\rho}(x_k,y_k) + \rho(x_k,y) < \varepsilon\ . \qquad \square$$

Proposition 1.2.19 ([4]). *Let (X,ρ) be a quasi-semimetric space and (x_n) a sequence in X.*

1. *If (x_n) is weakly left K-Cauchy, then $\{x_n : n \in \mathbb{N}\}$ is precompact.*
2. *If (x_n) is ρ-convergent, then it is outside precompact. If x is a limit of (x_n), then $\{x\} \cup \{x_n : n \in \mathbb{N}\}$ is precompact.*
3. *There exist ρ-convergent sequences which are not precompact.*

Proof. 1. For $\varepsilon > 0$ there exists k such that $\rho(x_k,x_n) < \varepsilon$, for every $n \geq k$, implying $\{x_n : n \in \mathbb{N}\} \subset \cup\{B_\rho(x_i,\varepsilon) : 1 \leq i \leq k\}$.

2. If x is a limit of (x_n), then for every $\varepsilon > 0$ there exists $k \in \mathbb{N}$ such that $p(x_n,x) < \varepsilon$, for all $n > k$, implying $\{x\} \cup \{x_n : n \in \mathbb{N}\} \subset B_\rho(x,\varepsilon) \cup \{B_\rho(x_i,\varepsilon) : 1 \leq i \leq k\}$.

If (x_n) is ρ-convergent to some $x \in X$, then the above reasoning shows that $\{x_n : n \in \mathbb{N}\}$ is outside precompact.

3. We shall present a counterexample in the space ℓ_∞ of all bounded real sequences equipped with the asymmetric norm $p(x) = \sup_i x_i^+$, for $x = (x_i)_{i=1}^\infty \in \ell_\infty$. Consider the sequence $x_n = (1,1,\ldots,\underset{n}{1},0,0,\ldots)$, $n \geq 1$. If $z = (1,1,\ldots)$, then $p(x_n - z) = p(0,\ldots,\underset{n}{0},-1,-1,\ldots) = 0$, for all n, so that (x_n) converges to z with respect to p. Let $\varepsilon = 1/2$ and let $n_1 < n_2 < \cdots < n_k$ be an arbitrary finite subset of $\mathbb{N}$. Then, for every $n > n_k$, $p(x_n - x_{n_i}) = 1$, $i = 1,\ldots,k$, showing that $\{x_n : n \in \mathbb{N}\}$ is not contained in $\cup\{B_\rho(x_{n_i},\varepsilon) : 1 \leq i \leq k\}$. $\square$

Proposition 1.2.20 ([27]). *A τ_ρ-compact quasi-semimetric space (X,ρ) is precompact and separable.*

Proof. The precompactness is obvious: for $\varepsilon > 0$, $\{B_\rho(x, \varepsilon) : x \in X\}$ is an τ_ρ-open cover of X, so there exists a finite subset Z_ε of X such that $X = \cup\{B_\rho(z, \varepsilon) : z \in Z_\varepsilon\}$.

To prove the separability, let $Z_k \subset X$ be a finite subset such that $X = \cup\{B_\rho(z, 1/k) : z \in Z_k\}$, $k \in \mathbb{N}$. Obviously, $Z = \cup_{k=1}^\infty Z_k$ is countable. For an arbitrary $\varepsilon > 0$ let $k \in \mathbb{N}$ be such that $1/k < \varepsilon$. Then for $x \in X = \cup\{B_\rho(z, 1/k) : z \in Z_k\}$ there exists $z \in Z_k \subset Z$ such that $x \in B_\rho(z, 1/k) \subset B_\rho(z, \varepsilon)$, proving the τ_ρ-density of Z in X. □

We mention also the following result from [196].

Proposition 1.2.21. *In a precompact quasi-semimetric space (X, ρ) every sequence admits a left ρ-Cauchy subsequence.*

If X is countable, then the converse assertion is also true.

Proof. Suppose that (X, ρ) is precompact and let (x_n) be a sequence in X. By the precompactness of X there exists a finite subset Z_1 of X such that $X = \cup\{B_\rho(z, 1/2) : z \in Z_1\}$. It follows that there exists $z_1 \in Z$ and an infinite subset $M_1 : n_1^1 < n_2^1 < \cdots$, of $\mathbb{N}$ such that $\rho(z_1, x_{n_1^k}) < 1/2$ for all $k \in \mathbb{N}$. Similarly, there exist an infinite subset $M_2 : n_1^2 < n_2^2 < \cdots$, of M_1 and $z_2 \in X$ such that $\rho(z_2, x_{n_2^k}) < 1/2^2$ for all $k \in \mathbb{N}$.

Continuing in this manner we obtain the infinite sets $\mathbb{N} \supset M_1 \supset M_2 \supset \cdots$, $M_k : n_1^k < n_2^k < \cdots$, and the points $z_k \in X$ such that $\rho(z_k, x_{n_i^k}) < 1/2^k$ for all $i \in \mathbb{N}$.

We show that the diagonal sequence $(x_{n_k^k})_{k\in\mathbb{N}}$ is left ρ-Cauchy. Indeed, given $\varepsilon > 0$ let $k \in \mathbb{N}$ such that $2^{-k} < \varepsilon$. Then for every $i > k$, $x_{n_i^i} \in B_\rho(z_k, 2^{-k}) \iff \rho(z_k, x_{n_i^i}) < 2^{-k} < \varepsilon$.

Suppose that the set X is countable, $X = \{x_n : n \in \mathbb{N}\}$. If X is not precompact then there exists $\varepsilon > 0$ such that for every finite subset Z of X there exists $x \in X$ with $\rho(z, x) \geq \varepsilon$ for all $z \in Z$. It follows that for every $n \in \mathbb{N}$ there exists $y_n \in X$ such that $\rho(x_k, y_n) \geq \varepsilon$ for all $k = 1, \ldots, n$. It is obvious that the sequence (y_n) does not contain any left ρ-Cauchy subsequence. □

The set Y is called *totally bounded* if for every $\varepsilon > 0$, Y can be covered by a finite family of sets of diameter less than ε, where the diameter of a set is defined by (1.2.7). Total boundedness implies precompactness. Indeed, if $Y \subset \cup\{A_i : 1 \leq i \leq n\}$ where $\operatorname{diam}(A_i) < \varepsilon$ and $A_i \cap Y \neq \emptyset$, for $i = 1, \ldots, n$, then, taking $z_i \in A_i \cap Y$, $1 \leq i \leq n$, it follows that $Y \subset \cup\{B_\rho(z_i, \varepsilon) : 1 \leq i \leq n\}$.

As it is known, in metric spaces the precompactness, the outside precompactness and the total boundedness are equivalent notions, a result that is no longer true in quasi-metric spaces, where outside precompactness is strictly weaker than precompactness, which, in its turn, is strictly weaker than total boundedness, see [143] or [153].

It is obvious that ρ^s-precompactness implies ρ-precompactness and $\bar{\rho}$-precompactness, but the converse is not true, as the following example shows.

If (X,p) is an asymmetric seminormed space, then the terms p-precompact, $\bar{p}$-precompact, etc are understood as ρ_p-precompact, $\rho_{\bar{p}}$-precompact, etc, where by ρ_q we denote the quasi-semimetric $\rho_q(x,y) = q(y-x)$, $x, y \in X$, corresponding to an asymmetric seminorm q. Total boundedness is defined similarly.

Example 1.2.22 ([4])**.** There exists a subset of an asymmetric normed space (X,p) that is both p- and $\bar{p}$-precompact, but not p^s-precompact.

Consider the space ℓ_∞ with the asymmetric norm $p(x) = \sup_i x_i^+$, $x = (x_i) \in \ell_\infty$. Then $p^s(x) = \sup_i |x_i|$ is the usual sup-norm on ℓ_∞. Let $x^0 = (1,1,\ldots)$. Because for $x \in B_{p^s}$, $x_i - 1 \le |x_i| - 1 \le 0$, it follows that $p(x - x^0) = 0$, so that $x \in x^0 + \varepsilon B_p$ for every $\varepsilon > 0$, showing that B_{p^s} is p-precompact. The relations $B_{p^s} = -B_{p^s} \subset -x^0 + \varepsilon(-B_p) = -x^0 + \varepsilon B_{\bar{p}}$ show that B_{p^s} is also $\bar{p}$-precompact.

Since any normed space with precompact unit ball is finite dimensional, it follows that B_{p^s} is not p^s-precompact.

It is also obvious that a precompact set is outside precompact, but the converse is not true, even in asymmetric normed spaces.

Example 1.2.23 ([4])**.** There exists a set that is outside precompact but not precompact.

In the same space consider the sequence $x_n = (1,1,\ldots,\underset{(n)}{1},0,\ldots)$, $n \in \mathbb{N}$ and let $x = (1,1,\ldots)$. Then $p(x_n - x) = p((0,\ldots,0,-1,-1,\ldots)) = 0$, that is $x_n \xrightarrow{p} x$. By Proposition 1.2.19.2, the set $A = \{x_n : n \in \mathbb{N}\}$ is outside precompact but not precompact. Indeed, let $\varepsilon = 1/2$. If $x_{n_1}, \ldots, x_{n_k} \in A$, where $N_1 < n_2 < \cdots < n_k$, then $p(x_{n_k+1} - x_{n_k}) = 1$, showing that A is not precompact.

Remark 1.2.24. The above example shows also that the set $B = A \cup \{x\}$ is precompact, but not hereditarily precompact, because its subset A is not precompact.

Recall the following general results.

Proposition 1.2.25 ([73])**.**

1. *Any sequentially compact topological space is countably compact.*
2. *If a countably compact topological space satisfies the first axiom of countability, then it is sequentially compact.*

From this proposition one obtains immediately the following consequence.

Corollary 1.2.26. *A quasi-semimetric space is sequentially compact if and only if it is countably compact.*

In contrast to the case of metric spaces, compactness and sequential compactness are different notions in quasi-metric spaces.

Example 1.2.27 ([130]). Let ω_1 be the first uncountable ordinal. The space $X = [1, \omega_1)$, equipped with the quasi-semimetric $d(x, y) = 1$ if $x < y$ and $d(x, y) = 0$ otherwise, is sequentially compact but not compact.

This space is also an example of a sequentially compact quasi-semimetric space that is not precompact.

Other examples are given in [75] and [214].

Indeed, if (x_n) is a sequence in X, then there exists $x \in X$ such that $x_n \leq x$ for all $n \in \mathbb{N}$, implying $d(x, x_n) = 0$ for all $n \in \mathbb{N}$, that is $x_n \xrightarrow{d} x$.

For each $x \in X$, $B(x, 1) = \{y \in X : d(x, y) = 0\} = \{y \in X : y \leq x\}$ is a countable set, and the family $B(x, 1)$, $x \in X$, covers X. The existence of a finite subset $x_1, \ldots, x_k$ of X such that $X = B(x_1, 1) \cup \cdots \cup B(x_k, 1)$ would imply that X is countable, a contradiction.

The sequential compactness of X plus precompactness would imply compactness (see Proposition 1.2.32), so X is not precompact.

The following condition for a sequentially compact quasi-semimetric space to be compact was given in [130].

Proposition 1.2.28. *A sequentially compact quasi-semimetric space (X, ρ) such that $\operatorname{cl}\{x\}$ is compact for every $x \in X$ is compact.*

Proof. If X is compact, then the closed subset $\operatorname{cl}\{x\}$ is compact for every $x \in X$.

Suppose now that X is sequentially compact and $\operatorname{cl}\{x\}$ is compact for every $x \in X$. Let $A = \{x \in X : \forall y \in \operatorname{cl}\{x\},\ \operatorname{cl}\{y\} = \operatorname{cl}\{x\}$ and let M be the subset of A obtained by choosing exactly one element in each set $\operatorname{cl}\{x\}$.

We have

$$y \in \operatorname{cl}\{x\} \iff \rho(y, x) = 0\,.$$

Observe first that $\rho|_M$ is a T_1 quasi-metric on M. Indeed, for $x, y \in M$, $\rho(x, y) = 0$ implies $x \in \operatorname{cl}\{y\}$, so that $\operatorname{cl}\{x\} = \operatorname{cl}\{y\}$, and, by the definition of the set M, $x = y$.

Let us show now that M is sequentially compact. If (x_n) is a sequence in M, then, by the sequential compactness of the space X, there exists a subsequence (x_{n_k}) of (x_n) converging to some $x \in X$, that is $\rho(x, x_{n_k}) \to 0$ as $k \to \infty$. Since $\operatorname{cl}\{x\}$ is compact, an application of Zorn's lemma shows that it contains a nonempty minimal closed subset C. If $y \in C$, then $\operatorname{cl}\{y\} \subset C$, so that, by the minimality of $C, \operatorname{cl}\{y\} = C$. It follows that $C \subset A$, so there exists $z \in M$ such that $\operatorname{cl}\{z\} = C$. If $y \in C \subset \operatorname{cl}\{x\}$, then $\rho(y, x) = 0$, and $z \in C = \operatorname{cl}\{y\}$ implies $\rho(z, y) = 0$. The relations

$$\rho(z, x_{n_k}) \leq \rho(z, y) + \rho(y, x) + \rho(z, x_{n_k}) = \rho(z, x_{n_k}) \to 0, \quad k \to \infty\,,$$

show that $x_{n_k} \xrightarrow{\rho} z \in M$.

As a sequentially compact quasi-metric space, M is compact. Let G_i, $i \in I$, be an open cover of the space X and let G_{i_k}, $1 \leq k \leq n$, be a finite cover of M.

Let $x \in X$. Since $\mathrm{cl}\{x\}$ is compact, there exists a nonempty minimal closed subset C of $\mathrm{cl}\{x\}$. Reasoning as above, it follows that there exists $z \in M$ such that $\mathrm{cl}\{z\} = C \subset \mathrm{cl}\{x\}$, so that $\rho(z, x) = 0$. Let $k \in \{1, \dots, n\}$ be such that $z \in G_{i_k}$ and let $r > 0$ such that $B_\rho(z, r) \subset G_{i_k}$. Then $x \in B_\rho(z, r) \subset G_{i_k}$, showing that $X = \cup_{k=1}^n G_{i_k}$. Therefore the space X is compact. □

Corollary 1.2.29. *A sequentially compact T_1 quasi-metric space is compact.*

Proof. Let (X, ρ) be a T_1 quasi-metric space. It follows that $\mathrm{cl}\{x\} = \{x\}$ for every $x \in X$, so that $\mathrm{cl}\{x\}$ is compact. By Proposition 1.2.28 the space X is $\tau(\rho)$-compact.

A direct proof. As an exercise we present the elegant direct proof of this result given by Künzi [137, Proposition 2.1.11]. Suppose that the T_1 quasi-metric space (X, ρ) is countably compact but not compact. Then there exists an open cover $\mathcal{G}$ of X without any countable subcover.

For every $x \in X$ choose $G_x \in G$ such that $x \in G_x$. By Proposition 1.2.13.4, $\tau(\rho) \subset \tau(\bar{\rho})$, so there exists $n_x \in \mathbb{N}$ with $B_{\bar{\rho}}(x, 2^{-n_x}) \subset G_x$.

Let ω and ω_1 be the first countable, respectively the first uncountable, ordinal number. Then $\mathbb{N}$ can be identified with $[1; \omega) = \{\alpha : 1 \leq \alpha < \omega\}$.

We shall construct inductively a transfinite sequence $\{x_\alpha\}_{\alpha<\omega_1}$ in the following way. Let $x_1 \in X$ arbitrary and $x_\alpha \in \complement(\cup_{\beta<\alpha} G_{x_\beta})$ for $2 \leq \alpha < \omega_1$, where for $Z \subset X$, $\complement(Z) = X \setminus Z$.

For $i \in \mathbb{N}$ let $B_i = \{\alpha < \omega_1 : 2^{-n_{x_\alpha}} = i\}$. Since $\cup_{i\in\mathbb{N}} B_i = [1; \omega_1)$, there exists $k \in \mathbb{N}$ such that the set B_k is uncountable. Take a strictly increasing sequence $\alpha_1 < \alpha_2 < \cdots$ in B_k. By the countable compactness of X, the sequence $(x_{\alpha_n})_{n\in\mathbb{N}}$ has a cluster point x. The sets $F_n = \complement(\cup_{i<n} G_{x_{\alpha_i}})$ are nonempty closed and $F_1 \supset F_2 \supset \cdots$. Since $x_{\alpha_i} \in F_n$ for every $i > n$ it follows that $x \in \cap_{n\in\mathbb{N}} F_n$.

But $x \in F_n$ implies $x \notin B_{\bar{\rho}}(x_{\alpha_i}, 2^{-k})$ for all $i < n$. It follows that

$$\bar{\rho}(x_{\alpha_n}, x) \geq 2^{-k} \iff \rho(x, x_{\alpha_n}) \geq 2^{-k},$$

for all $n \in \mathbb{N}$, in contradiction to the fact that x is a cluster point of the sequence (x_{α_n}). □

Proposition 1.2.30. *A sequentially compact quasi-semimetric space (X, ρ) is sequentially left K-complete and $\mathcal{U}_\rho$-complete.*

Proof. If (x_n) is a left K-Cauchy sequence in X, then it contains a subsequence (x_{n_k}) converging with respect to τ_ρ to some $x \in X$. By Proposition 1.2.4.2, the sequence (x_n) is τ_ρ-convergent to x.

Let $\mathcal{F}$ be a $\mathcal{U}_\rho$-Cauchy filter in X. Then for every $n \in \mathbb{N}$ there exists $x_n \in X$ such that $B(x_n, 2^{-n}) \in \mathcal{F}$. By the sequential compactness of the space X, there exists a subsequence (x_{n_k}) of (x_n) that is τ_ρ-convergent to some $x \in X$. Let us show that $\lim \mathcal{F} = x$ which is equivalent to

$$B(x, 2^{-n}) \in \mathcal{F},$$

for all $n \in \mathbb{N}$. For $n \in \mathbb{N}$ there exists $k \in \mathbb{N}$ such that $n_k > n$ and $x_{n_k} \in B(x, 2^{-(n+1)})$. The implication

$$\rho(x_{n_k}, y) < 2^{-n_k} \Rightarrow \rho(x,y) \le \rho(x, x_{n_k}) + \rho(x_{n_k}, y) < 2^{-(n+1)} + 2^{-n_k} \le 2^{-n}$$

shows that $B(x_{n_k}, 2^{-n_k}) \subset B(x, 2^{-n})$, so that $B(x, 2^{-n}) \in \mathcal{F}$. □

Corollary 1.2.31 ([185]). *A compact quasi-semimetric space is left K-complete.*

Proof. Let (X, ρ) be a τ_ρ-compact quasi-semimetric space. A compact topological space is countably compact, and a first countable countably compact topological space is sequentially compact (Proposition 1.2.25). Consequently (X, ρ) is τ_ρ-sequentially compact. If (x_n) is a left K-Cauchy sequence in X, then it contains a subsequence ρ-convergent to some $x \in X$. By Proposition 1.2.4, the sequence (x_n) is ρ-convergent to x. □

Based on Proposition 1.2.30 one can prove the following result.

Proposition 1.2.32. *A precompact countably compact quasi-semimetric space is compact.*

Proof. Let (X, ρ) be precompact countably compact quasi-semimetric space. The space X is compact if and only if any ultrafilter on X is τ_ρ-convergent. Let $\mathcal{F}$ be an ultrafilter on X. Show first that $\mathcal{F}$ is $\mathcal{U}_\rho$-Cauchy, which means that for every $\varepsilon > 0$ there exists $x \in X$ such that $B(x, \varepsilon) \in \mathcal{F}$. Since X is precompact, there exists a finite subset Z of X such that $X = \cup\{B(z, \varepsilon) : z \in Z\}$. Since $X \in \mathcal{F}$ and $\mathcal{F}$ is an ultrafilter, it follows that there exists $x \in Z$ such that $B(x, \varepsilon) \in \mathcal{F}$. By Proposition 1.2.30 the space X is $\mathcal{U}_\rho$-complete, so that the $\mathcal{U}_\rho$-Cauchy filter $\mathcal{F}$ is τ_ρ-convergent. □

We mention also the following result.

Proposition 1.2.33 ([131]). *A precompact left ρ-sequentially complete quasi-semimetric space (X, ρ) is compact.*

Proof. Let (x_n) be a sequence in X. By Proposition 1.2.21, the sequence (x_n) contains a left ρ-Cauchy subsequence (x_{n_k}), which, by hypothesis, is ρ-convergent to some $x \in X$. Consequently X is sequentially compact. Since for quasi-semimetric spaces, countable compactness and sequential compactness are equivalent (see Proposition 1.2.25), the conclusion follows from Proposition 1.2.32. □

The following theorem is the analogue of a well-known result in metric spaces. The necessity follows from Proposition 1.2.20 and Corollary 1.2.31. The sufficiency was proved in [138].

Theorem 1.2.34. *A quasi-semimetric space is compact if and only if it is precompact and left K-sequentially complete.*

For the proof of sufficiency we shall need the following result of König [118].

Lemma 1.2.35. *Let E_k, $k \in \mathbb{N}$, be nonempty finite sets, $E = \cup_{k=1}^{\infty} E_k$ and $R \subset E \times E$ a relation in E such that the condition*

$$\forall y \in E_{n+1}, \ \exists x \in E_n, \quad xRy\,, \tag{1.2.23}$$

holds for every $n \in \mathbb{N}$. Then there exists a sequence $a_k \in E_k$, $k \in \mathbb{N}$, such that $a_k R a_{k+1}$ for all $k \in \mathbb{N}$.

Proof. Consider the set $\mathcal{S}$ of all $s = (x_1, \ldots, x_n) \in E_1 \times \cdots \times E_n$ such that $x_i R x_{i+1}$, $1 \le i \le n-1$, where $n \ge 2$. Let $\#E_k = n_k$. Since, by condition (1.2.23), for every $k \ge 2$ there exist at least n_k distinct elements $(x_1, x_2, \ldots, x_k)$ in $\mathcal{S}$, it follows that the set $\mathcal{S}$ is infinite. For $x_1 \in E_1$ let

$$\mathcal{S}_1(x_1) = \{s \in \mathcal{S} : s = (x_1, x_2, \ldots, x_n),\ n \ge 2\}\,.$$

Since $\mathcal{S} = \cup\{\mathcal{S}_1(x_1) : x_1 \in E_1\}$, it follows that there exists $a_1 \in E_1$ such that the set $\mathcal{S}_1(a_1)$ is infinite. Let now

$$\mathcal{S}_2(a_1, x_2) = \{s \in \mathcal{S} : s = (a_1, x_2, x_3, \ldots, x_n),\ n \ge 2\}\,.$$

Since $\cup\{\mathcal{S}_1(a_1, x_2) : x_2 \in E_2\} = \mathcal{S}_1(a_1)$, it follows that there exists $a_2 \in E_2$ such that the set $\mathcal{S}_2(a_1, a_2)$ is infinite. Supposing, by induction, that there are given $a_i \in E_i$, $1 \le i \le n$, such that the sets

$$\mathcal{S}_k(a_1, \ldots, a_k) = \{s \in \mathcal{S} : s = (a_1, \ldots, a_k, x_{k+1}, \ldots, x_m),\ m \ge k+1\}$$

are infinite for all $k = 1, 2, \ldots, n$; put

$$\begin{aligned} &\mathcal{S}_{n+1}(a_1, \ldots, a_n, x_{n+1}) \\ &\quad = \{s \in \mathcal{S} : s = (a_1, \ldots, a_n, x_{n+1}, x_{n+2}, \ldots, x_m),\ m \ge n+1\}\,, \end{aligned}$$

then it follows that

$$\cup\{\mathcal{S}_{n+1}(a_1, \ldots, a_n, x_{n+1}) : x_{n+1} \in E_{n+1}\} = \mathcal{S}_n(a_1, \ldots, a_n) \setminus \{(a_1, \ldots, a_n)\},$$

so there exists $a_{n+1} \in E_{n+1}$ such that the set $\mathcal{S}_{n+1}(a_1, \ldots, a_n, a_{n+1})$ is infinite.

The sequence (a_n) obtained in this way satisfies $a_k R a_{k+1}$ for all $k \in \mathbb{N}$. □

Now we can prove Theorem 1.2.34.

Proof of Theorem 1.2.34. Since a compact space is countably compact, and in a quasi-semimetric space countable compactness and sequential compactness are equivalent, the sequential left K-completeness of X follows from Prop. 1.2.30.

We know (Proposition 1.2.20) that every compact quasi-semimetric space is precompact. For $\varepsilon > 0$, $X = \cup\{B(x, \varepsilon) : x \in X\}$, so there exists a finite subset Z of X such that $X = \cup\{B(z, \varepsilon) : z \in Z\}$, showing that the quasi-semimetric space X is precompact.

Conversely, suppose that the quasi-semimetric space (X, ρ) is precompact and left K-sequentially complete. For every $k \in \mathbb{N}$ there exists a finite subset Z_k of X such that $X = \cup\{B(z, 2^{-k}) : z \in Z_k\}$.

For $k \in \mathbb{N}$ let

$$E_k = \{z \in Z_k : x_m \in B(z, 2^{-k+1}) \text{ for infinitely many } m \in \mathbb{N}\} ,$$

and $E = \cup_{k=1}^{\infty} E_k$.

Define a relation R on E by the conditions

$$xRy \iff \exists k \in \mathbb{N}, \quad x \in E_k \wedge y \in E_{k+1} \wedge \rho(x, y) < 2^{-k} .$$

Let us show that the relation R satisfies the hypotheses of Lemma 1.2.35. If $y \in E_{k+1}$, then $y \in Z_{k+1}$ and $x_m \in B(y, 2^{-k})$ for infinitely many $m \in \mathbb{N}$. Let $x \in Z_k$ be such that $y \in B(x, 2^{-k})$. The implication

$$\rho(y, z) < 2^{-k} \Rightarrow \rho(x, z) \leq \rho(x, y) + \rho(y, z) < 2^{-k} + 2^{-k} = 2^{-k+1}$$

shows that $B(y, 2^{-k}) \subset B(y, 2^{-k+1})$. Consequently xRy.

Now, by Lemma 1.2.35 there exists the elements $a_k \in E_k$ such that $a_k R a_{k+1}$ for all $k \in \mathbb{N}$. It follows that each ball $B(a_k, 2^{-k+1})$ contains an infinity of terms x_m of the sequence (x_n). Let n_1 be such that $x_{n_1} \in B(a_1, 1)$ and $n_2 > n_1$ such that $x_{n_2} \in B(a_2, 2^{-1})$, and so on. One obtains a sequence $n_1 < n_2 \ldots$ such that $x_{n_k} \in B(a_k, 2^{-k+1})$, $k \in \mathbb{N}$. Since $\rho(a_k, a_{k+1}) < 2^{-k}$, $k \in \mathbb{N}$, $\sum_{k=1}^{\infty} \rho(a_k, a_{k+1}) < 1$, so that by Proposition 1.2.6.1, the sequence (a_k) is left K-Cauchy. Since X is left K-sequentially complete, there exists $x_0 \in X$ such that $a_k \xrightarrow{\rho} x_0$. It follows that $\rho(x_0, x_{n_k}) \leq \rho(x_0, a_k) + \rho(a_k, x_{n_k}) < \rho(x_0, a_k) + 2^{-k} \to 0$, that is $x_{n_k} \xrightarrow{\rho} x_0$, showing that the space X is sequentially compact. By Proposition 1.2.32 the space X is compact. □

In the metric case precompactness admits the following characterization: a semimetric space X is precompact if and only if every sequence in X has a Cauchy subsequence. In the quasi-metric case this holds only for hereditarily precompact spaces. A quasi-semimetric space (X, ρ) is called *hereditarily precompact* if every subset of X is precompact. Any subset of an outside precompact set is outside precompact, but there are examples of precompact sets that are not hereditarily precompact, see Remark 1.2.24.

Proposition 1.2.36 ([138]). *For a quasi-semimetric space (X, ρ) the following are equivalent.*

1. *The space X is hereditarily precompact.*
2. *Every countable subset of X is precompact.*
3. *Every sequence in X has a left K-Cauchy subsequence.*
4. *Every sequence in X has a weakly left K-Cauchy subsequence.*

Proof. Obviously, $1 \Rightarrow 2$.

$2 \Rightarrow 3$. For a sequence (x_n) in X put $A_0 = \{x_n : n \in \mathbb{N}\}$. By hypothesis, there exists a finite subset F_1 of A_0 such that $A_0 \subset \cup\{B(x,1) : x \in F_1\}$. It follows that there exists $n_1 \in \mathbb{N}$ such that $x_{n_1} \in F_1$ and the set $M_1 = \{n \in \mathbb{N} : n \geq n_1,\ x_n \in B(x_{n_1}, 1)\}$ is infinite. Put $A_1 = \{x_n : n \in M_1,\ n > n_1\}$ and let $n_2 \in M_1,\ n_2 > n_1$, be such that the set $M_2 = \{n \in M_1 : n \geq n_2,\ x_n \in B(x_{n_2}, 1/2)\}$ is infinite. Put $A_2 = \{x_n : n \in M_2,\ n > n_2\}$ and pick $n_3 \in M_2,\ n_3 > n_2$, such that the set $M_3 = \{n \in M_2 : n \geq n_3 \text{ and } x_n \in B(x_{n_3}, 1/3)\}$ is infinite. Continuing in this manner we find the infinite sets $M_1 \supset M_2 \supset \cdots$ and the indices $n_1 < n_2 < \cdots,\ n_k \in M_k$, such that $\rho(x_{n_k}, x_n) < 1/k$, for all $n \in M_k$. For $\varepsilon > 0$ let $k \in \mathbb{N}$ such that $1/k < \varepsilon$. For $k \leq i < j,\ n_j \in M_j \subset M_i$, so that $\rho(x_{n_i}, x_{n_j}) < 1/i \leq 1/k < \varepsilon$, showing that the subsequence (x_{n_k}) is left K-Cauchy.

The implication $3 \Rightarrow 4$ is again obvious (see Proposition 1.2.1).

$4 \Rightarrow 1$. Suppose that there exists a subset A of X which is not precompact. Then there exists $\varepsilon > 0$ such that $A \setminus \cup\{B(x,\varepsilon) : x \in F\} \neq \emptyset$ for every finite subset F of A. For $x_1 \in A$ let $x_2 \in A \setminus B(x_1, \varepsilon),\ x_3 \in A \setminus (B(x_1,\varepsilon) \cup B(x_2, \varepsilon)), \ldots, x_{n+1} \in A \setminus (B(x_1,\varepsilon) \cup \cdots \cup B(x_n, \varepsilon)),\ n \in \mathbb{N}$. Then $\rho(x_n, x_m) \geq \varepsilon$ for all $m, n \in \mathbb{N}$ with $n < m$, so that the sequence (x_n) does not contain any weakly left K-Cauchy subsequence. □

One says that a a subset Y of a quasi-semimetric space (X, ρ) has the *Lebesgue property* if for every τ_ρ-open cover $\mathcal{G}$ of Y there exists $\varepsilon > 0$ such that the family $\{B_\rho(x, r) : x \in X\}$ refines $\mathcal{G}$, that is

$$\forall x \in X, \quad \exists G \in \mathcal{G}, \quad B_\rho(x, r) \subset G\,. \tag{1.2.24}$$

The following result is an extension of a known result in metric spaces.

Proposition 1.2.37. *Any countably compact quasi-semimetric space has the Lebesgue property.*

Proof. Let $\mathcal{G}$ be a countable τ_ρ-open covering of the quasi-semimetric space (X, ρ). For every $x \in X$ there exists $G_x \in \mathcal{G}$ and $r_x > 0$ such that $x \in B_\rho(x, r_x) \subset G_x$. By the countable compactness of the space X there exists $x_1, \ldots, x_n \in X$ such that $X = \cup_{k=1}^n B_\rho(x_k, 2^{-1} r_k)$, where $r_k = r_{x_k}$, $k = 1, \ldots, n$.

Put $\varepsilon = \min\{2^{-1} r_k : 1 \leq k \leq n\}$ and show that this ε satisfies (1.2.24). For $x \in X$ let $k \in \{1, \ldots, n\}$ be such that $x \in B_\rho(x_k, 2^{-1} r_k)$. Then

$$B_\rho(x_k, \varepsilon) \subset B_\rho(x_k, r_k) \subset G_{x_k}\,.$$

Indeed, if $y \in B_\rho(x_k, \varepsilon)$, then

$$\rho(x_k, y) \leq \rho(x_k, x) + \rho(x, y) < \frac{r_k}{2} + \varepsilon \leq r_k\,. \qquad \square$$

Proposition 1.2.38 ([152]). *Every subset with the Lebesgue property of a quasi-semimetric space is weakly left K-sequentially complete.*

Proof. Suppose, by contradiction, that Y is a Lebesgue subset of a quasi-semimetric space (X, ρ) and that (x_n) is a weak left K-Cauchy sequence that does not converge to any $x \in A$. Then for every $x \in Y$ there exists $r_x > 0$ such that the set $\{n \in \mathbb{N} : x_n \notin B(x, r_x)\}$ is infinite. The family of open sets $\{B(x, r_x) : x \in Y\}$ is an open cover of Y, so that there exists $\varepsilon > 0$ such that for every $x \in Y$ there exists $y \in Y$ with $B(y, \varepsilon) \subset B(x, r_x)$. Since the sequence (x_n) is weakly left K-Cauchy, there exists $n_\varepsilon \in \mathbb{N}$ such that $\rho(x_{n_\varepsilon}, x_n) < \varepsilon$ for every $n > n_\varepsilon$, By the Lebesgue property of the set Y there exists $x \in Y$ such that $B(x_{n_\varepsilon}, \varepsilon) \subset B(x, r_x)$, leading to the contradiction $x_n \in B(x_{n_\varepsilon}, \varepsilon) \subset B(x, r_x)$ for all $n > n_\varepsilon$. □

By Proposition 1.2.2 a weakly left K-Cauchy sequentially complete quasi-semimetric space is weakly left K-Cauchy sequentially complete, so that

Corollary 1.2.39. *If (X, ρ) is a quasi-semimetric space, then every Lebesgue subset of X is left K-Cauchy sequentially complete.*

Combining Propositions 1.2.37, 1.2.38, 1.2.32 and Theorem 1.2.34 one obtains.

Theorem 1.2.40 ([152]). *Let Y be a subset of a quasi-semimetric space (X, ρ). The following are equivalent.*

1. *The set Y is compact.*
2. *The set Y is precompact and sequentially compact.*
3. *The set Y is precompact and has the Lebesgue property.*

A quasi-metric space (X, ρ) is called *equi-normal* if $\rho(A, B) := \inf\{\rho(a, b) : a \in A, b \in B\} > 0$ for every pair A, B of nonempty disjoint τ_ρ-closed subsets of X. As it is remarked in [194], every Lebesgue quasi-metric is equi-normal, and every equi-normal quasi-metric is point-symmetric (or strong, meaning that $\tau(\rho) \subset \tau(\bar{\rho})$). Other properties of Lebesgue, point-symmetric and equi-normal quasi-metrics are studied in [15, 189, 194]. The papers [27] and [186] contain results on the relations between compactness, completeness and precompactness as well as numerous examples and counterexamples.

Fixed point theorems in quasi-metric spaces were proved by Hicks, Huffman and Carlson [31, 103, 104], Romaguera [191] and Romaguera and Checa [195]. Chen et al. [34, 35] proved fixed point theorems using a slightly different notion of convergence (see [32]). In [33] some optimization problems in quasi-metric and in asymmetric normed spaces are discussed. A version of Ekeland variational principle in quasi-metric spaces with applications to fixed point theorems is given in [46]. Another version of Ekeland variational principle with applications to optimization problems was proved by Ume [233].

1.2.3 Baire category

As it is well known, the Baire category theorem plays a fundamental role in the proofs of two fundamental principles of functional analysis: the uniform boundedness principle (called also the Banach-Steinhaus theorem) and the open mapping theorem. With the aim to check for possible extensions of these principles to the asymmetric case, we shall present some Baire category results in quasi-metric spaces.

Recall that a subset S of a topological space (T,τ) is called

- *dense* in T if $\mathrm{cl}(S)=T$;
- *nowhere dense* if $\mathrm{int}(\mathrm{cl}(S))=\emptyset$;
- *of first Baire category* if S can be written as a countable union of nowhere dense sets;
- *of second Baire category* if it is not of first Baire category;
- *residual* if $T\setminus S$ is of first Baire category.

One says that T is a *Baire space* if every nonempty open subset of T is of second Baire category. For the convenience of the reader we include some results on Baire category.

Proposition 1.2.41. *Let T be a topological space and $S\subset T$.*

1. *If the set S is nowhere dense, then every subset of S is nowhere dense.*
2. *If the set S is of first Baire category, then every subset of S is of first Baire category.*
3. *The union of a countable family of first Baire category sets is a set of first Baire category.*
4. *The set S is nowhere dense if and only if $\overline{S}$ is nowhere dense.*
5. (i) *If the set S is nowhere dense, then $T\setminus S$ is dense in T.*
   (ii) *If S is closed, then S is nowhere dense if and only if $T\setminus S$ is dense in T.*

The following theorem contains some useful characterizations of Baire spaces.

Theorem 1.2.42 ([110]). *Let T be a topological space. The following are equivalent.*

1. *T is a Baire space.*
2. *For every family G_n, $n\in\mathbb{N}$, of open dense subsets of T, their intersection $\cap_{n=1}^{\infty}G_n$ is dense in T.*
3. *For every family F_n, $n\in\mathbb{N}$, of closed subsets such that $\mathrm{int}\,(\cup_{n=1}^{\infty}F_n)\neq\emptyset$ there exists $n\in\mathbb{N}$ such that $\mathrm{int}(F_n)\neq\emptyset$.*
4. *Any residual subset of T is dense in T.*
5. *Any first category subset of T has empty interior.*

The following proposition extends a well-known result for topological vector spaces.

Proposition 1.2.43. *A second category asymmetric LCS is a Baire space.*

Proof. Let (X, P) be an asymmetric LCS and suppose, by contradiction, that X contains a nonempty open set G that is of first Baire category in X. If $x_0 \in G$, then $U := -x_0 + G$ is an open neighborhood of 0 which is a set of first Baire category too. For $x \in X$ the sequence $(n^{-1}x)_{n\in\mathbb{N}}$ converges to 0 as $n \to \infty$, implying the existence of $n \in \mathbb{N}$ such that $n^{-1}x \in U \iff x \in nU$. It follows that $X = \cup_{n=1}^{\infty} nU$, so that the space X is of first Baire category. □

The first Baire type result for quasi-metric spaces was proved by Kelly [111] (see also [185]).

Theorem 1.2.44. *Let (X, ρ) be a quasi-semimetric space. If X is right K-$\bar{\rho}$-sequentially complete, then (X, τ_ρ) is of second category in itself.*

Proof. Let $X = \cup_{n=1}^{\infty} X_n$ where the sets X_n are τ_ρ-closed for all $n \in \mathbb{N}$, and suppose, by contradiction, that τ_ρ-int$(X_n) = \emptyset$ for all $n \in \mathbb{N}$. Then $X \setminus X_1$ is nonempty and τ_ρ-open, so there exist $x_1 \in X$ and $0 < r_1 < 1$ such that $B_\rho[x_1, r_1] \subset X \setminus X_1$. Similarly, the set $B_\rho(x_1, r_1) \setminus X_2$ is nonempty and τ_ρ-open so there exist $x_2 \in X$ and $0 < r_2 < 1/2$ such that $B_\rho[x_2, r_2] \subset B_\rho(x_1, r_1) \setminus X_2$. Continuing in this manner one obtains the points $x_n \in X$ and the numbers $0 < r_n < 1/n$ such that $B_\rho[x_n, r_n] \subset B_\rho(x_{n-1}, r_{n-1}) \setminus X_n$ for all $n \in \mathbb{N}$, where $B_\rho(x_0, r_0) = X$.

It follows

$$\bigcap_{n=1}^{\infty} B_\rho[x_n, r_n] \subset \bigcap_{n=1}^{\infty} (X \setminus X_n) = X \setminus \bigcup_{n=1}^{\infty} X_n = \emptyset,$$

in contradiction to Proposition 1.2.9. □

Remark 1.2.45. The right K-$\bar{\rho}$-completeness from Theorem 1.2.44 can be formulated in the equivalent form: every left K-ρ-Cauchy sequence is $\bar{\rho}$-convergent.

Other Baire type results were proved by Gregori and Ferrer [76] (rediscovered in [21]). A topological space (X, τ) is called *quasi-regular* if every nonempty open subset of X contains the closure of some nonempty open set, that is for every nonempty open subset V of X there exists a nonempty open set U such that

$$\overline{U} \subset V\,. \tag{1.2.25}$$

Theorem 1.2.46 ([76]). *If the quasi-semimetric space (X, ρ) is τ_ρ-quasi-regular and left ρ-sequentially complete, then it is a Baire space.*

Proof. The proof is similar to the proof of Theorem 1.2.44. Let G_n, $n \in \mathbb{N}$, be a family of dense open subsets of X, $G = \cap_{n=1}^{\infty} G_n$ and $U \subset X$ nonempty open. We have to show that $U \cap G \neq \emptyset$. Since G_1 is dense in X, $U \cap G_1 \neq \emptyset$, so that, by

(1.2.25), there exist $x_1 \in X$ and $0 < r_1 < 1$ such that τ_ρ-cl$(B_1) \subset U \cap G_1$, where $B_1 = B_\rho(x_1, r_1)$. Similarly $B_1 \cap G_2$ is nonempty open so there exist $x_2 \in X$ and $0 < r_2 < 1/2$ such that τ_ρ-cl$(B_2) \subset B_1 \cap G_2$, where $B_2 = B_\rho(x_2, r_2)$. Continuing in this manner one obtains the open balls $B_n = B_\rho(x_n, r_n)$ with $0 < r_n < 1/n$ such that τ_ρ-cl$(B_n) \subset B_{n-1} \cap G_n$, $n \in \mathbb{N}$, where $B_0 = U$. As in the proof of Theorem 1.2.44, the relations $x_n \in B_\rho(x_n, r_n) \subset B_\rho(x_m, r_m)$, valid for all $n, m \in \mathbb{N}$ with $m < n$, imply that the sequence (x_n) is left K-ρ-Cauchy, so that, by hypothesis, there exists $x \in X$ with $x_n \xrightarrow{\rho} x$ as $n \to \infty$. Since $x_m \in B_\rho(x_n, r_n) = B_n$ for all $m > n$, it follows that $x \in \tau_\rho$-cl$(B_n) \subset B_{n-1} \cap G_n \subset U \cap G_n$ for all $n \in \mathbb{N}$, implying $x \in \bigcap_{n \in \mathbb{N}} (U \cap G_n) = U \cap G$. □

Taking into account the implications from Proposition 1.2.2, we get the following

Corollary 1.2.47. *A quasi-regular quasi-semimetric space (X, ρ) is a Baire space in any of the following cases*

(a) *(X, ρ) is left ρ-sequentially complete;*

(b) *(X, ρ) is weakly left K-sequentially complete;*

(c) *(X, ρ) is left K-sequentially complete.*

1.2.4 Baire category in bitopological spaces

We shall present now, following the paper [7], pairwise Baire bitopological spaces and their properties.

Let (T, τ, ν) be a bitopological space. A subset S of T is called

- (τ, ν)-*nowhere dense* if ν-int$(\tau$-cl$(S)) = \emptyset$;
- of (τ, ν)-*first category* if it is the union of a countable family of (τ, ν)-nowhere dense sets;
- of (τ, ν)-*second category* if it is not of (τ, ν)-first category;
- (τ, ν)-*residual* if $T \setminus S$ is of (τ, ν)-first category.

The space (T, τ, ν) is called

- (τ, ν)-*Baire* if each nonempty τ-open subset of T is of (ν, τ)-second category;
- *pairwise Baire* if it is both (τ, ν)-Baire and (ν, τ)-Baire.

The following properties are the bitopological analogs of properties from Proposition 1.2.41.

Proposition 1.2.48. *Let (T, τ, ν) be a bitopological space and $S \subset T$.*

1. *If S is (τ, ν)-nowhere dense, then every subset of S is (τ, ν)-nowhere dense.*
2. *If S is of (τ, ν)-first category, then every subset of S is of (τ, ν)-first category.*

3. *The union of a countable family of* (τ,ν)*-first category sets, is of* (τ,ν)*-first category.*
4. *A subset* S *of a bitopological space* (T,τ,ν) *is* (τ,ν)*-nowhere dense if and only if* $\tau\text{-int}(T\setminus S)$ *is* ν*-dense in* T.

Proof. Only the equivalence from 4 needs a proof. Suppose that $\nu\text{-int}(\tau\text{-cl}(S))=\emptyset$. Then for any nonempty set $V\in\nu$, $V\cap(T\setminus\tau\text{-cl}(S))\neq\emptyset$. Since $T\setminus\tau\text{-cl}(S)\subset\tau\text{-int}(T\setminus S)$ it follows that $V\cap\tau\text{-int}(T\setminus S)\neq\emptyset$, that is $\tau\text{-int}(T\setminus S)$ is ν-dense in T.

To prove the converse, suppose that there exists $t\in\nu\text{-int}(\tau\text{-cl}(S))$. Then

$$\tau\text{-}\operatorname{cl}(S)\cap\tau\text{-}\operatorname{int}(T\setminus S)=\emptyset\,. \tag{1.2.26}$$

Indeed, if some $s\in T$ belongs to this intersection, then, as $\tau\text{-int}(T\setminus S)$ is a τ-neighborhood of s, it follows that $S\cap\tau\text{-int}(T\setminus S)\neq\emptyset$, yielding the contradiction $S\cap(T\setminus S)\neq\emptyset$.

Since $\tau\text{-cl}(S)$ is a ν-neighborhood of t the relation (1.2.26) shows that $t\notin\nu\text{-cl}(\tau\text{-int}(T\setminus S)$, that is $\tau\text{-int}(T\setminus S)$ is not ν-dense in T. □

The following theorem contains some characterizations of pairwise Baire spaces similar to those of Baire spaces (see Theorem 1.2.42) and with similar proofs.

Theorem 1.2.49 ([7]). *For a bitopological space* (T,τ,ν) *the following are equivalent.*

1. *The space* T *is* (τ,ν)*-Baire.*
2. *The intersection of each countable family of* τ*-dense* ν*-open subsets of* T *is* τ*-dense in* T.
3. $\tau\text{-int}\,(\cup_{n=1}^\infty F_n)=\emptyset$ *for every family* F_n, $n\in\mathbb{N}$, *of* ν*-closed sets with empty* τ*-interiors.*
4. *For every subset* M *of* T *of first* (ν,τ)*-category the set* $T\setminus M$ *is* τ*-dense in* T.

Proof. As an exercise of acquaintance with the notions we shall give the proofs of these equivalences.

1 ⇒ 2. Let G_n, $n\in\mathbb{N}$, be a family of ν-open τ-dense subsets of T and suppose that their intersection is not τ-dense in T, that is $G:=T\setminus\tau\text{-cl}(\cap_{n=1}^\infty G_n)\neq\emptyset$. Then $G\in\tau$ and $G\subset T\setminus\cap_{n=1}^\infty G_n=\cup_{n=1}^\infty(T\setminus G_n)$.

The set $F_n:=T\setminus G_n$ is ν-closed and $T\setminus F_n=G_n$ is τ-dense in T. By Proposition 1.2.48.4, F_n is (ν,τ)-nowhere dense, implying that the nonempty τ-open set G is of (ν,τ)-first category, that is T is not (τ,ν)-Baire.

2 ⇒ 1. To prove the converse, suppose that T is not (τ,ν)-Baire. Then there exists a nonempty set $G\in\tau$ such that $G=\cup_{n=1}^\infty S_n$, where each S_n is (ν,τ)-nowhere dense. By Proposition 1.2.48.4 this is equivalent to the fact that the ν-open set $G_n:=\nu\text{-int}(T\setminus S_n)$ is τ-dense in T. Since

$$\cap_{n=1}^\infty G_n\subset\cap_{n=1}^\infty(T\setminus S_n)=T\setminus\cup_{n=1}^\infty S_n=T\setminus G\,,$$

and $T \setminus G$ is τ-closed, this implies

$$\tau\text{-}(\cap_{n=1}^{\infty} G_n) \subset T \setminus G \neq T ,$$

that is $\cap_{n=1}^{\infty} G_n$ is not τ-dense in T.

1 $\Rightarrow$ 3. Let F_n be ν-closed with τ-int$(F_n) = \emptyset$ for all $n \in \mathbb{N}$. It follows that F_n is (ν, τ)-nowhere dense. If $G := \tau$-int$(\cup_n F_n) \neq \emptyset$, then $G \subset \cup_n F_n$ implies that G is a nonempty τ-open set which is of first (ν, τ)-category, in contradiction to the fact that the space T is (τ, ν)-Baire. Consequently τ-int$(\cup_n F_n = \emptyset$.

3 $\Rightarrow$ 4. Let $M = \cup_{n=1}^{\infty} S_n$, where τ-int$(\nu$-cl$(S_n)) = \emptyset$, $n \in \mathbb{N}$, and let $F_n = \nu$-cl(S_n), $n \in \mathbb{N}$. If $T \setminus M$ is not τ-dense in T, then the smaller set $T \setminus \cup_n F_n$ is not τ-dense in T too, so there exists a nonempty set $G \in \tau$ such that $G \cap (T \setminus \cup_n F_n) = \emptyset$. But then $G \subset \cup_n F_n$, in contradiction to the fact that τ-int $(\cup_{n=1}^{\infty} F_n) = \emptyset$.

4 $\Rightarrow$ 1. Suppose by contradiction that T is not (τ, ν)-Baire. Then there exists a nonempty set $G \in \tau$ which is of first (ν, τ)-category. Then $T \setminus G$ is τ-closed, so that τ-cl$(T \setminus G) = T \setminus G \neq T$, that is G is a set of first (ν, τ)-category that is not τ-dense in T. □

Fukutake [84] proved the following result.

Theorem 1.2.50. *If (T, τ, ν) is a bitopological space such that $\tau \subseteq \nu$ and ν is metrizable and complete, then T is τ-ν-Baire.*

To present the corresponding pairwise Baire result we need a new notion. The bitopological space (T, τ, ν) is called *pairwise fine* if every nonempty τ-open set contains a nonempty ν-open set, and every nonempty ν-open set contains a nonempty τ-open set.

Theorem 1.2.51 ([7]). *Let (T, τ, ν) be a pairwise fine bitopological space. Then T is pairwise Baire if and only if both (T, τ) and (T, ν) are Baire topological spaces.*

The notion of quasi-regularity can be also adapted to bitopological spaces, see [7]. A bitopological space (T, τ, ν) is called

- *τ-ν-quasi-regular* if each nonempty τ-open subset of T contains the ν-closure of a nonempty τ-open set;
- *pairwise quasi-regular* if it is both τ-ν-quasi-regular and ν-τ-quasi-regular.

A cover $\mathcal{U}$ of T is called *pairwise open* if $\mathcal{U} \subset \tau \cup \nu$ and both $\mathcal{U} \cap \tau$ and $\mathcal{U} \cap \nu$ contain at least one nonempty set.

Let (T, τ, ν) be a bitopological space. We say that

- T is *pairwise compact* if every pairwise open cover of T contains a finite subcover;
- the topology τ is *locally compact* with respect to ν if each point $t \in T$ admits a τ-open neighborhood U such that ν-cl(U) is ν-compact;
- the space T is *pairwise locally compact* if τ is locally compact with respect to ν and ν is locally compact with respect to τ, or, equivalently, if each point

$t \in T$ admits a τ-open neighborhood U and a ν-open neighborhood V such that both ν-cl(U) and τ-cl(V) are pairwise compact.

We mention the following results.

Proposition 1.2.52 ([145]). *Let (T, τ, ν) be a bitopological space.*

1. *If T is pairwise Hausdorff and ν is locally compact with respect to τ, then $\tau \subset \nu$.*
2. *If T is pairwise Hausdorff and (T, τ) compact, then $\tau \subset \nu$.*
3. *If (X, ρ) is a T_1 quasi-metric space and $\tau_{\bar{\rho}}$ is locally compact with respect to τ_ρ, then $\tau_\rho \subset \tau_{\bar{\rho}}$.*

Proof. Only the first assertion needs a proof, the others being direct consequences of it.

Let $t \in T$ and $U \in \tau$ such that $t \in U$. Since ν is locally compact with respect to τ, there exists $V \in \nu$ containing t such that the set $\overline{V}^\tau$ is τ-compact. Let $C = V \setminus U$. Then $\overline{C}^\tau \subset \overline{V}^\tau$, so that $\overline{C}^\tau$ is also τ-compact.

Observe that $t \notin \overline{C}^\tau$. Indeed, the hypothesis $t \in \overline{C}^\tau$ leads to the contradiction $\emptyset \neq U \cap C = U \cap (V \setminus U) = \emptyset$.

Since T is pairwise Hausdorff, for every $s \in \overline{C}^\tau$ there exists U_s, V_s such that $s \in U_s \in \tau$, $t \in V_s \in \nu$, and $U_s \cap V_s = \emptyset$.

The τ-compactness of $\overline{C}^\tau$ implies the existence of $s_1, \ldots, s_n \in T$ such that $\overline{C}^\tau \subset \cup_{i=1}^n U_{s_i}$.

Let $W := V \cap \cap_{i=1}^n V_{s_i} \in \nu$. Then $t \in W$ and $W \cap (\cup_{i=1}^n U_{s_i}) = \emptyset$, so that $W \cap C = \emptyset$.

If $s \in W$, then $s \in V$ and $s \notin C = V \cap \complement(U)$, so that $s \in \complement(\complement(V) \cap U) = U \cup \complement(V)$, implying $s \in U$. Consequently, $W \subset U$, showing that $U \in \mathcal{V}_\nu(t)$ and that $\tau \subset \nu$. □

We mention also the following result.

Theorem 1.2.53 ([144]). *A bitopological space (T, τ, ν) is quasi-uniformizable if and only if it is pairwise completely regular.*

It is well known that a quasi-regular locally compact topological space is a Baire space (see [102]). The following theorem contains the bitopological analog of this result.

Theorem 1.2.54 ([7]). *Every pairwise fine, pairwise normal and pairwise locally compact bitopological space is pairwise Baire.*

1.2.5 Completeness and compactness in quasi-uniform spaces

The completeness notions for quasi-metric spaces considered in the previous subsections can be extended to quasi-uniform spaces by replacing sequences by filters or by nets.

Let $(X,\mathcal{U})$ be a quasi-uniform space, $\mathcal{U}^{-1} = \{U^{-1} : U \in \mathcal{U}\}$ the conjugate quasi-uniformity on X, and $\mathcal{U}^s = \mathcal{U} \vee \mathcal{U}^{-1}$ the coarsest uniformity finer than $\mathcal{U}$ and $\mathcal{U}^{-1}$. The quasi-uniform space $(X,\mathcal{U})$ is called *bicomplete* if $(X,\mathcal{U}^s)$ is a complete uniform space. This notion is useful and easy to handle, because one can appeal to well-known results from the theory of uniform spaces, but it is not very appropriate for the study of the specific properties of quasi-uniform spaces.

Call a filter $\mathcal{F}$ on a quasi-uniform space $(X,\mathcal{U})$

- *left $\mathcal{U}$-Cauchy* (*right $\mathcal{U}$-Cauchy*) if for every $U \in \mathcal{U}$ there exists $x \in X$ such that $U(x) \in \mathcal{F}$ (respectively $U^{-1}(x) \in \mathcal{F}$);
- *left K-Cauchy* (*right K-Cauchy*) if for every $U \in \mathcal{U}$ there exists $F \in \mathcal{F}$ such that $U(x) \in \mathcal{F}$ (respectively $U^{-1}(x) \in \mathcal{F}$) for all $x \in F$.

The notions of left and right K-Cauchy filter were defined and studied by Romaguera in [190, 192].

A net $(x_i : i \in I)$ in $(X,\mathcal{U})$ is called

- *left $\mathcal{U}$-Cauchy* (*right $\mathcal{U}$-Cauchy*) if for every $U \in \mathcal{U}$ there exists $x \in X$ and $i_0 \in I$ such that $(x, x_i) \in U$ (respectively $(x_i, x) \in U$) for all $i \geq i_0$;
- *left K-Cauchy* if

$$\forall U \in \mathcal{U},\ \exists i_0 \in I,\ \forall i,j \in I,\ i_0 \leq i \leq j \Rightarrow (x_i, x_j) \in U\,. \tag{1.2.27}$$

- *right K-Cauchy* if

$$\forall U \in \mathcal{U},\ \exists i_0 \in I,\ \forall i,j \in I,\ i_0 \leq i \leq j \Rightarrow (x_j, x_i) \in U\,. \tag{1.2.28}$$

Observe that

$$(x_j, x_i) \in U \iff (x_i, x_j) \in U^{-1}\,,$$

so that a net is right K-Cauchy with respect to $\mathcal{U}$ if and only if it is left K-Cauchy with respect to $\mathcal{U}^{-1}$. A similar remark applies to $\mathcal{U}$-nets.

The quasi-uniform space $(X,\mathcal{U})$ is called

- *left $\mathcal{U}$-complete* (*left K-complete*) if every left $\mathcal{U}$-Cauchy (respectively, left K-Cauchy) filter in X is $\tau(\mathcal{U})$-convergent;
- *left $\mathcal{U}$-complete by nets* (*left K-complete by nets*) if every left $\mathcal{U}$-Cauchy (respectively, left K-Cauchy) net in X is $\tau(\mathcal{U})$-convergent;
- *Smyth left K-complete by nets* if every left K-Cauchy net in X is $\mathcal{U}^s$-convergent.

The notions of right completeness are defined similarly, by asking the $\tau(\mathcal{U})$-convergence of the corresponding right Cauchy filter (or net) with respect to the topology $\tau(\mathcal{U})$ (or with respect to $\tau(\mathcal{U}^s)$ in the case of Smyth completeness).

The notion of Smyth completeness (see [133] and [225], [229], [230]) has applications to computer science, see [216]. In fact, there are a lot of applications of quasi-metric spaces, asymmetric normed spaces and quasi-uniform spaces to computer science, abstract languages, the analysis of the complexity of programs, see, for instance, [89, 92, 171, 207, 208, 209, 210, 216].

Notice also that these notions of completeness can be considered within the framework of bitopological spaces in the sense of Kelly [111], since a quasi-uniform space is a bitopological space with respect to the topologies $\tau(\mathcal{U})$ and $\tau(\mathcal{U}^{-1})$. For this approach see the papers by Deák [57, 58].

The study of some bitopological and quasi-uniform structures on concrete spaces of semi-continuous or continuous functions was done in the papers [79] and [139], respectively.

The notions of total boundedness and precompactness can be also extended to quasi-uniform spaces.

A subset Y of a quasi-uniform space $(X,\mathcal{U})$ is called

- *precompact* if for every $U \in \mathcal{U}$ there exists a finite subset Z of Y such that $Y \subset U(Z)$;
- *outside precompact* if for every $U \in \mathcal{U}$ there exists a finite subset Z of X such that $Y \subset U(Z)$;
- *totally bounded* if for every $U \in \mathcal{U}$ there exists a finite family of subsets $Z_i,\ i = 1,\dots,n$, of X such that $Z_i \times Z_i \subset U,\ i = 1,\dots,n$, and $Y \subset \cup_{i=1}^n Z_i$, or, equivalently, if Y is totally bounded with respect to the associated uniformity $\mathcal{U}^s$.

The space $(X,\mathcal{U})$ is called *hereditarily precompact* if every subset of it is precompact.

We mention the following simple properties of Cauchy filters. The sequential versions of the assertions 2 and 3 are contained in Proposition 1.2.4.

Proposition 1.2.55. *Let $(X,\mathcal{U})$ be a quasi-uniform space.*

1. *Any left K-Cauchy filter is left $\mathcal{U}$-Cauchy. The same is true for right Cauchy filters.*
2. *If x is a cluster point of a left K-Cauchy filter $\mathcal{F}$, then $\mathcal{F}$ is $\tau(\mathcal{U})$-convergent to x.*
3. *If x is a cluster point of a right K-Cauchy filter $\mathcal{F}$, then $\mathcal{F}$ is $\tau(\mathcal{U})$-convergent to x.*

Proof. 1. Let $\mathcal{F}$ be a left K-Cauchy filter. For $U \in \mathcal{U}$ there exists $F \in \mathcal{F}$ such that $U(x) \in \mathcal{F}$ for all $x \in F$. Taking an element x in the nonempty set F it follows that $\mathcal{F}$ is left $\mathcal{U}$-Cauchy.

2. Let x be a $\tau(\mathcal{U})$-cluster point of a left K-Cauchy filter $\mathcal{F}$ on X, that is $x \in \cap\{\tau(\mathcal{U})\text{-cl}(F) : F \in \mathcal{F}\}$. For $U \in \mathcal{U}$ let $V \in \mathcal{U}$ such that $V^2 \subset U$. Let $F \in \mathcal{F}$ be such that $V(y) \in \mathcal{F}$ for all $y \in F$. By hypothesis, the set $F \cap V(x)$ is nonempty. Taking a point z in this intersection, then $V(z) \in \mathcal{F}$. Let $y \in V(z) \iff (z,y) \in V$. Since $(x,z) \in V$ it follows that $(x,y) \in V^2 \iff y \in V^2(x)$, that is $V(z) \subset V^2(x) \subset U(x)$. Consequently, $U(x) \in \mathcal{F}$ and $\mathcal{F}$ is $\tau(\mathcal{U})$-convergent to x.

3. Let x be a $\tau(\mathcal{U})$-cluster point of a right K-Cauchy filter $\mathcal{F}$ on X. For $U \in \mathcal{U}$ let $V \in \mathcal{U}$ be such that $V^2 \subset U$, and let $F \in \mathcal{F}$ be such that $V^{-1}(y) \in \mathcal{F}$ for all $y \in F$.

Let $y \in F$. Since $V^{-1}(y) \cap V(x) \neq \emptyset$, there exists $z \in V^{-1}(y) \cap V(x)$. It follows that $(z,y) \in V$ and $(x,z) \in V$, so that $(x,y) \in V^2 \iff y \in V^2(x)$. Consequently, $F \subset V^2(x) \subset U(x)$, showing that $U(x) \in \mathcal{F}$, that is $\mathcal{F}$ is $\tau(\mathcal{U})$-convergent to x. □

Now we shall present, following [218] a characterization of compactness in quasi-uniform spaces.

Proposition 1.2.56 ([218]). *Let $(X,\mathcal{U})$ be a quasi-uniform space.*

1. *The space $(X,\mathcal{U})$ is precompact if and only if every ultrafilter on X is left $\mathcal{U}$-Cauchy.*
2. *If $(X,\mathcal{U})$ is compact, then every left $\mathcal{U}$-Cauchy filter is $\tau(\mathcal{U})$-convergent.*

Proof. 1. Suppose that $(X,\mathcal{U})$ is precompact and let $\mathcal{F}$ be an ultrafilter on X. Then for every $U \in \mathcal{U}$ there exists a finite subset Z of X such that $X = \cup\{U(z) : z \in Z\}$. Since $\mathcal{F}$ is an ultrafilter there exists $z \in Z$ such that $U(z) \in \mathcal{F}$, proving that the ultrafilter $\mathcal{F}$ is left $\mathcal{U}$-Cauchy.

To prove the converse suppose that $(X,\mathcal{U})$ is not precompact. Then there exists $U \in \mathcal{U}$ such that for every finite subset Z of X, $X \setminus U(Z) = X \setminus \cup\{U(z) : z \in Z\} \neq \emptyset$. The family $\{X \setminus U(Z) : Z \subset X,\ Z \text{ finite nonempty}\}$ is the base of a filter $\mathcal{G}$ on X. Let $\mathcal{F}$ be an ultrafilter on X such that $\mathcal{G} \subset \mathcal{F}$. Then for every $x \in X$, $X \setminus U(x) \in \mathcal{G} \subset \mathcal{F}$. Consequently $U(x) \notin \mathcal{F}$ for all $x \in X$, so that the ultrafilter $\mathcal{F}$ is not left $\mathcal{U}$-Cauchy.

2. Suppose that X contains a left $\mathcal{U}$-Cauchy filter $\mathcal{F}$ that is not convergent. Then for every $x \in X$ there exists $U_x \in \mathcal{U}$ such that $U_x(x) \notin \mathcal{F}$. Let $V_x \in \mathcal{U}$ such that $V_x \circ V_x \subset U_x$, $x \in X$. Taking into account the compactness of X, there exists a finite subset $x_1, \dots, x_n$ of X such that $X = \cup_{k=1}^n V_{x_k}(x_k)$. Let $V = \cap_{k=1}^n V_{x_k} \in \mathcal{U}$. Since the filter $\mathcal{F}$ is left $\mathcal{U}$-Cauchy, there exists $z \in X$ such that $V(z) \in \mathcal{F}$. Let $k \in \{1, \dots, n\}$ be such that $z \in V_{x_k}(x_k) \iff (x_k, z) \in V_{x_k}$. If we show that $V(z) \subset U_{x_k}(x_k)$, then $U_{x_k}(x_k) \in \mathcal{F}$, in contradiction to the choice of the sets $U_x \in \mathcal{U}$. Indeed, $y \in V(z) \iff (z,y) \in V \subset V_{x_k}$ implies $(x_k, y) \in V_{x_k} \circ V_{x_k} \subset U_{x_k}$, so that $y \in U_{x_k}(x_k)$. □

The following characterization of compactness is the analog of a well-known result in uniform spaces.

Theorem 1.2.57 ([218]). *A quasi-uniform space $(X,\mathcal{U})$ is compact if and only if it is precompact and left $\mathcal{U}$-complete.*

Proof. It is clear that a compact quasi-uniform space is precompact. By Proposition 1.2.56.2 it is left $\mathcal{U}$-complete too.

Let $\mathcal{F}$ be an ultrafilter on X. By Proposition 1.2.56.1, $\mathcal{F}$ is left $\mathcal{U}$-Cauchy, so that it is $\tau(\mathcal{U})$-convergent, proving the $\tau(\mathcal{U})$-compactness of X. □

We shall present now following [229] the equivalence between completeness in terms of filters and in terms of nets.

A filter $\mathcal{F}$ in a quasi-uniform space $(X,\mathcal{U})$ is called

- *round* if
$$\forall A \in \mathcal{F},\ \exists B \in \mathcal{F},\ \exists U \in \mathcal{U}, \quad U(B) \subset A\,; \tag{1.2.29}$$
- S_u-*Cauchy* if
$$\forall U \in \mathcal{U},\ \forall A \in \mathcal{F},\ \exists x \in A, \quad U(x) \in \mathcal{F}\,. \tag{1.2.30}$$

For a net $\varphi = (x_i : i \in I)$ in $(X,\mathcal{U})$ let
$$E_i = \{x_j : j \in I,\ j \geq i\},\ i \in I, \quad \text{and} \quad F_{i,U} = U(E_i),\ (i,U) \in I \times \mathcal{U}\,. \tag{1.2.31}$$

Then $F_{i,U} = U(E_i),\ (i,U) \in I \times \mathcal{U}$ is the base of a filter on X denoted by $\Phi(\varphi)$ (or $\Phi(x_i)$).

Proposition 1.2.58 ([229]). *Let $(X,\mathcal{U})$ be a quasi-uniform space.*

1. *If $\varphi = (x_i : i \in I)$ is a left K-Cauchy net, then the associated filter $\Phi(\varphi)$ is round S_u-Cauchy and*
   (i) $x_i \xrightarrow{\mathcal{U}} x \implies \mathcal{V}_{\mathcal{U}}(x) \subset \Phi(\varphi)$ *(i.e., $\lim_{\mathcal{U}} \Phi(\varphi) = x$);*
   (ii) $x_i \xrightarrow{\mathcal{U}^{-1}} x \implies \Phi(\varphi) \subset \mathcal{V}_{\mathcal{U}}(x)$;
   (iii) $x_i \xrightarrow{\mathcal{U}^s} x \implies \Phi(\varphi) = \mathcal{V}_{\mathcal{U}}(x)$.
2. *To each round S_u-Cauchy filter $\mathcal{G}$ one can associate a left K-Cauchy net $\varphi = (x_\lambda : \lambda \in \Lambda)$ such that $\Phi(\varphi) = \mathcal{G}$.*

Proof. 1. Let $\Phi(\varphi)$ be the filter associated to the left K-Cauchy net $\varphi=(x_i : i\in I)$.

Claim I. The filter $\Phi(\varphi)$ is round.

For $(i_1, U) \in I \times U$ let $V \in \mathcal{U}$ such that $V^2 \subset U$ and let $i_0 \in I,\ i_0 \geq i_1$ be such that (1.2.27) holds for V. The conclusion will follow if we show that
$$U(F_{i_0,V}) \subset F_{i_1,U}\,. \tag{1.2.32}$$

We have
$$y \in U(F_{i_0,V}) \iff \exists z \in F_{i_0,V},\ (z,y) \in V,$$
$$z \in F_{i_0,V} \iff \exists i \geq i_0 \geq i_1,\ (x_i, z) \in V\,,$$
so that $(x_i, y) \in V^2 \subset U$, implying $y \in U(x_i) \subset U(E_{i_1}) = F_{i_1,U}$.

Claim II. The filter $\Phi(\varphi)$ is S_u-Cauchy.

Let again $F_{i_1,U} \in \Phi(\varphi)$, $V \in \mathcal{U}$ and $W \in \mathcal{U}$ such that $W^2 \subset V$. Since the net (x_i) is left K-Cauchy, there exists $i_0 \in I,\ i_0 \geq i_1$ such that

$$\forall i, j \in I,\ i_0 \leq i \leq j \Rightarrow (x_i, x_j) \in W\ . \tag{1.2.33}$$

Let us show that $F_{i_0,W} \subset V(x_{i_0})$, which implies $V(x_{i_0}) \in \Phi(\varphi)$, that is (1.2.30) holds for $x_{i_0} \in F_{i_1,U}$ (with V instead of U).

Observe first that $i_0 \geq i_1$ and $(x_{i_0}, x_{i_0}) \in U$ imply $x_{i_0} \in U(E_{i_1}) = F_{i_1,U}$ Also

$$y \in F_{i_0,W} \iff \exists i \geq i_0,\ (x_i, y) \in W\ .$$

By (1.2.33), $(x_{i_0}, x_i) \in W$, so that $(x_{i_0}, y) \in W^2 \subset V$, showing that $y \in V(x_{i_0})$.

The proof of (i). For $U \in \mathcal{U}$ let $V \in \mathcal{U}$ such that $V^2 \subset U$. By hypothesis there exists $i_0 \in I$ such that $E_{i_0} \subset V(x)$. If we show that $V(E_{i_0}) \subset U(x)$, then $U(x) \in \Phi(\varphi)$.

Again,

$$y \in V(E_{i_0}) \iff \exists i \geq i_0,\ (x_i, y) \in V\ .$$

By the choice of i_0, $(x, x_i) \in V$, so that $(x, y) \in V^2 \subset U$, that is $y \in U(x)$.

The proof of (ii). For $F_{i_1,U} \in \Phi$ let $V \in \mathcal{U}$ such that $V^2 \subset U$.

Let $i_0 \geq i_1$ such that

$$E_{i_0} \subset V^{-1} \iff \forall i \geq i_0, \quad (x_i, x) \in V\ .$$

Let $y \in V(x) \iff (x, y) \in V$. Since for every $i \geq i_0 \geq i_1$, $(x_i, x) \in V$, it follows that $(x_i, y) \in V^2 \subset U$, that is $V(x) \subset U(E_{i_1}) = F_{i_1,U}$, showing that $F_{i_1,U} \in \mathcal{V}_{\mathcal{U}}(x)$.

The assertion (iii) is a consequence of (i) and (ii).

2. Let now $\mathcal{G}$ be a round S_u-Cauchy filter in $(X, \mathcal{U})$. Consider the set

$$\Lambda = \{(U, A) : U \in \mathcal{U},\ A \in \mathcal{G}\}\ .$$

By the condition (1.2.30),

$$\forall (U, A) \in \mathcal{U} \times \mathcal{G},\ \exists x_{(U,A)} \in A, \quad U(x_{(U,A)}) \in \mathcal{G}\ . \tag{1.2.34}$$

Consider the elements $x_{(U,A)}$, $(U, A) \in \mathcal{U} \times \mathcal{G}$, given according to (1.2.34) and define the order on Λ by

$$(U, A) \leq (V, B) \iff V \subset U\ \wedge\ B \subset A \cap U(x_{(U,A)})\ . \tag{1.2.35}$$

It is clear that the so-defined relation is transitive. For $(U, A), (V, B) \in \Lambda$ let $W = U \cap V$ and $C = A \cap B \cap U(x_{(U,A)}) \cap V(x_{(V,B)})$. Then $(U, A) \leq (W, C)$ and $(V, B) \leq (W, C)$, showing that the set $(\Lambda, \leq)$ is directed.

Claim III. The net $\varphi = (x_\lambda : \lambda \in \Lambda)$ is left K-Cauchy.

For $U \in \mathcal{U}$ take $\lambda_0 = (U, X)$. If $(U, X) \le (V, B) \le (W, C)$, then

$$x_{(W,C)} \in C \subset V(x_{(V,B)}) \subset U(x_{(V,B)}) ,$$

that is $(x_{(V,B)}, x_{(W,C)}) \in U$.

Claim IV. $\mathcal{G} \subset \Phi(\varphi)$.

Let $A \in \mathcal{G}$. By (1.2.29), there exist $U \in \mathcal{U}$ and $B \in \mathcal{G}$ such that $U(B) \subset A$. Put $\lambda = (X \times X, B)$. Then

$$\mu = (W, C) \ge \lambda \Rightarrow C \subset B ,$$

so that for every $\mu \ge \lambda$, $x_\mu \in C \subset B$, that is $E_\lambda \subset B$, so that $U(E_\lambda) \subset U(B) \subset A$, implying $A \in \Phi(\varphi)$.

Claim V. $\Phi(\varphi) \subset \mathcal{G}$.

Let $U(E_\lambda) \in \Phi(\varphi)$ for some $\lambda = (V, B) \in \mathcal{U} \times \mathcal{G}$. If $(W, C) \in \Lambda$, is such that $(W, C) \ge (V, B)$ and $(W, C) \ge (U, X)$ then $W \subset V$ and $W \subset U$, implying

$$W(x_{(W,C)}) \subset U(x_{(W,C)}) \subset U(E_\lambda) .$$

Since, by (1.2.34), $W(x_{(W,C)}) \in \mathcal{G}$, it follows $U(E_\lambda) \in \mathcal{G}$. □

As a consequence of Proposition 1.2.58 one obtains.

Corollary 1.2.59 ([229]). *Let $(X, \mathcal{U})$ be a quasi-uniform space.*

1. *The space $(X, \mathcal{U})$ is left K-complete by nets if and only if every round S_u-Cauchy filter in X is $\mathcal{U}$-convergent.*
2. *The space $(X, \mathcal{U})$ is Smyth left K-complete by nets if and only if every round S_u-Cauchy filter in X is the neighborhood filter of some $x \in X$.*

Proof. 1. Suppose that every round S_u-Cauchy filter in X is $\mathcal{U}$-convergent. Let $\varphi = (x_i : i \in I)$ be a left K-complete by net in X. Then the associated filter $\Phi(\varphi)$ is round and S_u-Cauchy, so that it converges to some $x \in X$. Taking into account the definition of $\Phi(\varphi)$ (see (1.2.31)) it is easy to check that the net φ converges to x.

Suppose that X is left K-complete by nets and let $\mathcal{G}$ be a round S_u-Cauchy filter in X. By Proposition 1.2.58.2, one can associate to $\mathcal{G}$ a left K-Cauchy net $\varphi = (x_\lambda : \lambda \in \Lambda)$ such that $\Phi(\varphi) = \mathcal{G}$. By hypothesis the net φ is $\mathcal{U}$-convergent to some $x \in X$. By the assertion 1.(i) of the same proposition, the net $\mathcal{G} = \Phi(\varphi)$ is $\mathcal{U}$-convergent to x.

The proof of 2 is similar. □

Remark 1.2.60. 1. If $\mathcal{G}$ is a filter in a uniform space $(X,\mathcal{U})$, then one takes $\Lambda = \{(x,A) : A \in \mathcal{G},\ x \in A\}$ with the order

$$(x,A) \leq (y,B) \iff B \subset A\,.$$

The net φ associated to the filter $\mathcal{G}$ is defined by

$$\varphi(x,A) = x,\ (x,A) \in \Lambda\,.$$

One shows that if the filter $\mathcal{G}$ is Cauchy, then the associated net φ is Cauchy too, and if $\mathcal{G}$ is $\mathcal{U}$-convergent to some $x \in X$, then the net φ is also $\mathcal{U}$-convergent to x. As it is remarked in [229], in the quasi-uniform case this definition does not yield the equivalence between completeness with filters and nets.

Exploiting the ideas of Sünderhauf [229] (see Proposition 1.2.58), Künzi [133] (see also [137, Lemma 2.6.11]) was able to prove the equivalence between the left K-completeness and the left K-completeness by nets.

Theorem 1.2.61. *A quasi-uniform space $(X,\mathcal{U})$ is left K-complete if and only if every left K-Cauchy net in X is $\tau(\mathcal{U})$-convergent.*

Proof. The proof of an implication is standard. Suppose that $(X,\mathcal{U})$ is left K-complete and let $(x_i : i \in I)$ be a left K-Cauchy net in X. Then the sets $E_i = \{x_j : j \in I,\ i \leq j\},\ i \in I$, form the base of a left K-Cauchy filter $\mathcal{F}$ on X. Indeed, for $U \in \mathcal{U}$ let $i_0 \in I$ such that $i_0 \leq i \leq j$ implies $(x_i, x_j) \in U$. Then, for every fixed $i \geq i_0$, $(x_i, x_j) \in U$ for all $j \geq i$, that is $E_i \subset U(x_i)$ showing that $U(x) \in \mathcal{F}$ for every $x \in E_{i_0}$.

By hypothesis, $\mathcal{F}$ is $\tau(\mathcal{U})$-convergent to some $x \in X$. It is easy to check that the net (x_i) is also $\tau(\mathcal{U})$-convergent to x.

To prove the converse, let $\mathcal{F}$ be a left K-Cauchy filter on X. Then for every $U \in \mathcal{U}$ there exists $F_U \in \mathcal{F}$ such that $U(y) \in \mathcal{F}$ for all $y \in F_U$. For arbitrary $U \in \mathcal{U}$ and $A \in \mathcal{F}$, $A \cap F_U \neq \emptyset$, so there exists $x_{(U,A)} \in A$ such that $U(x_{(U,A)}) \in \mathcal{F}$. Put $\Lambda = \{(U,A) : U \in \mathcal{U},\ A \in \mathcal{F}\}$, define the order on Λ by (1.2.35) and the net $\varphi : \Lambda \to X$ by $\varphi(U,A) = x_{(U,A)},\ (U,A) \in \Lambda$.

As in the proof of Proposition 1.2.58 (Claim III), one shows that the net φ is left K-Cauchy. By hypothesis it is $\tau(\mathcal{U})$-convergent to some $x \in X$. Let us show that the filter $\mathcal{F}$ is $\tau(\mathcal{U})$-convergent to x. For $U \in \mathcal{U}$ let $V \in \mathcal{U}$ such that $V^2 \subset U$. Let $(U_0, A_0) \in \Lambda$ such that $x_{(U,A))} \in V(x)$ for all $(U,A) \in \Lambda$ with $(U,A) \geq (U_0, A_0)$. Defining B by

$$B := A_0 \cap F_V \cap U_0(x_{(U_0,A_0)}) \in \mathcal{F}\,,$$

it follows that $(U_0 \cap V, B) \in \Lambda$ and $(U_0 \cap V, B) \geq (U_0, B_0)$, so that

$$x_{(U_0 \cap V,A)} \in V(x) \iff (x, x_{(U_0 \cap V,A)}) \in V\,. \tag{1.2.36}$$

Let us prove now that $V(x_{(U_0\cap V,B)}) \subset V^2(x)$. Indeed,

$$y \in V(x_{(U_0\cap V,B)}) \iff (x_{(U_0\cap V,B)}, y) \in V ,$$

which, combined with (1.2.36), leads to

$$(x,y) \in V^2 \iff y \in V^2(x) .$$

The relations

$$\mathcal{F} \ni (U_0 \cap V)(x_{(U_0\cap V,B)}) \subset V(x_{(U_0\cap V,B)}) \subset V^2(x) \subset U(x) ,$$

show that $U(x) \in \mathcal{F}$. □

The following result appears in [138].

Proposition 1.2.62. *If $\mathcal{U}, \mathcal{V}$ are hereditarily precompact quasi-uniformities on a set X, then $\mathcal{U} \vee \mathcal{V}$ is also hereditarily precompact.*

Proof. We shall present the proof given in [137]. It is based on Ramsey theorem, see for instance [100]. Suppose that $\mathcal{U} \vee \mathcal{V}$ is not hereditarily precompact. Then, there exists an infinite subset W of X, $U \in \mathcal{U}$ and $V \in \mathcal{V}$ s.t $W \setminus (U \cap V)(Z) \neq \emptyset$ for every finite subset Z of W. Then for $x_1 \in W$ there exists $x_2 \in W \setminus (U \cap V)(x_1)$. Taking by induction $x_{n+1} \in W \setminus (U \cap V)(\{x_1, \dots, x_n\})$, then $x_n \notin (U \cap V)(x_k)$ for all $n, k \in \mathbb{N}$ with $n > k$. Let $S = \{x_n : n \in \mathbb{N}\}$ and denote by $[S]^2$ the set of all two-element subsets of S. Define a function $h : [S]^2 \to \{0, 1\}$ by $h(\{x_k, x_n\}) = 1$ if $k < n$ and $x_n \notin U(x_k)$ and $h(\{x_k, x_n\}) = 0$ if $k < n$ and $x_n \in U(x_k)$ (so that $x_n \notin V(x_k)$). By Ramsey Theorem, [100, Theorem A, p.19], there exists an infinite subset Y of S such that the map h is constant on $[Y]^2$. Since the subspace $(Y, \mathcal{U}|_Y)$ is precompact, h cannot be equal to 1 on $[Y]^2$. Since $(Y, \mathcal{V}|_Y)$ is also precompact, h cannot be equal to 0 on $[Y]^2$, a contradiction that shows that $(X, \mathcal{U} \vee \mathcal{V})$ must be hereditarily precompact. □

The above proposition has the following corollary.

Corollary 1.2.63 ([138, 137]). *A quasi-uniform space $(X, \mathcal{U})$ is totally bounded if and only if both $\mathcal{U}$ and $\mathcal{U}^{-1}$ are hereditarily precompact.*

A filter $\mathcal{F}$ on a quasi-uniform space $(X, \mathcal{U})$ is called *$\mathcal{U}$-stable* if for every $U \in \mathcal{U}$, $\cap\{U(F) : F \in \mathcal{F}\} \in \mathcal{F}$.

Proposition 1.2.64 ([137]). *A quasi-uniform space $(X, \mathcal{U})$ is hereditarily precompact if and only if every (ultra)filter on X is $\mathcal{U}^{-1}$-stable.*

Proof. Suppose that X contains a subset Y that is not precompact. Then there exists $U \in \mathcal{U}$ such that $Y \setminus U(Z) \neq \emptyset$, for every finite subset Z of Y. Since $(Y \setminus U(Z_1)) \cap (Y \setminus U(Z_2)) = Y \setminus U(Z_1 \cup Z_2)$, it follows that $\{Y \setminus U(Z) : Z \subset Y,\ Z \text{ finite}\}$ is a filter base on X. Let $\mathcal{F}$ be an (ultra)filter on X containing it.

Observe that $x \in U^{-1}(Y \setminus U(Z))$ implies the existence of $y \in Y \setminus U(Z)$ such that $(x, y) \in U$. But $y \in Y \setminus U(Z)$ is equivalent to $y \in Y$ and $(z, y) \notin U$ for all $z \in Z$. It follows that $x \in X \setminus Z$ and

$$\begin{aligned}&\cap\{U^{-1}(F) : F \in \mathcal{F}\} \subset \cap\{U^{-1}(Y \setminus U(Z)) : Z \subset Y,\ Z \text{ finite}\}\\ &\quad \subset \cap\{X \setminus Z : Z \subset Y,\ Z \text{ finite}\} = X \setminus \cup\{Z : Z \in Y,\ Z \text{ finite}\} = X \setminus Y\,.\end{aligned}$$

For fixed $y \in Y$, $\mathcal{F} \ni X \setminus U(y) \subset Y$, implying $Y \in \mathcal{F}$. It follows that $X \setminus Y \notin \mathcal{F}$, so that the filter $\mathcal{F}$ is not $\mathcal{U}^{-1}$-stable.

To prove the converse suppose that there exists a filter $\mathcal{F}$ on X which is not $\mathcal{U}^{-1}$-stable. It follows that there exists $U \in \mathcal{U}$ such that $\cap\{U^{-1}(F) : F \in \mathcal{F}\} \notin \mathcal{F}$.

Let $F_1 \in \mathcal{F}$. If for every $F \in \mathcal{F}$, $F_1 \setminus U^{-1}(F) = \emptyset$, then $F_1 \subset U^{-1}(F)$, so that $F_1 \subset \cap_F U^{-1}(F)$, implying $\cap_F U^{-1}(F) \in \mathcal{F}$, in contradiction to the choice of the set U. Consequently, there exists $A_2 \in \mathcal{F}$ such that $F_1 \setminus U^{-1}(A_2) \neq \emptyset$. Taking $F_2 = F_1 \cap A_2$, it follows that $F_2 \in \mathcal{F}$, $F_2 \subset F_1$ and $F_1 \setminus U^{-1}(F_2) \neq \emptyset$. By induction one obtains the sets $F_1 \supset F_2 \supset \cdots$ in $\mathcal{F}$ such that $F_n \setminus U^{-1}(F_{n+1}) \neq \emptyset$ for all $n \in \mathbb{N}$. Pick $x_n \in F_n \setminus U^{-1}(F_{n+1})$, $n \in \mathbb{N}$, and put $Y = \{x_n : n \in \mathbb{N}\}$. The proof will be done if we show that the set Y is not $\mathcal{U}$-precompact. For $n > k$ in $\mathbb{N}$, $x_n \in F_n \subset F_{k+1}$ and $x_k \in F_n \setminus U^{-1}(F_{k+1})$, so that

$$(x_n, x_k) \notin U^{-1} \iff (x_k, x_n) \notin U \iff x_n \notin U(x_k)\,.$$

It follows that the set Y is not $\mathcal{U}$-precompact. □

Our next aim is to prove the analog of Theorem 1.2.34 to quasi-uniform spaces: a quasi-uniform space is compact if and only if it is precompact and left K-complete, a result due to Künzi [133]. To do this we need some preliminary results.

Let $\mathcal{F}$ be a filter in a quasi-uniform space $(X, \mathcal{U})$. For a filter base (or subbase) $\mathcal{B}$ one denotes by Fil$(\mathcal{B})$ the filter generated by $\mathcal{B}$. Let also

$$\text{env}(\mathcal{B}) = \text{Fil}(\{U(B) : U \in \mathcal{U},\ B \in \mathcal{B},\}) \;-\; \text{the } \textit{envelope} \text{ of } \mathcal{B}\,, \tag{1.2.37}$$

and

$$\text{co-env}(\mathcal{B}) = \text{Fil}(\{U^{-1}(B) : U \in \mathcal{U},\ B \in \mathcal{B},\}) \;-\; \text{the } \textit{co-envelope} \text{ of } \mathcal{B}\,. \tag{1.2.38}$$

The filter $\mathcal{F}$ is called

- *round* if $\mathcal{F} = \text{env}(\mathcal{F})$;

and

- *co-round*if $\mathcal{F} = \text{co-env}(\mathcal{F})$.

Note that this definition of roundness is in concordance with that given in (1.2.29).

Recall also that the filter $\mathcal{F}$ is called

- *left S_u-Cauchy* provided that

$$\forall U \in \mathcal{U}, \ \ \forall F \in \mathcal{F}, \ \ \exists x \in F, \quad U(x) \in \mathcal{F}\,, \tag{1.2.39}$$

and

- *left K-Cauchy* provided that

$$\forall U \in \mathcal{U}, \ \ \exists F \in \mathcal{F}, \ \ \forall x \in F, \quad U(x) \in \mathcal{F}\,. \tag{1.2.40}$$

Proposition 1.2.65. *Let $(X, \mathcal{U})$ be a quasi-uniform space and $\mathcal{F}$ a filter on X.*

1. *If $\mathcal{F}$ is round and left S_u-Cauchy, then there exists a maximal round left S_u-Cauchy filter $\tilde{\mathcal{F}} \supset \mathcal{F}$.*
2. *If $\mathcal{F}$ is round and left S_u-Cauchy, then there exists a left K-Cauchy filter $\mathcal{K} \supset \mathcal{F}$.*
3. *The filter $\mathcal{F}$ is round left S_u-Cauchy if and only if there exists a left K-Cauchy filter $\mathcal{K} \supset \mathcal{F}$ such that* $\operatorname{env}(\mathcal{K}) \subset \mathcal{F}$.
4. *Any maximal co-round left S_u-Cauchy filter $\mathcal{F}$ is left K-Cauchy.*

Proof. The proof of 1 is obtained by an application of Zorn's Lemma.

2. For $U \in \mathcal{U}$ and $F \in \mathcal{F}$ put

$$M(U, F) = \{x \in F : U(x) \in \mathcal{F}\}\,. \tag{1.2.41}$$

It is easy to check that the sets $M(U, F)$ satisfy the conditions

$$\begin{array}{ll}
\text{(i)} & M(U,F) \neq \emptyset; \\
\text{(ii)} & U_1 \subset U_2 \ \Rightarrow M(U_1,F) \subset M(U_2,F); \\
 & F_1 \subset F_2 \ \Rightarrow M(U,F_1) \subset M(U,F_2); \\
\text{(iii)} & M(U_1,F) \cap M(U_2,F) = M(U_1 \cap U_2, F), \\
 & M(U,F_1) \cap M(U,F_2) = M(U, F_1 \cap F_2), \\
 & M(U_1,F_1) \cap M(U_2,F_2) \supset M(U_1 \cap U_2, F_1 \cap F_2)\,.
\end{array} \tag{1.2.42}$$

The assertion from (i) follows from the fact that $\mathcal{F}$ is left S_u-Cauchy. The last inclusion in (iii) needs also some motivation:

$$M(U_1,F_1) \cap M(U_2,F_2) \supset M(U_1,F_1 \cap F_2) \cap M(U_2,F_1 \cap F_2) = M(U_1 \cap U_2, F_1 \cap F_2)\,.$$

It follows that the family (1.2.41) is a filter base. Denote by $\mathcal{K}$ the filter generated by this family and show that $\mathcal{K}$ is left K-Cauchy and finer than $\mathcal{F}$.

For any $F \in \mathcal{F}$, the inclusion $M(U,F) \subset F$ implies $F \in \mathcal{K}$, that is $\mathcal{F} \subset \mathcal{K}$.

Let us show that $\mathcal{K}$ is left K-Cauchy. For $U \in \mathcal{U}$, $M(U,X) \in \mathcal{K}$ and $U(x) \in \mathcal{F} \subset \mathcal{K}$ for every $x \in M(U,X)$.

3. Suppose that the filter $\mathcal{F}$ is left S_u-Cauchy and let $\mathcal{K}$ the left K-Cauchy filter finer that $\mathcal{F}$ generated by the family (1.2.41). It remains to show that $\text{env}(\mathcal{K}) \subset \mathcal{F}$, where $\text{env}(\mathcal{K})$ is the filter generated by the family of sets

$$\{V(M(U,F)) : U,V \in \mathcal{U},\ F \in \mathcal{F}\} .$$

Let $U,V \in \mathcal{U}$ and $F \in \mathcal{F}$.

We have

$$x \in M(U \cap V, F) \iff x \in F \ \wedge\ (U \cap V)(x) \in \mathcal{F} .$$

The inclusions

$$(U \cap V)(x) \subset V(x) \subset V(M(U \cap V, F)) \subset V(M(U,F)) ,$$

imply $V(M(U,F)) \in \mathcal{F}$, so that $\text{env}(\mathcal{K}) \subset \mathcal{F}$.

Suppose now that $\mathcal{F}$ is a filter on X and there exists a left K-Cauchy filter $\mathcal{K} \supset \mathcal{F}$ such that $\text{env}(\mathcal{K}) \subset \mathcal{F}$.

For $U \in \mathcal{U}$ and $F \in \mathcal{F}$ let $V \in \mathcal{U}$ such that $V^2 \subset U$.

Because $\mathcal{K}$ is left K-Cauchy, there exists $K \in \mathcal{K}$ such that $V(x) \in \mathcal{K}$ for all $x \in K$. Since $\mathcal{F} \subset \mathcal{K}$, it follows that $F \cap K \neq \emptyset$, so that there exists $x \in F$ with $V(x) \in \mathcal{K}$.

Since $V^2(x) \subset U(x)$ and $V^2(x) = V(V(x)) \in \text{env}(\mathcal{K})$, it follows that $U(x) \in \mathcal{F}$, that is $\mathcal{F}$ is left S_u-Cauchy.

4. Let $\mathcal{F}$ be a maximal co-round left S_u-Cauchy filter on X. Let $\mathcal{K}$ be the left K-Cauchy filter finer than $\mathcal{F}$ associated to $\mathcal{F}$ according to the construction given in the proof of the assertion 2. If we show that $\mathcal{K}$ is co-round, then, by the maximality of $\mathcal{F}$, we must have $\mathcal{F} = \mathcal{K}$, implying that $\mathcal{F}$ is left K-Cauchy.

For $U \in \mathcal{U}$ and $E \in \mathcal{F}$ let $V \in \mathcal{U}$ and $F \in \mathcal{F}$ such that

$$V^2 \subset U \quad \text{and} \quad V^{-1}(F) \subset E .$$

To prove that $\mathcal{K}$ is co-round it is sufficient to show that $V^{-1}(M(V,F)) \subset M(U,E)$.

Let $y \in V^{-1}(M(V,F))$. We have

$$y \in V^{-1}(M(V,F)) \iff \exists x \in M(V,F),\ \ (y,x) \in V .$$

But, according to (1.2.41),

$$x \in M(V,F) \iff x \in F \quad \text{and} \quad V(x) \in \mathcal{F} .$$

We have also $V(x) \subset U(y)$. Indeed, $z \in V(x) \iff (x,y) \in V$. Since $(y,x) \in V$, it follows that $(y,z) \in V^2 \subset U$, that is $z \in U(y)$.

It follows that $U(y) \in \mathcal{F}$, and since $x \in F$,

$$(y,x) \in V \ \Rightarrow\ y \in V^{-1}(x) \subset V^{-1}(F) \subset E .$$

Consequently, $y \in E$ and $U(y) \in \mathcal{F}$ which is equivalent to $y \in M(U,E)$. $\square$

The following result will be used in the proof of the characterization of compactness.

Proposition 1.2.66 ([133]). *A quasi-uniform space is precompact if and only if every maximal co-round filter is left K-Cauchy.*

Proof. Let $(X,\mathcal{U})$ be a quasi-uniform space. Supposing that $(X,\mathcal{U})$ is not precompact, there exists $U \in \mathcal{U}$ such that $X \setminus U(Z) \neq \emptyset$ for every finite subset Z of X. It follows that the family $\{X \setminus U(x) : x \in X\}$ is the subbase of a filter on X. Denote by $\mathcal{F}$ the co-envelope of this filter and let $\mathcal{G} \supset \mathcal{F}$ be a maximal co-round filter on X.

For $U \in \mathcal{U}$ let $V \in \mathcal{U}$ be such that $V^2 \subset U$.

For $x \in X$ arbitrary, $V^{-1}(X \setminus V^2(x)) \cap V(x) = \emptyset$. Indeed, if there is $y \in V(x) \cap V^{-1}(X \setminus V^2(x))$, then $(x,y) \in V$ and

$$y \in V^{-1}(X \setminus V^2(x)) \iff \exists z \in X \setminus V^2(x),\ (y,z) \in V\,.$$

It follows that

$$(x,z) \in V^2 \iff z \in V^2(x)\,,$$

in contradiction to the choice of the element z.

Because $V^{-1}(X \setminus V^2(x)) \in \mathcal{F} \subset \mathcal{G}$ and $V(x) \cap V^{-1}(X \setminus V^2(x)) = \emptyset$ it follows that $V(x) \notin \mathcal{G}$, so that the maximal co-round filter $\mathcal{G}$ is not left S_u-Cauchy.

To prove the converse, suppose that $(X,\mathcal{U})$ is precompact and let $\mathcal{F}$ be a maximal co-round filter on X.

First we show that $\mathcal{F}$ is left S_u-Cauchy. For $U \in \mathcal{U}$ and $E \in \mathcal{F}$, let $V \in \mathcal{U}$ and $F \in \mathcal{F}$ such that

$$V^2 \subset U \quad \text{and} \quad V^{-1}(F) \subset E\,.$$

By the precompactness of $(X,\mathcal{U})$ there exists a finite subset Z of X such that $V(z) \cap F \neq \emptyset$ for every $z \in Z$ and

$$F \subset V(Z)\,.$$

It follows that

$$Z \subset V^{-1}(F) \subset E\,.$$

Indeed, for $z \in Z$ there exists $x \in F$ such that

$$(x,z) \in V \iff (z,x) \in V^{-1} \iff z \in V^{-1}(x) \subset V^{-1}(F) \subset E\,.$$

Suppose that, for a fixed $z \in Z$,

$$(X \setminus U(z)) \cap B \neq \emptyset\,,$$

for all $B \in \mathcal{F}$. It follows that the filter $\mathcal{G}(z)$ generated by the family $\{W^{-1}([X \setminus U(z)] \cap B) : W \in \mathcal{U},\ B \in \mathcal{F}\}$ is co-round.

From

$$W^{-1}([X \setminus U(z)] \cap B) \subset W^{-1}(B) ,$$

it follows that $W^{-1}(B) \in \mathcal{G}(z)$, so that $\mathcal{F} \subset \mathcal{G}(z)$. Since $\mathcal{F}$ is maximal, it follows that $\mathcal{F} = \mathcal{G}(z)$.

Taking $B = X \in \mathcal{F}$ and $W = V$ one obtains

$$V^{-1}(X \setminus U(z)) \in \mathcal{F} . \tag{1.2.43}$$

Let us show that

$$V^{-1}(X \setminus U(z)) \subset X \setminus V(z) . \tag{1.2.44}$$

Indeed, suppose that there exists $y \in V(z) \cap V^{-1}(X \setminus U(z))$. Then $(z, y) \in V$ and

$$y \in V^{-1}(X \setminus U(z)) \iff \exists x \in X \setminus U(z),\ (y, x) \in V .$$

But $(z, y) \in V$ and $(y, x) \in V$ implies

$$(z, x) \in V^2 \iff x \in V^2(z) \subset U(z) ,$$

in contradiction to the choice of the element x.

Now, by (1.2.43) and (1.2.44), $X \setminus V(z) \in \mathcal{F}$.

But $F \subset V(Z)$ implies $(X \setminus V(Z)) \cap F = \emptyset$, so that there must exist $z \in Z$ such that $X \setminus V(z) \notin \mathcal{F}$.

Consequently, there exist $z_0 \in Z$ and $B_0 \in \mathcal{F}$ such that

$$(X \setminus U(z_0)) \cap B_0 = \emptyset \iff B_0 \subset U(z_0) ,$$

implying $U(z_0) \in \mathcal{F}$. Consequently the filter $\mathcal{F}$ is left S_u-Cauchy.

Since it is co-round and maximal, by Proposition 1.2.65.4 it is left K-Cauchy. □

Now we can prove the characterization of compactness in quasi-uniform spaces.

Theorem 1.2.67. *A quasi-uniform space is compact if and only if it is precompact and left K-complete.*

Proof. Suppose that the quasi-uniform space $(X, \mathcal{U})$ is precompact and left K-complete. To prove the compactness it suffices to show that every ultrafilter on X is convergent.

Let $\mathcal{G}$ be an ultrafilter on X. For $U \in \mathcal{U}$ put

$$M(U) = \{x \in X : U(x) \in \mathcal{G}\} .$$

Since X is precompact, for every $U \in \mathcal{U}$ there exists a finite subset $\{z_1, \ldots, z_n\}$ of X such that

$$U(z_1) \cup \cdots \cup U(z_n) = X \in \mathcal{G} .$$

Since $\mathcal{G}$ is maximal, there exists k, $1 \le k \le n$, such that $U(z_k) \in \mathcal{G}$. It follows that the family $\{M(U) : U \in \mathcal{U}\}$ is the base of a filter $\mathcal{F}$ on X.

Let us show that $\mathcal{F}$ is co-round. If for $U \in \mathcal{U}$, $V \in \mathcal{U}$ is such that $V^2 \subset U$, then

$$V^{-1}(M(V)) \subset M(U) \,. \tag{1.2.45}$$

Indeed,

$$y \in V^{-1}(M(V)) \iff \exists x \in M(V),\ (y,x) \in V \,.$$

Let us show that

$$V(x) \subset U(y) \,. \tag{1.2.46}$$

Indeed, $x \in M(V) \iff V(x) \in \mathcal{G}$. But then $z \in V(x) \iff (x,z) \in V$ implies

$$(y,z) \in V^2 \subset U \Rightarrow z \in U(y) \,.$$

By the inclusion (1.2.46), $U(y) \in \mathcal{G}$ or, equivalently, $y \in M(U)$. The inclusion (1.2.45) shows that $\mathcal{F}$ is co-round. Let now $\mathcal{H} \supset \mathcal{F}$ be a maximal co-round filter on X. By Proposition 1.2.66, $\mathcal{H}$ is left K-Cauchy, so that, by hypothesis, it converges to some $x \in X$.

Let us show that the ultrafilter $\mathcal{G}$ converges also to x. For $U \in \mathcal{U}$ let $V \in \mathcal{U}$ be such that $V^2 \subset U$. Since $M(V) \in \mathcal{F} \subset \mathcal{H}$ and $V(x) \in \mathcal{H}$ there exists $y \in M(V) \cap V(x)$. Since $y \in M(V)$, $V(y) \in \mathcal{G}$. If we show that

$$V(y) \subset U(x) \,, \tag{1.2.47}$$

then $U(x) \in \mathcal{G}$, proving that $\mathcal{G}$ converges to x.

To prove (1.2.47), observe that $y \in V(x) \iff (x,y) \in V$, so that $z \in V(y)$ implies $(y,z) \in V$ and $(x,z) \in V^2 \subset U$. Consequently, $z \in U(x)$. □

The Lebesgue property (see (1.2.24)) can be extended from quasi-metric spaces to quasi-uniform spaces. We present some properties related to this notion following [147]. One says that a quasi-uniform space $(X,\mathcal{U})$ has the *Lebesgue property* if for every $\tau(\mathcal{U})$-open cover $\mathcal{G}$ of the space X there exists $U \in \mathcal{U}$ such that $\{U(x) : x \in X\}$ refines $\mathcal{G}$, that is

$$\forall x \in X,\ \exists G \in \mathcal{G},\quad U(x) \subset G \,. \tag{1.2.48}$$

The analog of Proposition 1.2.37 holds in the quasi-uniform case too.

Proposition 1.2.68. *Any compact quasi-uniform space has the Lebesgue property.*

Proof. Let $\mathcal{G}$ be a $\tau(\mathcal{U})$-open cover of the quasi-uniform space $(X,\mathcal{U})$. For every $x \in X$ choose s $G_x \in \mathcal{G}$ and $U_x \in \mathcal{U}$ such that $U_x^2(x) \subset G_x$. If V_x denotes the $\tau(\mathcal{U})$-interior of the set U_x, then the $\tau(\mathcal{U})$-open cover $\{V_x(x) : x \in X\}$ of X contains a finite subcover $X = \cup\{V_z(z) : z \in F\}$, where $F \subset X$ is nonempty and finite. Put

$U = \cap\{U_z : z \in F\}$ and show that U satisfies (1.2.48). For $x \in X$ let $z \in F$ be such that $x \in V_z(z) \subset U_z(z)$.

Then

$$U(x) \subset U_z^2(z) \subset G_z \,.$$

Indeed, $y \in U(x) \iff (x, y) \in U$ and $x \in U_z(z) \iff (z, x) \in U_z$, so that $(z, y) \in U_z \circ U \subset U^2$, that is $y \in U_z^2(z)$. □

A quasi-uniform space $(X, \mathcal{U})$ is called *point-symmetric* provided that

$$\tau(\mathcal{U}) \subset \tau(\mathcal{U}^{-1}) \,. \tag{1.2.49}$$

It follows that

$$\tau(\mathcal{U}^s) = \tau(\mathcal{U}^{-1}) \,. \tag{1.2.50}$$

The following result is the quasi-uniform analog of Proposition 1.2.13.4.

Proposition 1.2.69. *If the topology $\tau(\mathcal{U})$ of a quasi-uniform space $(X, \mathcal{U})$ is T_1 and compact, then $\mathcal{U}$ is point-symmetric.*

Proof. Suppose on the contrary that there exists $G \in \tau(\mathcal{U}) \setminus \tau(\mathcal{U}^{-1})$, meaning that

(i) $\forall x \in G,\ \exists U_x \in \mathcal{U},\quad U_x(x) \subset G,$

(ii) $\exists x_0 \in G,\ \forall U \in \mathcal{U},\quad U^{-1}(x_0) \cap \complement(G) \neq \emptyset\,.$

The family $\{U^{-1}(x_0) \cap \complement(G) : U \in \mathcal{U}\}$ is the base of a filter $\mathcal{F}$ on X, which, by the compactness of X, has a cluster point y. Since $\complement(G)$ is closed $y \in \complement(G)$, so that $y \neq x_0$. As the topology $\tau(\mathcal{U})$ is T_1 there exists $U_0 \in \mathcal{U}$ such that $x_0 \notin U_0(y)$.

Also

$$\forall U \in \mathcal{U},\quad U(y) \cap U^{-1}(x_0) \cap \complement(G) \neq \emptyset\,.$$

Let $V \in \mathcal{U}$ such that $V^2 \subset U_0$. Since $V \subset U_0$ it follows that $x_0 \notin V(y)$.

Choosing an element $z \in V(y) \cap V^{-1}(x_0)$ it follows that $(y, z) \in V$ and $(z, x_0) \in V$, so that $(y, x_0) \in V^2 \subset U_0$, leading to the contradiction $x_0 \in U_0(y)$. □

The following proposition is an extension of Theorem 1.1.57.

Proposition 1.2.70. *Let $(X, \mathcal{U})$, $(Y, \mathcal{V})$ be quasi-uniform spaces such that $\mathcal{U}$ has the Lebesgue property and $\mathcal{V}^{-1}$ is point-symmetric. Then any continuous mapping $f : (X, \mathcal{U}) \to (Y, \mathcal{V})$ is quasi-uniformly continuous (in fact, $(\mathcal{U}, \mathcal{V}^s)$-quasi-uniformly continuous).*

Proof. For $V \in \mathcal{V}$ let $W \in \mathcal{V}$ such that $W^2 \subset V$. The condition on $\mathcal{V}$ means that $\tau(\mathcal{V}^{-1}) \subset \tau(\mathcal{V})$ so that $\tau(\mathcal{V}) = \tau(\mathcal{V}^s)$. Putting $W^s = W \cap W^{-1}$ it follows that for every $x \in X$, there exists $W_x \in \mathcal{V}$ such that $W_x(f(x)) \subset W^s(f(x))$. Since $\{f^{-1}(\tau(\mathcal{V})\text{-}\operatorname{int}(W_x(f(x))) : x \in X\}$ is a $\tau(\mathcal{U})$-open cover of X, the Lebesgue property of $\mathcal{U}$ yields an entourage $U \in \mathcal{U}$ such that

$$\forall x \in X,\ \exists y \in X,\quad U(x) \subset f^{-1}\left(W_y(f(y))\right)\ ,$$

implying

$$\forall x \in X, \ \exists y \in X, \quad f(U(x)) \subset W_y(f(y) \subset W^s(f(y)))\,.$$

The proof will be done if we show that $(f(x), f(z)) \in V$ for every $(x, z) \in U$. Since $x \in U(x)$ it follows that $f(x) \in f(U(x)) \subset W^s(f(y)$, and so $(f(y), f(x)) \in W^{-1} \iff (f(x), f(y)) \in W$.

Also $(x, z) \in U \iff z \in U(x)$ implies $f(z) \in f(U(x)) \subset W^s(f(y))$ so that $(f(y), f(z)) \in W$ and $(f(x), f(z)) \in W^2 \subset V$. □

Other results on the relations between completeness, compactness, precompactness, total boundedness and other related notions in quasi-metric and quasi-uniform spaces can be found in the papers [10, 101, 131, 138, 139, 173, 174, 185, 190, 196, 214].

1.2.6 Completions of quasi-metric and quasi-uniform spaces

Two difficult problems in the study of quasi-metric and quasi-uniform spaces are those of completion and compactification for such spaces. As it is mentioned in the paper [185], the authors did not succeed to obtain a satisfactory theory for these problems. Some progress in this direction was obtained by Alemany and Romaguera [9], Doitchinov [62, 66], Gregori and Romaguera [101], Romaguera and Sánchez-Granero [199]. We shall discuss in Section 2.1 the existence of a bicompletion for an asymmetric normed space, following the paper [88]. The existence of a bicompletion of a normed cone was proved by Oltra and Valero [170].

As it is well known, every metric space (X, ρ) admits a *completion*, meaning a complete metric space (X^*, ρ^*) and an isometric embedding $\varphi : X \to X^*$ such that $\varphi(X)$ is dense in X. The standard construction (see, e.g., [73, Exercise 4.5.6]) consists in considering X^* as the space of equivalence classes of Cauchy sequences in X. Two Cauchy sequences $(x_n), (y_n)$ in X are considered equivalent if $\lim_n \rho(x_n, y_n) = 0$. For two equivalence classes $\xi, \eta \in X^*$ one puts $d^*(\xi, \eta) = \lim_n \rho(x_n, y_n)$, where $(x_n) \in \xi$ and $(y_n) \in \eta$. One shows that (X^*, d^*) is a complete metric space fulfilling all the requirements of a completion for (X, d). Furthermore, the completion of a metric space is unique up to an isometric isomorphism.

It turns out that these ideas are not easy to transpose to quasi-metric spaces. Call a mapping f between two quasi-metric spaces (X, ρ) and (Y, d) a *quasi-isometry* if

$$d(f(x), f(y)) = \rho(x, y)\,, \tag{1.2.51}$$

for all $x, y \in X$. Then it is natural to call a *completion* of a quasi-metric space (X, ρ) a complete quasi-metric space (X^*, ρ^*) such that there exists a quasi-isometry $\varphi : X \to X^*$ with $\varphi(X)$ dense in (X^*, τ_{ρ^*}).

First of all, the definition depends on the kind of completeness we consider (see Subsection 1.2.1), but as it is shown by some examples a completion of an arbitrary quasi-metric space does not exist in general.

An approach similar to that in the metric case, proposed by Doitchinov [63], will be briefly described in what follows. Let (X, ρ) be a quasi-metric space. A sequence (x_n) in X is called a *D-Cauchy sequence* if there exists another sequence (y_n) in X such that

$$\lim_{m,n} \rho(y_m, x_n) = 0 . \tag{1.2.52}$$

The quasi-metric space (X, ρ) is called *D-complete* if every D-Cauchy sequence converges. A sequence (y_n) satisfying (1.2.52) is called a cosequence to (x_n).

The D-completeness of spaces of semi-Lipschitz functions will be discussed in Theorem 2.6.20 from Section 2.6.

Some of the advantages of this definition are mentioned in the following proposition.

Proposition 1.2.71. *Let (X, ρ) be a quasi-metric space. Then*

1. *Every convergent sequence is D-Cauchy.*
2. *Every subsequence of a D-Cauchy sequence is D-Cauchy.*
3. *If (X, ρ) is a metric space, then the notion of D-Cauchy sequence agrees with that of a usual Cauchy sequence.*

As a major drawback, Doitchinov [63] mentions the fact that a D-complete subspace of a quasi-metric space (X, ρ) need not to be $\tau(\rho)$-closed.

The construction proposed by Doitchinov [63] follows the ideas from the metric case, but it is more complicated due to the nature of quasi-metric spaces. Let (X, ρ) be a balanced quasi-metric space (see (1.1.17)). Two D-Cauchy sequences $(x_n), (y_n)$ are called *equivalent* if they have the same cosequences.

The following proposition emphasizes the steps needed for the construction of the completion. The proofs are based on the properties of balanced quasi-metric spaces, see Proposition 1.1.11.

Proposition 1.2.72. *Let (X, ρ) be a balanced quasi-metric space.*

1. *Let $x_n \to x$ (i.e., $\lim_n \rho(x, x_n) = 0$). Then (y_m) is a cosequence to (x_n) if and only if $\lim_m (y_m, x) = 0$.*
2. *If $(x'_n), (x''_n)$ are two D-Cauchy sequences with a common cosequence (y_m), then $x'_n \to x$ implies $x''_n \to x$.*
3. *If $(x'_n), (x''_n)$ are two equivalent D-Cauchy sequences, then $x'_n \to x$ implies $x''_n \to x$.*
4. *If $(x'_n), (x''_n)$ are two D-Cauchy sequences with a common cosequence (y_m), then they are equivalent.*
5. *A D-Cauchy sequence is equivalent with any of its subsequences.*
6. *The collection of all sequences converging to a point x forms a class of equivalent sequences.*

One denotes by X^* the set of all equivalence classes of D-Cauchy sequences and one puts

$$\rho^*(\xi, \eta) = \lim_{m,n} \rho(y_m, x_n) , \tag{1.2.53}$$

for $\xi, \eta \in X^*$, where (x_n) is a D-Cauchy sequence in the class ξ and (y_m) is a cosequence to the class η. One shows that the so-defined mapping $\rho^* : X^* \times X^* \to \mathbb{R}_+$ is well defined, ρ^* is a balanced quasi-metric on X^*, and the quasi-metric space (X^*, ρ^*) is D-complete.

Furthermore, the application $\varphi_X : X \to X^*$ given by

$$\varphi_X(x) = \{(x_n) : x_n \to x\}, \quad x \in X , \tag{1.2.54}$$

is a quasi-metric embedding of (X, ρ) into (X^*, ρ^*) and the image $\varphi_X(X)$ of X by φ_X is dense in X^*. The construction given above is called the *standard D-completion* of a quasi-metric space (X, ρ). We mention the following properties of this completion.

Theorem 1.2.73. *Let (X, ρ) be a balanced quasi-metric space.*

1. *If (X, ρ) is D-complete, then it coincides (up to a quasi-isometry) with its standard D-completion.*
2. *If (X, ρ) is a metric space, then its standard D-completion coincides with the usual metric completion of X.*
3. *The standard completion is the minimal D-complete balanced quasi-metric space containing (X, ρ), in the sense that if (Y, d) is a D-complete balanced quasi-metric space such that $(X, \rho) \subset (Y, d)$, then $(X^*, \rho^*) \subset (Y, d)$. (All the inclusions are understood as quasi-metric embeddings.)*
4. *If $(X, \rho) \subset (Y, d)$ are balanced quasi-metric space, then their standard D-completions satisfy $(X^*, \rho^*) \subset (Y^*, d^*)$.*
5. *(The extension property.) Any quasi-uniformly continuous mapping*

   $$f : (X, \rho) \to (Y, d)$$

   between two balanced quasi-metric spaces has a unique quasi-uniformly continuous extension $f^* : (X^*, rho^*) \to (Y^*, d^*)$, in the sense that $f^* \circ \varphi_X = \varphi_Y \circ f$, where $\varphi_X : X \to X^*$ and $\varphi_Y : Y \to Y^*$ are the standard embeddings given by (1.2.54).

A natural question is what happens with the conjugate quasi-metric. As it was shown by Doitchinov [63], there is no difference between right and left D-completeness.

Proposition 1.2.74. *Let (X, ρ) be a quasi-metric space and $\bar{\rho}$ the conjugate quasi-metric.*

1. *The quasi-metric ρ is balanced if and only if $\bar{\rho}$ is balanced.*
2. *The quasi-metric space (X, ρ) is D-complete if and only if $(X, \bar{\rho})$ is D-complete.*

3. *The standard D-completions* (X^*, ρ^*) *of* (X, ρ) *and* $(\bar{X}^*, (\bar{\rho})^*)$ *of* $(X, \bar{\rho})$ *are related by the equality*
$$(\bar{X}^*, (\bar{\rho})^*) = (X^*, \overline{\rho^*}) . \tag{1.2.55}$$

Replacing sequences by nets Doitchinov [65] extended this method to quasi-uniform spaces. A *completion* of a quasi-uniform space $(X,\mathcal{U})$ is a complete quasi-uniform space $(X^*,\mathcal{U}^*)$ such that there exists a quasi-uniform embedding $\varphi_X : X \to X^*$ with $\varphi_X(X)$ dense in X^*.

A net $(x_i : i \in I)$ in $(X,\mathcal{U})$ is called *D-Cauchy* if there exists another net $(y_j : j \in J)$ such that
$$\lim_{i,j} \rho(y_j, x_i) = 0 . \tag{1.2.56}$$

The condition (1.2.56) means that
$$\forall U \in \mathcal{U},\ \exists i_U \in I,\ \exists j_U \in J,\ \ \forall j \geq j_u,\ \forall i \geq i_U, \quad (y_j, x_i) \in U .$$

A net $(y_j : j \in J)$ satisfying (1.2.56) is called a conet to (x_i). A quasi-uniform space $(X,\mathcal{U})$ is called *D-complete* if every D-Cauchy net in X is convergent.

As we have seen, standard D-completions were obtained for the class of balanced quasi-metric spaces. In this case again the construction is possible only for a restricted class of quasi-uniform spaces, called quiet quasi-uniform spaces. A quasi-uniform space $(X,\mathcal{U})$ is called *quiet* if for every $U \in \mathcal{U}$ there exists $V \in \mathcal{U}$ such that
$$(\forall i \in I,\ (x, x_i) \in V)\ \wedge\ (\forall j \in J,\ (y_j, y) \in V)\ \wedge\ \lim_{i,j}(y_j, x_i) = 0\ \Rightarrow (x, y) \in U, \tag{1.2.57}$$
for all pairs of filters $(x_i : i \in I)$ and $(y_j : j \in J)$ in X and all $x, y \in X$.

Adapting the construction with sequences given in quasi-metric spaces to nets in quasi-uniform spaces, Doitchinov was able to prove the existence of a D-completion of a quasi-uniform space having properties similar to those of the standard D-completion of a quasi-metric space.

The notion of D-completeness and the construction of D-completion can be also treated through filters, see Doitchinov [61, 66]. A filter $\mathcal{F}$ in a quasi-uniform space $(X,\mathcal{U})$ is called *D-Cauchy* provided there exists a so-called co-filter $\mathcal{G}$ in X such that for every $U \in \mathcal{U}$ there are $G \in \mathcal{G}$ and $F \in \mathcal{F}$ with $G \times F \subset U$.

The quasi-uniform space $(X,\mathcal{U})$ is called *D-complete* provided that every D-Cauchy filter converges.

A related notion of completeness was considered by Andrikopoulos [13]. For a comparative study of the completeness notions defined by filters and nets see also Andrikopoulos [14], Deák [56, 57, 58]. Stoltenberg [226] proved the existence of a left S_t-completion for every quasi-uniform space.

Call a quasi-uniform space $(X,\mathcal{U})$ *Smyth complete* if every round S_u-Cauchy filter is the neighborhood filter of a unique point $x \in X$.

Theorem 1.2.75 ([228]). *A quasi-uniform space $(X,\mathcal{U})$ admits a Smyth completion if and only if every round S_u-Cauchy filter on X is stable.*

For other results on this line, see also Sünderhauf [229, 230].

A quasi-uniform space $(X,\mathcal{U})$ is called half-complete if every $\mathcal{U}^s$-Cauchy filter on X is $\tau(\mathcal{U})$-convergent. The existence of half-completions of quasi-uniform spaces is discussed by Romaguera in [9], the existence of left K-completions of quasi-metric and quasi-uniform spaces in [190] and that of right K-completions in [192].

Romaguera [193] considered another class of quasi-uniform spaces, called fitting quasi-uniform spaces. He showed that every quiet-quasi-uniform space is fitting, the bicompletion of a fitting quasi-uniform space is again fitting and a totally bounded fitting quasi-uniform space is a uniform space.

The existence of a compact completion of a quasi-semimetric spaces is given in [185].

Theorem 1.2.76. *Any quasi-semimetric space has a compact (left, right) completion.*

Proof. For a quasi-semimetric space (X,ρ) let $z \notin X$ and $X^* = X \cup \{z\}$. Without loss of generality we can suppose $\rho(x,y) \le 1$ for all $x, y \in X$. Define $\rho^* : X^* \times X^* \to [0;\infty)$ by

$$\rho^*(x,y) = \begin{cases} \rho(x,y) & \text{if} \quad x, y \in X, \\ 0 & \text{if} \quad x = z, \ y \in X^*, \\ 1 & \text{if} \quad x \in X, \ y = z\ . \end{cases}$$

Then ρ^* is a quasi-semimetric on X^*, $B_{\rho^*}(z,\varepsilon) = X^*$ and $B_{\rho^*}(x,\varepsilon) = B_\rho(x,\varepsilon)$ for every $x \in X$ and $0 < \varepsilon \le 1$.

It follows that X is dense in X^*, $\rho^*|_{X^* \times X^*} = \rho$ and the identity map from (X,ρ) to (X^*,ρ^*) is a quasi-isometry. Since every sequence in X^* is convergent to z it follows that (X^*,ρ^*) is complete in all senses of completeness definitions. Any open cover of X^* contains X^*, so that X^* is compact. □

The completion given in Theorem 1.2.76 is the smallest one-point compactification of (X,ρ). A disadvantage is that (X^*,ρ^*) is never T_1 because $\rho^*(z,y) = 0$ for all $y \in X^*$. Another disadvantage is that this completion is not unique, in general. A related question is that of the quasi-semimetrizability of the one-point Alexandrov compactification of a quasi-semimetric space.

In [185] it is shown also the existence of a complete metric space completion. The proof is based on the following remark. For a topological space T denote by $U(T)$ the space of all bounded usc functions with the metric $d(f,g) = \sup\{|f(t) - g(t)| : t \in T\}$.

The proof of the following result is based on standard methods of mathematical analysis.

Theorem 1.2.77. *The space $(U(T), d)$ is a complete metric space.*

Based on this result one can prove the following.

Theorem 1.2.78. *Any quasi-semimetric space has a quasi-completion.*

Proof. Let (X, ρ) be a quasi-semimetric space and $z \in X$ fixed. Assume again $\rho(x, y) < 1$ for all $x, y \in X$ and define $i : X \to U(X)$ by $i(x)(t) = \rho(x, t) + \rho(z, t)$, for $t \in X$, $x \in X$. Then $|i(x)(t)| < 2$ for all $x, t \in X$. By Proposition 1.1.8.4, the functions $\rho(x, \cdot)$ and $\rho(z, \cdot)$ are τ_ρ-lsc, so that $i(x) \in U(X)$ for every $x \in X$.

One shows that

$$\begin{aligned} d(i(x), i(y)) &:= \sup\{|i(x)(t) - i(y)(t)| : t \in X\} \\ &= \max\{\rho(x, y), \rho(y, x)\} = \rho^s(x, y) \, . \end{aligned}$$

The desired completion of (X, ρ) is $X^* = d\text{-cl}(i(X))$. □

Remark 1.2.79. Here a quasi-isometry is an application f between two quasi-semimetric spaces (X, ρ_1) and (X, ρ_2) such that $\rho_2(f(x), f(y)) = \rho_1^s(x, y)$. Also by a quasi-completion of a quasi-semimetric space (X, ρ) one understands a complete metric space (Y, d) and a quasi-isometry $i : (X, \rho) \to (Y, d)$ with $i(X)$ dense in Y.

Chapter 2

Asymmetric Functional Analysis

An asymmetric seminorm is a positive sublinear functional p on a real vector space X. If $p(x) = p(-x) = 0$ implies $x = 0$, then p is called an asymmetric norm. The conjugate asymmetric seminorm is given by $\bar{p}(x) = p(-x)$, $x \in X$, and $p^s = p \vee \bar{p}$ is a seminorm, respectively a norm on X if p is an asymmetric norm. An important example is the asymmetric norm u on $\mathbb{R}$ given by $u(t) = t^+$, $t \in \mathbb{R}$, generating the upper topology on $\mathbb{R}$. In this case $\bar{u}(t) = t^-$ and $u^s(t) = |t|$, $t \in \mathbb{R}$. The dual $X_p^\flat$ of an asymmetric normed space (X, p) is formed by all upper semi-continuous linear functionals from (X, p) to $(\mathbb{R}, |\cdot|)$, or equivalently, by all continuous linear functionals from (X, p) to $(\mathbb{R}, u)$. In contrast to the usual case, $X_p^\flat$ is not a linear space but merely a cone contained in the dual X^* of the normed space space (X, p^s). The aim of this chapter is to present some basic results on asymmetric normed spaces, their duals and on continuous linear operators acting between them. Applications are given to best approximation in asymmetric normed spaces. As important examples one considers asymmetric norms on normed lattices and spaces of semi-Lipschitz functions on quasi-metric spaces. Asymmetric locally convex spaces are considered as well.

2.1 Continuous linear operators between asymmetric normed spaces

The basic objects of functional analysis are normed spaces and locally convex spaces and spaces of continuous linear operators acting between them, with special emphasis on continuous linear functionals and the duals of these spaces. The situation is quite different in the asymmetric case, mainly due to the fact that the dual of an asymmetric normed space or of an asymmetric LCS X, meaning the set of all lower semi-continuous linear functionals on X, is not a linear space but merely a cone in the space of all continuous linear functionals on X. In spite of the existing differences, some results from the symmetric case have their counterparts in the asymmetric one, a study that was initiated in [90]. As an application one considers the important case of asymmetric norms on normed lattices.

2.1.1 The asymmetric norm of a continuous linear operator

Let (X,p) and (Y,q) be two asymmetric seminormed spaces. Denote by $L_a(X,Y)$ the space of all linear operators from X to Y. A linear operator $A : X \to Y$ is called (p,q)-continuous if it is continuous with respect to the topologies τ_p on X and τ_q on Y. The set of all (p,q)-continuous linear operators from X to Y is denoted by $L_{p,q}(X,Y)$. For $\mu \in \{p, \bar{p}, p^s\}$ and $\nu \in \{q, \bar{q}, q^s\}$, the (μ,ν)-continuity and the set $L_{\mu,\nu}(X,Y)$ are defined similarly. The space of all continuous linear operators between the associated seminormed spaces (X,p^s) and (Y,q^s) is denoted by $L(X,Y)$.

In the case of linear functionals, i.e., when $Y=(\mathbb{R},u)$, we put $X_p^\flat = L_{p,u}(X,\mathbb{R})$ and $X^* = L((X,p^s),(\mathbb{R},|\cdot|))$. The meaning of $X_{\bar{p}}^\flat$ is clear. A linear operator $A : (X,p) \to (Y,q)$ is called *(p,q)-semi-Lipschitz* (or *(p,q)-bounded*) if there exists a number $\beta \geq 0$ such that

$$q(Ax) \leq \beta\, p(x)\,, \tag{2.1.1}$$

for all $x \in X$.

The characterizations of the continuity of linear mappings will be based on the following proposition.

Proposition 2.1.1. *Let X be a real vector space, $f, g : X \to \mathbb{R}$ sublinear functionals and $\alpha, \beta > 0$.*

Then the following conditions are equivalent:

$$\forall x \in X, \quad g(x) \leq \beta \;\Rightarrow\; f(x) \leq \alpha, \tag{2.1.2}$$

and

$$\forall x \in X, \quad u(f(x)) \leq \frac{\alpha}{\beta} u(g(x)). \tag{2.1.3}$$

If $g(x) \geq 0$ for all $x \in X$, then these two conditions are also equivalent to

$$\forall x \in X, \quad f(x) \leq \frac{\alpha}{\beta} g(x). \tag{2.1.4}$$

Proof. (2.1.2) $\Rightarrow$ (2.1.3) Let $x \in X$. If $g(x) \leq 0$, then $g(nx) = ng(x) \leq 0 < \alpha,\ n \in \mathbb{N}$, so that $nf(x) = f(nx) \leq \beta,\ n \in \mathbb{N}$, implying $f(x) \leq 0$ and

$$u(f(x)) = 0 = \frac{\alpha}{\beta} u(g(x))\,.$$

If $g(x) > 0$, then $g\left(\frac{\beta}{g(x)} x\right) = \beta$, so that

$$f\left(\frac{\beta}{g(x)} x\right) \leq \alpha \iff f(x) \leq \frac{\alpha}{\beta} g(x) \iff u(f(x)) \leq \frac{\alpha}{\beta} u(g(x))\,.$$

(2.1.3) $\Rightarrow$ (2.1.2) Let $x \in X$. If $g(x) \leq 0 < \beta$, then $u(g(x)) = 0$, so that

$$f(x) \leq u(f(x)) \leq \frac{\alpha}{\beta} u(g(x)) = \frac{\alpha}{\beta} g(x) \leq \alpha\,.$$

If $g(x) > 0$, then, by hypothesis,

$$f(x) \le u(f(x)) \le \frac{\alpha}{\beta} u(g(x)) = \frac{\alpha}{\beta} g(x) \le \alpha \, .$$

Since $g(x) \ge 0$, $x \in X$, implies $u(g(x)) = g(x)$, $x \in X$, the equivalence (2.1.3) $\iff$ (2.1.4) is obvious. □

The following proposition contains some characterizations of continuity, similar to those known in the symmetric case.

Proposition 2.1.2. *For a linear operator A between two asymmetric seminormed spaces (X, p), (Y, q), the following are equivalent.*

1. *The operator A is continuous on X.*
2. *The operator A is continuous at $0 \in X$ (or at an arbitrary point $x_0 \in X$).*
3. *The operator A is (p, q)-semi-Lipschitz.*
4. *The operator A is $(\mathcal{U}_p, \mathcal{U}_q)$-quasi-uniformly continuous on X.*

Proof. The equivalence 1 $\iff$ 2 holds for any additive operator (see Proposition 1.1.42).

2 $\Rightarrow$ 3 By the continuity of A at $0 \in X$ there exists $r > 0$ such that $A(B_p[0, r]) \subset B_q[0, 1]$, meaning that

$$p(x) \le r \;\Rightarrow\; q(Ax) \le 1 \, ,$$

for every $x \in X$. Applying Proposition 2.1.1 to the sublinear functionals $f(x) = q(Ax)$ and $g(x) = p(x)$, it follows that

$$q(Ax) \le \frac{1}{r} p(x),$$

for all $x \in X$.

3 $\Rightarrow$ 4 Suppose that (2.1.1) holds with some $\beta > 0$ and let $\varepsilon > 0$. For $V = \{(y_1, y_2) \in Y \times Y : q(y_2 - y_1) < \varepsilon\} \in \mathcal{U}_q$, let $U := \{(x_1, x_2) \in X \times X : p(x_2 - x_1) < \varepsilon/\beta\} \in \mathcal{U}_p$. Then for every $(x_1, x_2) \in U$, $q(Ax_2 - Ax_1) = q(A(x_2 - x_1)) \le \beta p(x_2 - x_1) < \varepsilon$, showing that $(Ax_1, Ax_2) \in V$.

The implication 4 $\Rightarrow$ 1 is a general result: any quasi-uniformly continuous mapping between two quasi-uniform spaces is continuous with respect to the topologies induced by the quasi-uniformities. □

Based on these properties one can introduce an asymmetric seminorm on the cone $L_{p,q}(X, Y)$ by

$$\|A|_{p,q} = \sup\{q(Ax) : x \in X, \, p(x) \le 1\} \, , \tag{2.1.5}$$

for every $A \in L_{p,q}(X, Y)$.

The seminorm $\|\cdot\| := \|\cdot|_{p^s,q^s}$ is the usual operator seminorm on the space $L(X,Y) = L_{p^s,q^s}(X,Y)$ given for $A \in L(X,Y)$ by

$$\|A\| = \sup\{q^s(Ax) : x \in X,\, p^s(x) \le 1\}\,. \tag{2.1.6}$$

We mention the following results, whose proofs are similar to those for normed spaces.

Proposition 2.1.3. *Let (X,p) and (Y,q) be asymmetric seminormed spaces and $A \in L_{p,q}(X,Y)$. Then the number $\|A|_{p,q}$ is the smallest semi-Lipschitz constant for A and $\|\cdot|_{p,q}$ is an asymmetric seminorm on the cone $L_{p,q}(X,Y)$, which is an asymmetric norm if q is an asymmetric norm.*

The asymmetric seminorm $\|A|_{p,q}$ can be calculated also by the formula

$$\|A|_{p,q} = \sup\{q(Ax)/p(x) : x \in X,\, p(x) > 0\}\,. \tag{2.1.7}$$

A *cone* (in fact, a *convex cone*) is a subset Z of a linear space X such that $w+z \in Z$ and $\lambda z \in Z$ for all $w, z \in Z$ and $\lambda \ge 0$. A cone is called also a *semilinear space.* In order to study spaces of linear operators between asymmetric normed spaces and the duals of such spaces, we shall consider asymmetric norms on cones.

Remark 2.1.4. In fact, one can define an abstract notion of cone as a set K with two operations, addition $+$ which is supposed to be commutative, associative and having a neutral element denoted by 0, and multiplication by nonnegative scalars (denoted by $\cdot$), satisfying the properties

$$\begin{aligned} &\text{(i) } (\lambda\mu)a = \lambda(\mu a), \quad \text{(ii) } \lambda(a+b) = \lambda a + \lambda b, \quad \text{(iii)} (\lambda+\mu)a = \lambda a + \mu a, \\ &\text{(iv) } 1\cdot a = 1, \quad \text{and} \quad \text{(v) } 0 \cdot a = 0\,. \end{aligned} \tag{2.1.8}$$

The cone X is called *cancellative* if $a+c = b+c \Rightarrow a = b$ for all $a, b, c \in X$. A cone X is cancellative if and only if it can be embedded in a vector space.

The theory of locally convex cones, with applications to Korovkin type approximation theory for positive operators and to vector-measure theory, is developed in the books by Keimel and Roth [109] and Roth [212], respectively. A recent paper by Galanis [85] discusses Gâteaux and Hukuhara differentiability on topological cones (called by him topological semilinear spaces and meaning cones for which the operations of addition and multiplication by positive scalars are continuous).

The following proposition shows that continuous linear operators between two asymmetric normed spaces are continuous with respect to the associated norm topologies. Also the set of all these continuous linear operators is a cone (a cancellative one).

Proposition 2.1.5. *Let (X,p) and (Y,q) be asymmetric seminormed spaces. Any (p,q)-continuous linear operator $A : X \to Y$ is also (p^s,q^s)-continuous and the set $L_{p,q}(X,Y)$ is a convex cone in $L(X,Y)$. Also*

$$L_{p,q}(X,Y) = L_{\bar{p},\bar{q}}(X,Y)\,, \tag{2.1.9}$$

and

$$\|A|_{p,q} = \|A|_{\bar{p},\bar{q}} \geq \|A|_{p^s,q^s}, \tag{2.1.10}$$

for every $A \in L_{p,q}(X,Y)$.

In particular, every (p,u)*-continuous linear functional is* $(p^s, |\cdot|)$*-continuous and* $X_p^\flat$ *is a cone in the space* X^* *of all continuous linear functionals on the normed space* (X, p^s).

Proof. Observe that

$$\forall x \in X,\ q(Ax) \leq \beta p(x) \iff \forall x \in X,\ \bar{q}(Ax) \leq \beta \bar{p}(x),$$

proving the equality (2.1.9). Taking into account Proposition 2.1.3 (the fact that the seminorm is the smallest semi-Lipschitz constant), this equivalence implies also the equality $\|A|_{p,q} = \|A|_{\bar{p},\bar{q}}$.

Also, $q(Ax) \leq \beta p(x) \leq \beta p^s(x)$ and $\bar{q}(Ax) \leq \beta \bar{p}(x) \leq \beta p^s(x)$, implies $q^s(Ax) \leq \beta p^s(x)$, proving the inclusion $L_{p,q}(X,Y) \subset L_{p^s,q^s}(X,Y)$ and the inequality $\|A|_{p^s,q^s} \leq \|A|_{p,q}$. □

The following example shows that $L_{p,q}(X,Y)$ is not a subspace of $L(X,Y)$.

Example 2.1.6. On the space $X = C_0[0;1]$ from Example 1.1.43 consider the functional $\varphi(f) = f(1)$, $f \in C_0[0;1]$. Then φ is continuous on (X,p), but $-\varphi$ is not continuous.

Another example is furnished by the functional $\mathrm{id} : (\mathbb{R}, u) \to (\mathbb{R}, u)$ which is (p,u)-continuous, but $-\mathrm{id}$ is not.

Indeed, $\varphi(f) = f(1) \leq \max f([0;1]) = p(f)$, $f \in C_0[0;1]$, so that φ is (p,u)-continuous. Taking $f_n(t) = 1-nt$, $t \in [0;1]$, it follows that $p(f_n) = 1$ and $-\varphi(f_n) = -f_n(1) = n-1$, so that $-\varphi$ is not bounded on the unit ball of (X,p), so it is not continuous.

2.1.2 Continuous linear functionals on an asymmetric seminormed space

In this subsection we consider an asymmetric seminormed space (X,p) with non-trivial seminorm p (that is $p \neq 0$) with conjugate seminorm $\bar{p}$ and the (symmetric) seminorm p^s. Note that in this case the fact that a linear functional $\varphi : (X,p) \to \mathbb{R}$ is (p,u)-semi-Lipschitz is equivalent to

$$\varphi(x) \leq \beta p(x)\ , \tag{2.1.11}$$

for all $x \in X$.

Note also that the continuity of a function f from an asymmetric normed space (X,p) to $(\mathbb{R},u)$ is equivalent to its upper semicontinuity from (X,p) to $(\mathbb{R},|\cdot|)$.

Denote by $X_p^\flat$ and $X_{\bar{p}}^\flat$ the cones of p-continuous, respectively $\bar{p}$-continuous, linear functionals on X and let $X^* = (X,p^s)^*$ be the dual of the seminormed space (X,p^s). When there is no danger of confusion the space $X_p^\flat$ will be denoted simply by $X^\flat$ and we shall call it *the asymmetric dual* of the space (X,p).

Let

$$B_p = \{x \in X : p(x) \le 1\} \quad \text{and} \quad B'_p = \{x \in X : p(x) < 1\}$$

be the closed, respectively open, unit ball of X.

The functionals given by

$$\|\varphi|_p = \sup \varphi(B_p),\ \varphi \in X_p^\flat, \quad \text{and} \quad \|\psi|_{\bar{p}} = \sup \psi(B_{\bar{p}}),\ \psi \in X_{\bar{p}}^\flat, \tag{2.1.12}$$

are asymmetric norms on $X_p^\flat$ and $X_{\bar{p}}^\flat$, respectively.

The functional defined by

$$\|x^*\| = \sup x^*(B_{p^s}) = \sup\{x^*(x) : x \in X,\ p^s(x) \le 1\},\ x^* \in X^*\ , \tag{2.1.13}$$

is a norm on the space $X^* = (X,p^s)^*$ and the space X^* is complete with respect to this norm, i.e., it is a Banach space.

We mention the following properties of continuous linear functionals.

Proposition 2.1.7. *Let (X,p) be a space with asymmetric seminorm.*

1. *The functionals given by* (2.1.12) *are asymmetric norms on $X_p^\flat$, and $X_{\bar{p}}^\flat$, respectively, satisfying*

$$\varphi = 0 \iff \|\varphi|_p = 0 \quad \text{and} \quad \psi = 0 \iff \|\psi|_{\bar{p}} = 0\ . \tag{2.1.14}$$

   Also, if $\varphi \neq 0$, then

$$\varphi(x_0) = \|\varphi|_p \Rightarrow p(x_0) = 1\ , \tag{2.1.15}$$

   for any $x_0 \in B_p$.

2. *The norm $\|\varphi|_p$ of a functional $\varphi \in X_p^\flat$ is the smallest semi-Lipschitz constant for φ and it can be also calculated by the formula*

$$\|\varphi|_p = \sup\{\varphi(x)/p(x) : x \in X,\ p(x) > 0\}\ .$$

3. *The cones $X_p^\flat$ and $X_{\bar{p}}^\flat$ are contained in X^* and*

$$\|\varphi\| \le \|\varphi|_p \quad \text{and} \quad \|\psi\| \le \|\psi|_{\bar{p}}\ ,$$

   for every $\varphi \in X_p^\flat$ and every $\psi \in X_{\bar{p}}^\flat$.

4. *The following hold:*

$$\varphi \in X_p^\flat \iff -\varphi \in X_{\bar{p}}^\flat \quad and \quad \|\varphi|_p = \|-\varphi|_{\bar{p}}\,.$$

Consequently, $X_p = -X_{\bar{p}}$ *and the linear spans of* $X_p^\flat$ *and* $X_{\bar{p}}^\flat$ *agree, being given by*

$$\operatorname{sp}(X_p^\flat) = \operatorname{sp}(X_{\bar{p}}^\flat) = X_p^\flat + X_{\bar{p}}^\flat.\,.$$

Proof. All of the assertions from 1 and 2, excepting (2.1.14) and (2.1.15), are consequences of Propositions 2.1.2 and 2.1.3.

To prove (2.1.14), observe that if $\varphi \neq 0$, then there exists $x_0 \in X$ such that $\varphi(x_0) > 0$, implying

$$0 < \varphi(x_0) \le \|\varphi|_p\, p(x_0)\,.$$

If there exists $x_0 \in X$ with $p(x_0) < 1$ such that $\varphi(x_0) = \|\varphi|_p > 0$, then $x_1 = [p(x_0)]^{-1}x_0$ satisfies $p(x_1) = 1$ and $\varphi(x_1) = \varphi(x_0)/p(x_0) > \|\varphi|_p$, a contradiction, showing that (2.1.15) holds, too.

3. Let $\varphi \in X_p^\flat$. The inequalities

$$\varphi(x) \le \|\varphi|_p p(x) \le \|\varphi|_p p^s(x),\ x \in X\,,$$

imply $\varphi \in X^*$ and $\|\varphi\| \le \|\varphi|_p$. The situation for the conjugate seminorm $\bar{p}$ is similar.

4. Let now $\varphi : X \to \mathbb{R}$ be a linear functional. The assertion will be a consequence of the following equalities:

$$\|-\varphi|_{\bar{p}} = \sup\{-\varphi(x) : \bar{p}(x) \le 1\} = \sup\{\varphi(-x) : p(-x) \le 1\} = \|\varphi|_p\,.$$

It follows that $X_p^\flat = -X_{\bar{p}}^\flat$, so that $\operatorname{sp}(X_p^\flat) = X_p^\flat - X_p^\flat = X_p^\flat + X_{\bar{p}}^\flat = \operatorname{sp}(X_{\bar{p}}^\flat)$. □

In the following proposition we collect some simple properties of the norm $\|\cdot|_p$ that we shall need in the proofs of the separation theorems.

Proposition 2.1.8. *If* φ *is a continuous linear functional on a space with asymmetric seminorm* (X,p), $p \neq 0$, *then the following assertions hold.*

1. *We have*

$$\begin{aligned}\|\varphi|_p &= \sup\{\varphi(x) : x \in X,\ p(x) < 1\}\\ &= \sup\{\varphi(x) : x \in X,\ p(x) = 1\}\end{aligned} \tag{2.1.16}$$

and

$$\begin{aligned}-\|\varphi|_{\bar{p}} &= \inf\{\varphi(x) : x \in X,\ p(x) < 1\}\\ &= \inf\{\varphi(x) : x \in X,\ p(x) = 1\}\,.\end{aligned} \tag{2.1.17}$$

2. *If* φ *is both* p- *and* $\bar{p}$-*continuous (i.e.,* $\varphi \in X_p^\flat \cap X_{\bar{p}}^\flat$*), then*

$$\varphi(B_p') = (-\|\varphi|_{\bar{p}}; \|\varphi|_p) \quad and \quad \varphi(B_{\bar{p}}') = (-\|\varphi|_p; \|\varphi|_{\bar{p}})\,.$$

3. *If* $\varphi \in X^\flat_p \setminus X^\flat_{\bar{p}}$ *and* $\psi \in X^\flat_{\bar{p}} \setminus X^\flat_p$, *then*

$$\varphi(B'_p) = (-\infty, \|\varphi|_p) \quad \text{and} \quad \psi(B'_{\bar{p}}) = (-\|\psi|_{\bar{p}}; \infty)\,.$$

Proof. 1. We can suppose $\varphi \neq 0$. Then $c := \sup\{\varphi(x) : p(x) < 1\} \le \|\varphi|_p$. If $x \in X$ is such that $p(x) = 1$, then $p(n(n+1)^{-1}x) = n(n+1)^{-1}p(x) < 1$, so that $n(n+1)^{-1}\varphi(x) = \varphi(n(n+1)^{-1}x) \le c$, for all $n \in \mathbb{N}$. Letting $n \to \infty$, it follows that $\varphi(x) \le c$, implying $\|\varphi|_p = \sup \varphi(B_p) \le c$, that is $\|\varphi|_p = c$.

By Proposition 2.1.7,

$$\|\varphi|_p = \sup\{\varphi(x/p(x)) : x \in X,\, p(x) > 0\} = \sup\{\varphi(y) : y \in X,\, p(y) = 1\}.$$

By (2.1.16) applied to $\bar{p}$, $\|\varphi|_{\bar{p}} = \sup \varphi(B'_{\bar{p}})$, so that

$$\inf\{\varphi(x):p(x)<1\}=\inf\{\varphi(-x):p(-x)<1\}=-\sup\{\varphi(x):\bar{p}(x)<1\}=-\|\varphi|_{\bar{p}},$$

proving the first equality in (2.1.17).

The second equality is proved similarly.

2. By the first assertion of the proposition,

$$\sup \varphi(B'_p) = \|\varphi|_p \quad \text{and} \quad \inf \varphi(B'_p) = -\|\varphi|_{\bar{p}}\,.$$

By Proposition 2.1.7.1, $-\|\varphi|_{\bar{p}} < \varphi(x) < \|\varphi|_p$ for every $x \in B'_p$.

The convexity of B'_p and the linearity of φ imply that $\varphi(B'_p)$ is convex, that is it is an interval in $\mathbb{R}$, so that

$$\varphi(B'_p) = (\inf \varphi(B'_p); \sup \varphi(B'_p)) = (-\|\varphi|_{\bar{p}}; \|\varphi|_p)\,.$$

3. If $\varphi \in X^\flat_p \setminus X^\flat_{\bar{p}}$, then

$$\infty = \sup\{\varphi(x) : p(-x) < 1\} = \sup\{-\varphi(x) : p(x) < 1\} = -\inf\{\varphi(x) : p(x) < 1\}\,,$$

that is $\inf \varphi(B'_p) = -\infty$, and $\sup \varphi(B'_p) = \|\varphi|_p$.

Reasoning as above, it follows that $\varphi(B'_p) = (-\infty; \|\varphi|_p)$.

For the second equality, note that $\sup \varphi(B'_p) = \infty$ and, by (2.1.17), $\inf \varphi(B'_p) = -\|\varphi|_{\bar{p}}$, implying $\varphi(B'_{\bar{p}}) = (-\|\varphi|_{\bar{p}}; \infty)$. □

2.1.3 Continuous linear mappings between asymmetric locally convex spaces

Let (X, P), (Y, Q) be two asymmetric locally convex spaces with the topologies τ_P and τ_Q generated by the families P and Q of asymmetric seminorms on X and Y, respectively. In the following when we say that (X, P) is an asymmetric locally convex space, we understand that X is a real vector space, P is a family of asymmetric seminorms on X and τ_P is the asymmetric locally convex topology associated to P.

A linear mapping $A : X \to Y$ is called (P, Q)-*bounded* if for every $q \in Q$ there exist $F \in \mathcal{F}(P)$ (the family of all nonempty finite subsets of P) and $\beta \geq 0$ such that

$$\forall x \in X,\ q(Ax) \leq \beta \max\{p(x) : p \in F\}. \tag{2.1.18}$$

If the family P is directed, then the (P, Q)-boundedness of A is equivalent to the condition: for every $q \in Q$ there exist $p \in P$ and $\beta \geq 0$ such that

$$\forall x \in X,\ q(Ax) \leq \beta p(x). \tag{2.1.19}$$

The continuity of the mapping A from (X, τ_P) to (Y, τ_Q) is called (τ_P, τ_Q)-*continuity.* We shall use also the terms (P, Q)-continuity for this property, and (P, u)-continuity in the case of (τ_P, τ_u)-continuous linear functionals, where u is the quasi-metric on $\mathbb{R}$ given in Example 1.1.3.

Because both of the topologies τ_P and τ_Q are translation invariant, we have the following result.

Proposition 2.1.9. *Let (X, P) and (Y, Q) be asymmetric locally convex spaces and $A : X \to Y$ a linear mapping. The following conditions are equivalent.*

1. *The mapping A is (P, Q)-continuous on X.*
2. *The mapping A is continuous at $0 \in X$.*
3. *The mapping A is continuous at some point $x_0 \in X$.*

The following proposition emphasizes the equivalence of continuity and boundedness for linear mappings.

Proposition 2.1.10. *Let (X, P) and (Y, Q) be two asymmetric locally convex spaces and $A : X \to Y$ a linear mapping. The following assertions are equivalent.*

1. *The mapping A is (P, Q)-continuous on X.*
2. *The mapping A is continuous at $0 \in X$.*
3. *The mapping A is (P, Q)-bounded.*
4. *The mapping A is $(\mathcal{U}_P, \mathcal{U}_Q)$-quasi-uniformly continuous on X.*

Proof. The equivalence 1 $\iff$ 2 follows from the preceding proposition.

Suppose the families P and Q to be directed.

2 $\Rightarrow$ 3. For $q \in Q$ consider the τ_Q-neighborhood $V = B_q(0, 1)$ of $A(0) = 0 \in Y$, and let U be a neighborhood of $0 \in X$ such that $A(U) \subset V$. If $p \in P$ and $r > 0$ are such that $B_p(0, r) \subset U$, then

$$\forall x \in X, \quad p(x) \leq r \ \Rightarrow\ q(Ax) \leq 1\ .$$

By Proposition 2.1.1 applied to f(x) = q(Ax) and $g(x) = p(x)$, this relation implies

$$\forall x \in X, \quad q(Ax) \leq \frac{1}{r} p(x)\ .$$

$3 \Rightarrow 4$. For $W \in \mathcal{U}_Q$ there exists $q \in Q$ and $\varepsilon > 0$ such that $W_{q,\varepsilon} = \{(y, y') \in Y \times Y : q(y' - y) < \varepsilon\} \subset W$. If $p \in P$ and $\beta > 0$ are such that (2.1.19) holds, then $(f(x), f(y)) \in W_{q,\varepsilon}$ for every $(x, y) \in U_{p,r}$, where $r := \varepsilon/\beta$.

The implication $4 \Rightarrow 1$ holds for arbitrary quasi-uniform spaces. □

In the case of linear functionals on an asymmetric locally convex space we have the following characterization of continuity, where u is as in Example 1.1.3.

Proposition 2.1.11. *Let (X, P) be an asymmetric locally convex space, where P is a directed family of asymmetric seminorms on X, and $\varphi : X \to \mathbb{R}$ a linear functional. The following assertions are equivalent.*

1. *The functional φ is (P, u)-continuous at $0 \in X$.*
2. *The functional φ is (P, u)-continuous on X.*
3. *The functional φ is upper semi-continuous from (X, τ_P) to $(\mathbb{R}, |\cdot|)$.*
4. *There exist $p \in P$ and $\beta \geq 0$ such that*

$$\forall x \in X, \quad \varphi(x) \leq \beta p(x). \tag{2.1.20}$$

5. *The functional φ is (P, u)-quasi-uniformly continuous on X.*

Using Proposition 2.1.1 and the inequality

$$f(y) - f(x) \leq f(y - x), \; x, y \in X, \tag{2.1.21}$$

valid for any sublinear functional on a vector space X, it is easy to check that these results hold for the slightly more general case of sublinear functionals.

Proposition 2.1.12. *Let (X, P) be an asymmetric locally convex space, where P is a directed family of asymmetric seminorms on X, and let $f : X \to \mathbb{R}$ be a sublinear functional. The following assertions are equivalent.*

1. *The functional f is (P, u)-continuous at $0 \in X$.*
2. *The functional f is (P, u)-continuous on X.*
3. *The functional f is upper semi-continuous from (X, τ_P) to $(\mathbb{R}, |\cdot|)$.*
4. *There exist $p \in P$ and $\beta > 0$ such that*

$$\forall x \in X, \quad f(x) \leq \beta p(x) \; .$$

5. *The functional f is (P, u)-quasi-uniformly continuous on X.*

Proof. $1 \Rightarrow 2$ Let $x \in X$. For $\varepsilon > 0$ there exist $p \in P$ and $r > 0$ such that

$$p(z) < r \;\Rightarrow\; f(z) < \varepsilon.$$

Then, for any $y \in X$ such that $p(y - x) < r$ we have

$$f(y) - f(x) \leq f(y - x) < \varepsilon,$$

proving the (P, u)-continuity of f at x.

The equivalence 2 $\iff$ 3 is true for an arbitrary function from X to $\mathbb{R}$.

2 $\Rightarrow$ 4. Since $(-\infty; 1]$ is a τ_u-neighborhood of $f(0) = 0 \in \mathbb{R}$, there exist $p \in P$ and $r > 0$ such that $f(B_p[0, r]) \subset (-\infty; 1]$, i.e.,

$$\forall x \in X, \quad p(x) \le r \Rightarrow f(x) \le 1 .$$

By Proposition 2.1.1 this implies

$$\forall x \in X, \quad f(x) \le \frac{1}{r} p(x) .$$

4 $\Rightarrow$ 5. For $\varepsilon > 0$ let $U_{u,\varepsilon} = \{(s, t) \in \mathbb{R}^2 : u(t - s) < \varepsilon\}$. If $p \in P$ and $\beta > 0$ are given by 4, let $V_{p,\varepsilon/\beta} = \{(x, y) \in X^2 : p(y - x) < \varepsilon/\beta\}$.

The inequalities

$$f(y) - f(x) \le f(y - x) \le \beta p(y - x) < \varepsilon$$

show that $f(V_{p,\varepsilon/\beta}) \subset U_{u,\varepsilon}$.

The implication 5 $\Rightarrow$ 1 is a general topological property. □

The above proposition has the following useful corollary.

Corollary 2.1.13. *Let (X, P) be an asymmetric LCS.*

1. *Let f, g be sublinear functionals defined on an asymmetric locally convex space (X, P). If $f \le g$ and g is (P, u)-continuous, then f is (P, u)-continuous too. In particular the result is true when f is linear.*
2. *Every asymmetric seminorm $p \in P$ is quasi-uniformly (P, u)-continuous.*

Proof. By Proposition 2.1.12, there exist $p \in P$ and $\beta \ge 0$ such that $\forall x \in X,\ g(x) \le \beta p(x)$. It follows that $\forall x \in X,\ f(x) \le g(x) \le \beta p(x)$, which, by the same proposition, implies the continuity of f.

The second assertion is a consequence of the first one and of the translation invariance of the topology. □

Remark 2.1.14. If the family P is not directed, then the family $\widetilde{P} = \{p_F : F \in \mathcal{F}(P)\}$, where $p_F(x) = \max\{p(x) : p \in F\}$, is a directed family of seminorms generating the same topology as P.

Consequently, the (P, u)-continuity of a functional φ (or f) is equivalent to the condition: there exist $F \in \mathcal{F}(P)$ and $\beta \ge 0$ such that

$$\forall x \in X, \quad \varphi(x) \le \beta p_F(x) = \beta \max\{p(x) : p \in F\}. \tag{2.1.22}$$

As in the case of asymmetric normed spaces, see Subsection 2.1.4, the set $L_{P,Q}(X,Y)$ of all (P,Q)-continuous mappings between two asymmetric LCS (X,P) and (Y, Q) is a convex cone in the space $L_a(X, Y)$ of all linear operators between X and Y. In fact it is contained in the linear space $L(X, Y) = L((X, P^s), (Y, Q^s))$

of all continuous linear operators between the locally convex spaces (X, P^s) and (Y, Q^s). Indeed, supposing P, Q directed, then for $A \in L_{P,Q}(X, Y)$ and $q \in Q$ there exist $p \in P$ and $\beta \geq 0$ such that

$$\forall x \in X, \quad q(Ax) \leq \beta p(x) \leq \beta p^s(x) .$$

Since

$$\bar{q}(Ax) = q(-Ax) = q(A(-x)) \leq \beta p(-x) \leq \beta p^s(x) ,$$

it follows that

$$\forall x \in X, \quad q^s(Ax) \leq \beta p^s(x) ,$$

showing that $A \in L(X, Y)$.

Here for $p \in P$, $\bar{p}(x) = p(-x)$, $p^s(x) = \max\{p(x), p(-x)\}$, $x \in X$, and $P^s = \{p^s : p \in P\}$ with similar definitions for the family Q.

For an asymmetric locally convex space (X, P) denote by $X^\flat = X^\flat_P$ the set of all linear (P, u)-continuous functionals. It follows that $X^\flat_P$ is a convex cone in $X^\#$ the algebraic dual space of X, i.e., the space of all linear functionals on X. In fact, by the above remark, it is contained in the dual space $X^* = L((X, P^s), (\mathbb{R}, |\cdot|))$.

Remark 2.1.15. It is easy to check that a linear functional $\varphi(t) = at$, $t \in \mathbb{R}$, is (τ_u, τ_u)-continuous if and only if $a \geq 0$. Indeed if $a \geq 0$, then $\varphi(t) = at \leq u(at) = au(t)$, $t \in \mathbb{R}$. If $a < 0$, then, reasoning as above, one concludes that φ fails to be continuous.

2.1.4 Completeness properties of the normed cone of continuous linear operators

Following [90], we can consider an extended asymmetric norm on the space $L(X, Y)$ of all linear continuous operators from (X, p^s) to (Y, q^s), defined by the same formula:

$$\|A|^*_{p,q} = \sup\{q(Ax) : x \in X,\, p(x) \leq 1\} = \sup\{q(Ax) : x \in B_p\} , \tag{2.1.23}$$

for every $A \in L(X, Y)$. If $A \in L_{p,q}(X.Y)$, then $A \in L(X, Y)$, and so $-A \in L(X, Y)$, but, as the above examples show, it is possible that $\| - A|^*_{p,q} = \infty$, so that $\| \cdot |^*_{p,q}$ could be effectively an extended asymmetric norm.

With the asymmetric norm $\| \cdot |^*_{p,q}$ one associates a symmetric extended norm on $L(X, Y)$ defined by

$$\|A\|^*_{p,q} = \|A|^*_{p,q} \vee \| - A|^*_{p,q} . \tag{2.1.24}$$

Since

$$\begin{aligned} \| - A|^*_{p,q} &= \sup\{q(-Ax) : p(x) \leq 1\} = \sup\{q(Ax) : p(-x) \leq 1\} \\ &= \sup\{q(Ax) : x \in B_{\bar{p}}\} , \end{aligned}$$

it follows that

$$\|A\|^*_{p,q} = \sup\{q(Ax) : x \in B_p \cup B_{\bar{p}}\} . \tag{2.1.25}$$

The norm $\|A\|^*_{p,q}$ can be calculated also by the formula

$$\|A\|^*_{p,q} = \sup\{q^s(Ax) : x \in B_p\} . \tag{2.1.26}$$

Indeed, denoting by λ the right side member of the above equality, we have $q(Ax) \le q^s(Ax) \le \lambda$ for every $x \in B_p$, so that $\|A|^*_{p,q} \le \lambda$. Similarly $q(-Ax) \le q^s(Ax) \le \lambda$ for every $x \in B_p$ implies $\| - A|^*_{p,q} \le \lambda$, so that $\|A\|^*_{p,q} \le \lambda$. Also, $q(Ax) \le \|A|^*_{p,q} \le \|A\|^*_{p,q}$ and $q(-Ax) \le \| - A|^*_{p,q} \le \| - A\|^*_{p,q} = \|A\|^*_{p,q}$ for every $x \in B_p$, implies $\lambda \le \|A\|^*_{p,q}$.

Recall that an asymmetric normed space (X, p) is called biBanach if the associated normed space (X, p^s) is a Banach space (i.e., a complete normed space).

Proposition 2.1.16 ([90]). *Let (X, p) and (Y, q) be asymmetric normed spaces.*

1. *The functional $\| \cdot |^*_{p,q}$ given by* (2.1.23) *is an extended asymmetric norm on the space $L(X, Y)$ and $\|A\| \le \|A\|^*_{p,q}$ for all $A \in L(X, Y)$. An operator $A \in L(X, Y)$ belongs to $L_{p,q}(X, Y)$ if and only if $\|A|^*_{p,q} < \infty$. Also $\|A|^*_{p,q} = \|A|_{p,q}$ for $A \in L_{p,q}(X, Y)$.*
2. *If the space (Y, q) is biBanach, then the space $(L(X, Y), \| \cdot \|^*)$ is complete.*
3. *The set $L_{p,q}(X, Y)$ is closed in $(L(X, Y), \|\cdot\|^*_{p,q})$, so it is complete with respect to the restriction of the extended norm $\| \cdot \|^*_{p,q}$ to $L_{p,q}(X, Y)$.*

Proof. We shall omit the subscripts p, q in what follows.

1. It is easy to check that $\| \cdot |^*$ is an extended asymmetric norm on $L(X, Y)$. We can suppose $\|A\|^* < \infty$. Then

$$q(Ax) \le \|A|p(x) \le \|A\|^* p^s(x), \quad \text{and}$$
$$q(-Ax) \le \| - A|p(x) \le \|A\|^* p^s(x) ,$$

so that $q^s(Ax) \le \|A\|^* p^s(x)$, for all $x \in X$, implying $\|A\| \le \|A\|^*$.

2. Let (A_n) be a $\| \cdot \|^*$-Cauchy sequence in $L(X, Y)$, that is for every $\varepsilon > 0$ there exists $n_0 \in \mathbb{N}$ such that

$$\|A_m - A_n\|^* \le \varepsilon \tag{2.1.27}$$

holds for all $m, n \ge n_0$. By the first point of the theorem, $\|A_m - A_n\| \le \|A_m - A_n\|^*$, so that (A_n) is a Cauchy sequence in the Banach space $(L(X, Y), \| \cdot \|)$, and so (A_n) has a $\| \cdot \|$-limit $A \in L(X, Y)$.

It remains to show that (A_n) converges to A with respect to the norm $\| \cdot \|^*$. The inequality $q^s(A_n x - Ax) \le \|A_n - A\|\, p^s(x)$ implies

$$\lim_{n\to\infty} q^s(A_n x - Ax) = 0 , \tag{2.1.28}$$

that is the sequence $(A_n x)$ converges to Ax in the normed space (Y, q^s), for every $x \in X$.

For $\varepsilon > 0$ let n_0 be such that (2.1.27) holds. Then, by (2.1.26), for every $x \in B_p$,

$$q^s(A_m x - A_n x) \le \|A_m - A_n\|^* \le \varepsilon \,.$$

By (2.1.28), the inequality $q^s(A_m x - A_n x) \le \varepsilon$ yields for $n \to \infty$,

$$q^s(A_m x - Ax) \le \varepsilon \,,$$

for every $x \in B_p$ and $m \ge n_0$. Taking into account (2.1.26), it follows that $\|A_m - A\|^* \le \varepsilon$ for every $m \ge n_0$.

3. To show that $L_{p,q}(X,Y)$ is closed in $(L(X,Y), \|\cdot\|^*)$ let (A_n) be a sequence in $L_{p,q}(X,Y)$ which is $\|\cdot\|^*$-convergent to some $A \in L(X,Y)$. For $\varepsilon = 1$ let $n_0 \in \mathbb{N}$ be such that $\|A_n - A\|^* \le 1$ for all $n \ge n_0$. Then $\|A\|^* \le \|A - A_{n_0}\|^* + \|A_{n_0}\|^* \le 1 + \|A_{n_0}\|^*$, which implies that both A and $-A$ belong to $L_{p,q}(X,Y)$. □

Remark 2.1.17. 1. If a sequence (A_n) in $L_{p,q}(X,Y)$ converges to $A \in L(X,Y)$ with respect to the conjugate norm $(\|\cdot|^*_{p,q})^{-1}$ of $\|\cdot|^*_{p,q}$, then $A \in L_{p,q}(X,Y)$.

2. As it is known, in the classical case, the completeness of $(L(X,Y), \|\cdot\|)$ implies the completeness of the normed space Y. We do not know if a similar result holds for the extended norm $\|\cdot\|^*$.

Let n_0 be such that $\|A - A_n|^* \le 1$ for all $n \ge n_0$. Then $\|A|^* \le \|A - A_{n_0}|^* + \|A_{n_0}|^* \le 1 + \|A_{n_0}|^* < \infty$, shows that $A \in L_{p,q}(X,Y)$, proving the validity of the assertion from 1.

2.1.5 The bicompletion of an asymmetric normed space

As it is well known, any normed space $(X, \|\cdot\|)$ has a completion, meaning that there exists a Banach space $(\widetilde{X}, \|\cdot\|^\sim)$ such that $(X, \|\cdot\|)$ is isometrically isomorphic to a dense subspace of $(\widetilde{X}, \|\cdot\|^\sim)$. The Banach space $(\widetilde{X}, \|\cdot\|^\sim)$, called the *completion* of X is uniquely determined, in the sense that any Banach space Z such that X is isometrically isomorphic to a dense subspace of Z is isometrically isomorphic to $(\widetilde{X}, \|\cdot\|^\sim)$.

A *bicompletion* of an asymmetric normed space (X,p) is a bicomplete asymmetric normed space (Y,q) such that X is isometrically isomorphic to a τ_{q^s}-dense subspace of Y. An *isometry* between two asymmetric normed spaces $(X,p), (Y,q)$ is a mapping $T : X \to Y$ such that

$$q(Tx - Ty) = p(x-y), \quad \text{for all} \quad x, y \in X \,. \tag{2.1.29}$$

If T is linear, then (2.1.29) is equivalent to

$$q(Tx) = p(x), \quad \text{for all} \quad x \in X \,. \tag{2.1.30}$$

Note that, as defined, the isometry T is in fact an isometry between the associated normed spaces $(X, p^s), (Y, q^s)$, because

$$q^s(Tx - Ty) = q(Tx - Ty) \vee q(Ty - Tx) = p(x - y) \vee p(y - x) = p^s(x - y) ,$$

for all $x, y \in X$. The construction of the bicompletion of an asymmetric normed space was done in [88] (see [170] for the case of normed cones), following ideas from the normed case, adapted to the asymmetric one. We only sketch the construction, referring for details to the mentioned papers.

Let (X, p) be an asymmetric normed space. In the set of all p^s-Cauchy sequences in X define an equivalence relation by

$$(x_n) \sim (y_n) \iff \lim_{n\to\infty} p^s(x_n - y_n) = 0 . \tag{2.1.31}$$

Lemma 2.1.18.

1. *The relation $\sim$ is an equivalence relation on the set of all p^s-Cauchy sequences in X.*
2. *If (x_n) is a p^s-Cauchy sequence then the sequence $(p(x_n))$ is convergent and $\lim_n p(x_n) = \lim_n p(y_n)$ whenever (y_n) is a p^s-Cauchy sequence equivalent to (x_n).*

Proof. The verification of 1 is routine.

2. The inequalities $p(x_n) - p(x_m) \le p(x_n - x_m) \le p^s(x_n - x_m)$, valid for all $m, n \in \mathbb{N}$, and the fact that (x_n) is p^s-Cauchy, imply that $(p(x_n))$ is a Cauchy sequence in $\mathbb{R}$, so it converges to some $\alpha \in \mathbb{R}$. If (y_n) is another p^s-Cauchy sequence equivalent to (x_n), then the inequalities

$$\begin{aligned} &p(x_n) - p(y_n) \le p(x_n - y_n) \le p^s(x_n - y_n), \\ &p(y_n) - p(x_n) \le p(y_n - x_n) \le p^s(y_n - x_n) = p^s(x_n - y_n) \end{aligned}$$

and the condition $\lim_n p^s(x_n - y_n) = 0$ imply $\lim_n p(x_n) = \lim_n p(y_n)$. □

Denote by $\tilde{X}$ the linear space of all equivalence classes of p^s-Cauchy sequences with addition and multiplication by scalars defined, as usual, by $[(x_n)] + [(y_n)] = [(x_n + y_n)]$ and $\lambda[(x_n)] = [(\lambda x_n)]$, where $[(x_n)]$ denotes the equivalence class containing the p^s-Cauchy sequence (x_n). Based on Lemma 2.1.18, one can define on the space $\tilde{X}$ an asymmetric norm $\tilde{p}$ by

$$\tilde{p}([(x_n)]) = \lim_n p(x_n) , \tag{2.1.32}$$

for any p^s-Cauchy sequence (x_n) in X.

As it is known, equipped with the norm

$$\widetilde{p^s}([(x_n)]) = \lim_n p^s(x_n) , \tag{2.1.33}$$

the space $(\tilde{X}, \widetilde{p^s})$ is a Banach space and the mapping $i : X \to \tilde{X}$ defined by

$$i(x) = [(x_n)] \quad \text{where} \quad x_n = x, \forall n \in \mathbb{N} , \tag{2.1.34}$$

is a linear isometry of X into $(\tilde{X}, \widetilde{p^s})$ such that $i(X)$ is $\widetilde{p^s}$-dense in $\tilde{X}$. Hence, the fact that (X, p) is biBanach, meaning that $(\tilde{X}, (\tilde{p})^s)$ is Banach, will follow once we prove the equality

$$(\tilde{p})^s = \widetilde{p^s} . \tag{2.1.35}$$

For a p^s-Cauchy sequence (x_n) in X, let $\alpha_n = p(x_n)$ and $\beta_n = p(-x_n)$, $n \in \mathbb{N}$. Then

$$\begin{aligned}\widetilde{p^s}([x_n]) &= \lim_n p^s(x_n) = \lim_n (\alpha_n \vee \beta_n) = (\lim_n \alpha_n) \vee (\lim_n \beta_n) \\ &= \tilde{p}([(x_n)]) \vee \tilde{p}(-[(x_n)]) = (\tilde{p})^s([(x_n)]) .\end{aligned}$$

Remark 2.1.19. The fact that $\alpha_n \to \alpha$ and $\beta_n \to \beta$ implies $\alpha_n \vee \beta_n \to \alpha \vee \beta$ follows from the relations

$$\alpha_n \vee \beta_n = \frac{\alpha_n + \beta_n + |\alpha_n - \beta_n|}{2} \to \frac{\alpha + \beta + |\alpha - \beta|}{2} = \alpha \vee \beta .$$

We summarize the results in the following theorem.

Theorem 2.1.20. *Let (X, p) be an asymmetric normed space, $\tilde{X}$ the space constructed above and $\tilde{p}$, $\widetilde{p^s}$ the norms on $\tilde{X}$ given by* (2.1.32) *and* (2.1.33)*, respectively.*

1. *The space $(\tilde{X}, \tilde{p})$ is biBanach, or, equivalently, $(\tilde{X}, (\tilde{p})^s)$ is a Banach space.*
2. *The mapping $i : X \to \tilde{X}$, defined by* (2.1.34)*, is a linear isometry of (X, p) into $(\tilde{X}, \tilde{p})$ and the space $i(X)$ is $\tilde{p}^s$-dense in $\tilde{X}$.*
3. *If (Y, q) is an asymmetric biBanach space such that (X, p) is isometrically isomorphic to a q^s-dense subspace of Y, then (Y, q) is isometrically isomorphic to $(\tilde{X}, \tilde{p})$.*

2.1.6 Asymmetric topologies on normed lattices

Alegre, Ferrer and Gregori, [77] and [6, 8], introduced an asymmetric norm on a normed lattice and studied the properties of the induced quasi-uniformity and topology, in connection with the usual properties of normed lattices.

An *ordered vector space* is real vector space X equipped with a partial order relation such that

$$x \le y \ \Rightarrow \ x + z \le y + z \quad \text{and} \quad tx \le ty , \tag{2.1.36}$$

for all $z \in X$ and $t \ge 0$. Denoting by X_+ the cone of positive elements, $X_+ = \{x \in X; x \ge 0\}$, it follows that

$$x \le y \iff y - x \in X_+ .$$

The ordered vector space $(X, \le)$ is called a *vector lattice* if every pair x, y of elements in X admits a lowest upper bound $x \vee y$. Since

$$x \le y \iff -y \le -x ,$$

it follows that

$$x \wedge y = -\left((-x) \vee (-y)\right) ,$$

so that every pair of elements $x, y \in X$ has a greatest lower bound.

Put

$$x^+ = x \vee 0, \quad x^- = (-x) \vee 0 = -(x \wedge 0) \quad \text{and} \quad |x| = x^+ + x^- .$$

It follows that $x = x^+ - x^-$.

A norm $\|\cdot\|$ on an ordered vector space $(X, \le)$ is called a *lattice norm* if it satisfies one of the following equivalent conditions:

$$\begin{aligned} &\text{(i)} \quad |x| \le |y| \ \Rightarrow \ \|x\| \le \|y\|, \\ &\text{(ii)} \quad 1^\circ. \ \ \||x|\| = \|x\|, \ \text{and} \ 2^\circ. \ \ 0 \le x \le y \ \Rightarrow \ \|x\| \le \|y\| , \end{aligned} \tag{2.1.37}$$

for all $x, y \in X$. An ordered vector space equipped with a lattice norm is called a *normed lattice* and is denoted by $(X, \|\cdot\|, \le)$. If, in addition, $(X, \|\cdot\|)$ is a Banach space, then $(X, \|\cdot\|, \le)$ is called a *Banach lattice.*

A normed lattice $(X, \|\cdot\|, \le)$ is called an L-space, M-space or an E-space, provided that

$$\begin{aligned} &\text{(L)} && \|x + y\| = \|x\| + \|y\|, \\ &\text{(M)} && \|x \vee y\| = \|x\| \vee \|y\|, \\ &\text{(E)} && \|x + y\|^2 + \|x - y\|^2 = 2\|x\|^2 + 2\|y\|^2 , \end{aligned}$$

for all positive $x, y \in X$.

To an asymmetric norm p on a vector space X one can associate the following norms, defined for $x \in X$ by the equalities:

$$\begin{aligned} &p^s_L(x) = p^s_L(x) = p(x) + p(-x), \qquad p^s_M(x) = p^s(x) = p(x) \vee p(-x), \\ &p^s_E(x) = \left(p(x)^2 + p(-x)^2\right)^{1/2} . \end{aligned} \tag{2.1.38}$$

The norm p^s_M is the usual norm p^s we have associated to an asymmetric norm p and all the norms given in (2.1.38) are equivalent.

A subset Z of a normed lattice X is called *increasing* if for every $y, z \in X$, $z \in Z$ and $z \le y$ implies $y \in Z$. It is called *decreasing* if for every $y, z \in X$, $z \in Z$ and $y \le z$ implies $y \in Z$.

For a normed lattice $(X, \|\cdot\|, \le)$ consider the functional

$$p(x) = \|x \vee 0\|, \quad x \in X . \tag{2.1.39}$$

Proposition 2.1.21. *The functional p given by (2.1.39) is an asymmetric norm on X with conjugate*

$$\bar{p}(x) = \|x^-\|, \quad x \in X . \tag{2.1.40}$$

The topology τ_p ($\tau_{\bar{p}}$) generated by p ($\bar{p}$) is T_0 but not T_1.

Proof. If $p(x) = p(-x) = 0$, then $\|x^+\| = 0$ implies $x^+ = 0$ and $\|(-x)^+\| = 0$ implies $x^- = (-x)^+ = 0$, so that $x = x^+ - x^- = 0$.

Also, $0 \le (x+y)^+ \le x^+ + y^+$ implies

$$p(x+y) = \|(x+y)^+\| \le \|x^+ + y^+\| \le \|x^+\| + \|y^+\| = p(x) + p(y) .$$

The positive homogeneity is obvious.

The conjugate $\bar{p}$ of p satisfies the equalities

$$\bar{p}(x) = p(-x) = \|(-x)^+\| = \|x^-\| .$$

Since p is an asymmetric norm the topology τ_p is T_0. The topology is T_1 if and only if $p(x) > 0$ for every $x \neq 0$ (see Proposition 1.1.8.3). But $p(x) = \|x^+\| = 0$ is equivalent to $x^+ = 0$, that is $x \le 0$. That is, excepting the trivial case $X_+ = X$, there are non-null elements $x \in X$ with $x \le 0$. □

The remark concerning the separation properties is taken from [51], where some properties of convergent sequences were also proved. Recall that we denote by $L_\rho((x_n))$ the set of all ρ-limits of a sequence (x_n) in a quasi-semimetric space (X, ρ), see (1.1.8).

Proposition 2.1.22 ([51]). *Let $(X, \|\cdot\|)$ be a normed lattice, p the asymmetric norm given by (2.1.39) and (x_n) a sequence in X.*

1. *If (x_n) is p-convergent, then $L_\rho((x_n))$ is increasing.*
2. *The sequence (x_n) is p-convergent to every $z \in X$ such that $x_n \le z$ for all $n \in \mathbb{N}$.*

   Similar results hold for $\bar{p}$-convergence.
3. *If (x_n) is $\bar{p}$-convergent, then $L_{\bar{p}}((x_n))$ is decreasing.*
4. *The sequence (x_n) is $\bar{p}$-convergent to every $y \in X$ such that $y \le x_n$ for all $n \in \mathbb{N}$.*

The following proposition contains some properties of this asymmetric norm and of the corresponding topology and quasi-uniformity.

Proposition 2.1.23. *Let $(X, \|\cdot\|, \le)$ be a normed lattice and p the asymmetric norm given by (2.1.39).*

1. *The norms p^s_M, p^s_L and p^s_E are mutually equivalent norms on X which are also equivalent to the original norm. Further, if X is an M-space, an L-space, or an E-space, then p^s_M, p^s_L, respectively p^s_E agree with the original norm $\|\cdot\|$.*

2. *The quasi-uniformity* $\mathcal{U}_p$ determines *the normed lattice structure in the sense that* $\tau_{\|\cdot\|} = \tau(\mathcal{U}_p^s)$ *and* $\text{Graph}(\leq) = \bigcap \mathcal{U}_{\bar{p}}$, *where* $\text{Graph}(\leq) = \{(x, y) \in X \times X : x \leq y\}$.
3. *A linear functional* $\varphi : X \to \mathbb{R}$ *is* (p, u)*-continuous if and only if it is* $(\|\cdot\|, |\cdot|)$*-continuous and positive, that is* $\varphi(x) \geq 0$ *whenever* $x \geq 0$.
4. *A subset* Y *of* X *is* p*-open* ($\bar{p}$*-open*) *if and only if it is* $\|\cdot\|$*-open and decreasing (resp. increasing).*

Proof. 1. Since the norms p_L^s, p_L^s, p_L^s are mutually equivalent, it remains to show their equivalence with $\|\cdot\|$. From (2.1.40)

$$\|x\| = \|x^+ - x^-\| \leq \|x^+\| + \|x^-\| = p(x) + p(-x) = p_L^s(x) .$$

On the other side, by (2.1.37), $x^+ \leq |x|$, $x^- \leq |x|$ implies $\|x^+\| \leq \|x\|$ and $\|x^-\| \leq \|x\|$, so that

$$p_L^s(x) \leq \|x^+\| + \|x^-\| \leq 2\|x\| .$$

If X is an L-space, then

$$p_L^s(x) = \|x^+\| + \|x^-\| = \|x^+ + x^-\| = \||x|\| = \|x\| .$$

The case when X is an M-space follows similarly using the equality $|x| = x^+ \vee x^-$.

If X is an E-space, then

$$\begin{aligned} 2\left(p_E^s(x)\right)^2 &= 2\left(\|x^+\|^2 + \|x^-\|^2\right) \\ &= \|x^+ + x^-\|^2 + \|x^+ - x^-\|^2 = \||x|\|^2 + \|x\|^2 = 2\|x\|^2 . \end{aligned}$$

2. The quasi-uniformity $\mathcal{U}_p$ is generated by the set

$$U_\varepsilon = \{(x, y) \in X \times X : p(y - x) < \varepsilon\} ,$$

and the uniformity $\mathcal{U}_p^s$ by the sets

$$U_\varepsilon^s = \{(x, y) \in X \times X : p_M^s(y - x) < \varepsilon\} .$$

Since the norm p_M^s is equivalent to the norm $\|\cdot\|$, it follows that it generates the same topology as $\|\cdot\|$.

To prove the equality $\text{Graph}(\leq) = \bigcap \mathcal{U}_{\bar{p}}$ observe that

$$\begin{aligned} (x, y) \in \bigcap \mathcal{U}_{\bar{p}} &\iff \forall \varepsilon,\ \bar{p}(y - x) < \varepsilon \iff \bar{p}(y - x) = 0 \iff \|(x - y)^+\| = 0 \\ &\iff (x - y)^+ = 0 \iff x - y \leq 0 \iff x \leq y . \end{aligned}$$

3. Let $f : (X, p) \to (\mathbb{R}, u)$ be linear and continuous. Then f is $(p_M^s, |\cdot|)$-continuous, and since the norm p_M^s is equivalent to $\|\cdot\|$, f is $(\|\cdot\|, |\cdot|)$-continuous.

The (p,u)-continuity of f implies the existence of $\beta \geq 0$ such that

$$\forall z \in X, \quad f(z) \leq \beta p(z) \, .$$

If $f(x) < 0$ for some $x \geq 0$, then $y = x/f(x) \leq 0$, and so $p(y) = \|y^+\| = 0$, that leads to the contradiction

$$1 = f(y) \leq \beta p(y) = 0 \, .$$

Conversely, suppose that $f : (X, \|\cdot\|) \to \mathbb{R}, |\cdot|)$ is linear, continuous and positive. Then there exists $\beta \geq 0$ such that

$$f(x) \leq \beta \|x\|, \; x \in X \, ,$$

implying

$$f(x) = f(x^+) - f(x^-) \leq f(x^+) \leq \beta \|x^+\| = \beta p(x) \, ,$$

for all $x \in X$, proving the (p,u)-continuity of the functional f.

4. Suppose that $G \subset X$ is p-open. For $x \in G$ there exists $r > 0$ such that $B_p(x,r) \subset G$ Then $B_{\|} \cdot \|(x,r) \subset B_p(x,r)$, showing that G is $\|\cdot\|$-open, and $y \leq x$ implies $(y-x)^+ = 0$ and $p(y-x) = \|(y-x)^+\| = 0 < r$.

Conversely, suppose that G is $\|\cdot\|$-open and decreasing and let $x \in G$. Then there exists $r > 0$ such that $B_{\|} \cdot \|(x,r) \subset G$. We shall show that $B_p(x,r) \subset G$. Indeed, let $y \in B_p(x,r)$. If $y \leq x$, then $y \in G$, because G is decreasing. If $y > x$, then $(y-x)^+ = y - x$, so that $\|y-x\| = \|(y-x)^+\| = p(y-x) < r$, showing that $y \in B_{\|} \cdot \|(x,r) \subset G$.

The case of $\bar{p}$-open sets can be treated similarly. □

Remark 2.1.24. Properties 1–3 are taken from [77] and 4 from [6].

In all examples given below of asymmetric normed lattices the order is the pointwise order

$$x \leq y \iff \forall k \in \mathbb{N}, \; x_k \leq x_k,$$

if $x = (x_k)$ and $y = (y_k)$ are sequences (possibly finite) of real numbers, and

$$f \leq g \iff \forall t \in T, \; f(t) \leq g(t),$$

if f, g are real-valued functions defined on a set T.

Consider the Banach lattices ℓ^p, $1 \leq p < \infty$, of all real sequences $x = (x_k)$ such that

$$\|x\|_p := \Big(\sum_k |x_k|^p \Big)^{1,p}.$$

By ℓ^∞ one denotes the Banach lattice of all bounded sequences with the sup-norm $\|\cdot\|_\infty$ and by c_0 its subspace formed by all converging to 0 sequences.

If T is a Hausdorff compact topological space, then $C(T)$ denotes the Banach lattice of all real-valued continuous functions on T with the sup-norm $\|\cdot\|_\infty$.

In all of these cases we shall use the notation

$$\|x\|^+ = \|x^+\| \quad \text{and} \quad \|x\|^- = \|x^-\|$$

for the asymmetric norm and its conjugate in a normed lattice $(X, \|\cdot\|)$.

In $\mathbb{R}^m$ one considers the Euclidean norm $\|\cdot\|_2$, but the results hold for any other lattice norm on $\mathbb{R}^m$, because for two equivalent lattice norms $\|\cdot\|$, $|||\cdot|||$ on $\mathbb{R}^m$, the corresponding asymmetric norms $\|\cdot\|^+$, $|||\cdot|||^-$ are also equivalent ([51, Proposition 2.5].

We mention the following characterizations of convergence in some of these asymmetric normed lattices obtained in [51].

Proposition 2.1.25.

1. *If a sequence $x_n = (x_{n,i})_{i\in\mathbb{N}}$, $n \in \mathbb{N}$, in $X = \ell^p$, $1 \le p \le \infty$, or in $X = c_0$, is $\|\cdot\|_p^+$-convergent to $x = (x_i)_{i\in\mathbb{N}} \in X$, then*

$$\forall i \in \mathbb{N}, \quad x_{n,i} \xrightarrow{u} x_i \quad \text{as} \quad n \to \infty. \tag{2.1.41}$$

2. *If $1 \le p < \infty$, then $x_n \xrightarrow{\|\cdot\|_p^+} x$ if and only if (2.1.41) holds and*

$$\sup_{n\in\mathbb{N}} \sum_{i=m}^{\infty} (x_{n,i}^+)^p \to 0 \quad \text{as} \quad m \to \infty. \tag{2.1.42}$$

3. *If $X = c_0$, then $x_n \xrightarrow{\|\cdot\|_\infty^+} x$ if and only if (2.1.41) holds and*

$$\sup_{n\in\mathbb{N}} \sup_{i\ge m} x_{n,i}^+ \to 0 \quad \text{as} \quad m \to \infty. \tag{2.1.43}$$

The following completeness results for these concrete Banach lattices were obtained also in [51], as consequences of some more general results concerning completeness in asymmetric normed lattices.

Theorem 2.1.26. *The asymmetric normed lattices $(\mathbb{R}^m, \|\cdot\|_2^\pm)$, $(C(T), \|\cdot\|_\infty^\pm)$, $(\ell^\infty, \|\cdot\|_\infty^\pm)$, $(c_0, \|\cdot\|_\infty^+)$ and $(\ell^p, \|\cdot\|_p^+)$, $1 \le p < \infty$, are all left K-sequentially complete.*

Now we shall present some Baire properties of the asymmetric topology. As it is remarked in [6, Proposition 1] the asymmetric topology τ_p of a normed lattice is never Baire.

Proposition 2.1.27. *Let $(X, \|\cdot\|, \le)$ be a normed lattice.*

1. *Every nonempty p-open subset of X is p-dense in X.*
2. *Any p-dense increasing subset of X is $\|\cdot\|$-dense in X. Similarly, a $\bar{p}$-dense decreasing subset of X is $\|\cdot\|$-dense in X.*
3. *The associated asymmetric normed space (X, p) is never Baire.*

Proof. 1. Let G be a nonempty p-open subset of X. For an arbitrary nonempty $U \in \tau_p$ let $x \in G$ and $y \in U$. Since both G and U are decreasing it follows that $z = x \wedge y \in G \cap U$, so that G is p-dense in X.

2. Suppose that $Y \subset X$ is p-dense and increasing. For $x \in X$ and $\varepsilon > 0$ there exists $y \in Y$ such that $p(y - x) < \varepsilon$. Then

$$y = x + (y - x) \le x + (y - x)^+$$

implies $z := x + (y - x)^+ \in Y$. Since

$$\|z - x\| = \|(y - x)^+\| = p(y - x) < \varepsilon ,$$

it follows that Y is $\|\cdot\|$-dense in X.

The second case can be treated similarly.

3. For $x < 0$ in X consider the family $G_n = B_p(nx, 1),\ n \in \mathbb{N}$, of p-open sets and show that $\cap_n G_n = \emptyset$. Indeed, if $y \in \cap_n G_n = \emptyset$, then

$$\begin{aligned} n\|x\| &= \|nx\| = \|-nx\| = \|(-nx)^+\| = \|(y - nx - y)^+\| \\ &\le \|(y - nx)^+\| + \|(-y)^+\| < 1 + \|(-y)^+\| , \end{aligned}$$

for all $n \in \mathbb{N}$, leading to the contradiction $\|(-y)^+\| = \infty$.

Consequently, $G_n,\ n \in \mathbb{N}$, is a family of p-open p-dense subsets of X whose intersection is not p-dense in X, showing that X is not a Baire space. □

For this reason the authors defined in *loc. cit.* another property: a normed lattice is called *quasi-Baire* if the intersection of any sequence of monotonic (all of the same kind) $\|\cdot\|$-dense sets is $\|\cdot\|$-dense. By a monotonic set one understands a set that is increasing or decreasing. Recall that a bitopological space (T, τ, ν) is called *pairwise Baire* provided the intersection of any sequence of τ-open ν-dense sets is ν-dense, and the intersection of any sequence of ν-open τ-dense sets is τ-dense (see Subsection 1.2.4).

Let $(X, \|\cdot\|, \le)$ be a normed lattice and X^* the dual space of $(X, \|\cdot\|)$. A subset $F \subset X^*$ is called *order determining* if

$$x \le y \iff \forall \varphi \in F,\ \varphi(x) \le \varphi(y) .$$

The following proposition puts in evidence the relevance of this notion for the quasi-Baire property.

Proposition 2.1.28. *Let $(X, \|\cdot\|, \le)$ be a normed lattice and F an order determining subset of X^*. If*

$$\sup_{x \in X} \inf_{\varphi \in F} \varphi(x) > 0 ,$$

then X is a quasi-Baire space.

In fact, the proof given in [6] shows that X does not contain decreasing proper dense subsets, so it is quasi-Baire in a trivial way.

Based on this proposition one can give the following example of a quasi-Baire lattice.

Example 2.1.29 ([6]). Let $\mathcal{A}$ be an algebra of subsets of a nonempty set T. Consider the linear space $\ell_0^\infty(T, \mathcal{A})$ generated by the set $\{\chi_A : A \in \mathcal{A}\}$ of characteristic functions of sets in $\mathcal{A}$, equipped with the norm $\|x\| = \sup\{|x(t)| : t \in T\}$ and the pointwise order.

Then $\ell_0^\infty(T, \mathcal{A})$ is a quasi-Baire space.

Let $X = \ell_0^\infty(T, \mathcal{A})$. It is well known that the dual of X is the space $\mathrm{ba}(T, \mathcal{A})$ of all finitely additive bounded measures on $\mathcal{A}$. It is obvious that the Dirac measures $\delta_t,\ t \in T$, defined by $\delta_t(x) = x(t),\ x \in X$, determines the order in X. Because $\delta_t(\chi_T) = 1$, for all $t \in T$, it follows that

$$\sup\{\inf_{t\in T} \delta_t(x) : x \in X\} \geq \inf_{t\in T} \delta_t(\chi_T) = 1 ,$$

so that, By Proposition 2.1.28, X is quasi-Baire.

Theorem 2.1.30 ([6]). *A normed lattice $(X, \|\cdot\|, \leq)$ is quasi-Baire if and only if the associated bitopological space $(X, \tau_p, \tau_{\bar{p}})$ is pairwise Baire.*

Proof. Suppose that X is quasi-Baire and let $G_n,\ n \in \mathbb{N}$, be a family of p-open $\bar{p}$-dense subsets of X. By Proposition 2.1.27.1, each G_n is decreasing, so that, by the second assertion of the same proposition, G_n is $\|\cdot\|$-dense in X. Since X is quasi-Baire, it follows that $\cap_n G_n$ is $\|\cdot\|$-dense in X, and so $\bar{p}$-dense too.

Similarly, if $G_n,\ n \in \mathbb{N}$, is a family of $\bar{p}$-open p-dense subsets of X, then $\cap_n G_n$ is p-dense in X.

Consequently, X is pairwise Baire.

Conversely, suppose that X is pairwise Baire and let $G_n,\ n \in \mathbb{N}$, be a family of decreasing $\|\cdot\|$-open and $\|\cdot\|$-dense subsets of X. By Proposition 2.1.23.4, each G_n is p-open and $\bar{p}$-dense in X, so that their intersection is $\bar{p}$-dense in X. By Proposition 2.1.27.2, $\cap_n G_n$ is $\|\cdot\|$-dense in X.

If the sets $G_n,\ n \in \mathbb{N}$, are increasing $\|\cdot\|$-open and $\|\cdot\|$-dense in X, one proceeds similarly. □

We shall present now an example given by Alegre [2] of an asymmetric dual of a normed lattice.

Consider the real Banach space ℓ^1 with the usual norm

$$\|x\|_1 = \sum_{i=1}^{\infty} |x_i|, \quad \text{for} \quad x = (x_i) \in \ell^1 ,$$

and the pointwise order

$$x \leq y \iff \forall i \in \mathbb{N}, \quad x_i \leq y_i .$$

In this case

$$x^+ = (x_i^+)_{i\in\mathbb{N}} \, .$$

The dual of ℓ^1 is the space ℓ^∞ of all bounded sequences of real numbers with the supremum norm

$$\|x\|_\infty = \sup\{|x_i| : i \in \mathbb{N}\}, \quad \text{for} \quad x = (x_i) \in \ell^\infty \, .$$

The mapping $y \mapsto f_y$, where for $y \in \ell^\infty$,

$$f_y(x) = \sum_{i=1}^{\infty} x_i y_i, \quad \text{for} \quad x = (x_i) \in \ell^1 \, ,$$

is an isometric isomorphism between $(\ell^1)^*$ and ℓ^∞.

Let

$$X = \{x = (x_n) \in \ell^1 : x_1 + x_2 = 0\} \quad \text{and} \quad H = \{y = (y_n) \in \ell^\infty : y_1 + y_2 = 0\} \, .$$

Consider on X the induced norm $\|\cdot\|_1$ and the associated asymmetric norm $p(x) = \|x^+\|_1$, $x \in X$, and consider H equipped with the norm $\|\cdot\|_\infty$. Let $X_p^\flat = (X,p)^\flat$ be the asymmetric dual of X and let $\mathrm{sp}(X_p^\flat) = X_p^\flat - X_p^\flat$ be the linear subspace of $X^* = (X, \|\cdot\|_1)^*$ generated by $X_p^\flat$ equipped with the norm

$$\|f\|_1^* = \sup\{|f(x)| : x \in X, \ \|x\|_1 \le 1\} \, .$$

Denote by (e_n) the canonical Schauder basis of ℓ^1, where for each $n \in \mathbb{N}$,

$$e_{n,i} = \delta_{n,i}, \quad i \in \mathbb{N} \, ,$$

($\delta_{n,i}$ is the Kronecker symbol).

Proposition 2.1.31. *The mapping Φ defined on $(\mathrm{sp}(X_p^\flat), \|\cdot\|_1^*)$ by*

$$\Phi(f) = \left(\frac{1}{2}(e_1 - e_2), \frac{1}{2}(e_2 - e_1), f(e_3), \dots\right), \quad f \in \mathrm{sp}(X_p^\flat) \, , \tag{2.1.44}$$

is an isometrical isomorphism between the spaces $(\mathrm{sp}(X_p^\flat), \|\cdot\|_1^*)$ and $(H, \|\cdot\|_\infty)$.

Proof. Let $f \in \mathrm{sp}(X_p^\flat) \subset X^*$. Then there exists $\beta \ge 0$ such that

$$|f(x)| \le \beta \|x\|_1, \quad x \in X \, ,$$

which shows that $\Phi(f) \in H$. It is obvious that $\Phi : X \to H$ is linear.

If $\Phi(f) = 0$, then $f(e_1 - e_2) = 0, f(e_3) = 0, \dots$. Since every $x \in X$ can be written as $x = x_1(e_1 - e_2) + x_3 e_3 + \cdots$ it follows that $f(x) = 0$, that is f is the null functional on X, showing that Φ is injective.

To show that Φ is surjective, for $y = (y_n) \in H$ put

$$\varphi(x) = 2x_1 y_1^+ + \sum_{i=3}^{\infty} x_i y_i^+ \quad \text{and} \quad \psi(x) = 2x_1 y_1^- + \sum_{i=3}^{\infty} x_i y_i^- .$$

Since $\varphi, \psi : X \to \mathbb{R}$ are positive and $(\|\cdot\|_1, |\cdot|)$-continuous, it follows that $\varphi, \psi \in X_p^\flat$, and $f = \varphi - \psi \in \mathrm{sp}(X_p^\flat)$. Since $\Phi(f) = y$, the mapping Φ is surjective, so bijective, and

$$\Phi^{-1}(y)(x) = 2x_1 y_1 + \sum_{i \geq 3} x_i y_i ,$$

for any $x = (x_1, -x_1, x_3, \dots) \in X$.

The functional $\Phi^{-1}(y)$ acts on ℓ^1 by the rule

$$\Phi^{-1}(y)(z) = z_1 y_1 - z_2 y_2 + \sum_{i \geq 3} z_i y_i ,$$

for any $z = (z_1, z_2, \dots) \in \ell^1$.

It remained to prove that Φ is an isometry.

For $z \in \ell^1$ with $\|z\|_1 \leq 1$, the element x given by

$$x = \left(\frac{1}{2}(z_1 - z_2), \frac{1}{2}(z_2 - z_1), z_3, \dots \right)$$

belongs to X, $\|x\|_1 \leq \|z\|_1 \leq 1$ and $\Phi^{-1}(y)(x) = \Phi^{-1}(y)(z)$.

It follows that

$$|\Phi^{-1}(y)(z)| = |\Phi^{-1}(y)(x)| \leq \|\Phi^{-1}(y)\|_1^* .$$

Since the dual of $(\ell^1, \|\cdot\|_1)$ is $(\ell^\infty, \|\cdot\|_\infty)$,

$$\sup\{|\Phi^{-1}(y)(z)| : z \in \ell^1, \ \|z\|_1 \leq 1\} = \|y\|_\infty ,$$

so that $\|y\|_\infty \leq \|\Phi^{-1}(y)\|_1^*$.

On the other side, for every $x \in X$,

$$|\Phi^{-1}(y)(x)| = |2x_1 y_1 + \sum_{i \geq 3} x_i y_i| \leq \|y\|_\infty \ \|x\|_1 ,$$

implying $\|\Phi^{-1}(y)\|_1^* \leq \|y\|_\infty$.

Consequently, $\|\Phi^{-1}(y)\|_1^* = \|y\|_\infty$. □

2.2 Hahn-Banach type theorems and the separation of convex sets

One of the fundamental principles of functional analysis is the Hahn-Banach extension theorem for a linear functional dominated by a sublinear functional. Based on this theorem one can prove extension results for lsc linear functionals on asymmetric normed spaces and on asymmetric LCS. Some separation results for convex subsets of asymmetric LCS, relying on the properties of the Minkowski gauge functional and on the extension results, are also proved. As an application, an asymmetric version of the Krein-Milman theorem is proved.

2.2.1 Hahn-Banach type theorems

Let X be a real vector space. Recall that a functional $p : X \to \mathbb{R}$ is called *sublinear* if

$$\text{(i) } p(\lambda x) = \lambda p(x) \quad \text{and} \quad \text{(ii)} \quad p(x+y) \leq p(x) + p(y) ,$$

for all $x, y \in X$ and $\lambda \geq 0$.

Notice that, as defined, a sublinear functional need not be positive. A positive sublinear functional is an asymmetric seminorm.

A sublinear functional is called a *seminorm* if instead of (i) it satisfies

$$p(\lambda x) = |\lambda| p(x) ,$$

for all $x \in X$ and $\lambda \in \mathbb{R}$. A seminorm is necessarily positive, that is $p(x) \geq 0$ for all $x \in X$. A seminorm is called a norm if

$$\text{(iii)} \quad p(x) = 0 \iff x = 0 .$$

A function $f : X \to \mathbb{R}$ is said to be *dominated* by a function $g : X \to \mathbb{R}$ if

$$f(x) \leq g(x) ,$$

for all $x \in X$. If the above inequality holds only for x in a subset Y of X, then we say that f is dominated by g on Y.

Theorem 2.2.1 (Hahn-Banach Extension Theorem). *Let X be a real vector space and $p : X \to \mathbb{R}$ a sublinear functional. If Y is a subspace of X and $f : Y \to \mathbb{R}$ is a linear functional dominated by p on Y then there exists a linear functional $F : X \to \mathbb{R}$ dominated by p on X such that $F|_Y = f$.*

We present now the extension results in the asymmetric case.

Theorem 2.2.2. *Let (X, p) be a space with asymmetric seminorm.*

1. *If Y is a subspace of X and $\varphi_0 : Y \to \mathbb{R}$ is a continuous linear functional on the asymmetric seminormed space $(Y, p|_Y)$, then there exists a continuous linear functional $\varphi : X \to \mathbb{R}$ such that*

$$\varphi|_Y = \varphi_0 \quad \textit{and} \quad \|\varphi|_p = \|\varphi_0|_p \, .$$

2. *If x_0 is a point in X with $p(x_0) > 0$, then there exists a continuous linear functional $\varphi : X \to \mathbb{R}$ such that*

$$\|\varphi|_p = 1 \quad \textit{and} \quad \varphi(x_0) = p(x_0) \, . \tag{2.2.1}$$

Proof. 1. The inequality

$$\varphi_0(y) \le \|\varphi_0|p(y) \, ,$$

valid for all $y \in Y$, shows that the linear functional φ_0 is dominated by the sublinear functional $q(\cdot) = \|\varphi_0|p(\cdot)$. By the Hahn-Banach Theorem (Theorem 2.2.1), it has a linear extension $\varphi : X \to \mathbb{R}$ dominated by q, that is

$$\varphi(x) \le \|\varphi_0|p(x) \, ,$$

for all $x \in X$, implying $\|\varphi| \le \|\varphi_0|$.

The relations

$$\begin{aligned} \|\varphi| &= \sup\{\varphi(x) : x \in X, \; p(x) \le 1\} \ge \sup\{\varphi(y) : y \in X, \; p(y) \le 1\} \\ &= \sup\{\varphi(x) : x \in X, \; p(x) \le 1\} \, , \end{aligned}$$

prove the reverse inequality, so that $\|\varphi| = \|\varphi_0|$.

2. Let $Y = \mathbb{R}x_0$ the one-dimensional subspace generated by x_0. Define $\varphi_0 : Y \to \mathbb{R}$ by $\varphi_0(tx_0) = tp(x_0)$, $t \in \mathbb{R}$.

Since for $t \ge 0$, $\varphi_0(tx_0) = tp(x_0) = p(tx_0)$ and $\varphi_0(tx_0) = tp(x_0) \le 0 \le p(tx_0)$ for $t < 0$, it follows that the linear functional φ_0 is dominated by p on Y, and $\|\varphi_0| \le 1$. The equality $\varphi(x_0/p(x_0)) = 1$ implies $\|\varphi_0| = 1$. □

Remark 2.2.3. 1. The conditions (2.2.1) are equivalent to

$$\varphi(x_0) = p(x_0) \quad \text{and} \quad \forall x \in X, \;\; \varphi(x) \le p(x) \, . \tag{2.2.2}$$

2. Taking $\psi = (1/p(x_0))\, \varphi$ it follows that ψ satisfies the conditions

$$\psi(x_0) = 1 \quad \text{and} \quad \forall x \in X, \;\; \psi(x) \le \frac{1}{p(x_0)}\, p(x) \, ,$$

or, equivalently,

$$\|\psi|_p = \frac{1}{p(x_0)} \quad \text{and} \quad \psi(x_0) = 1 \, .$$

We agree to call a functional φ satisfying the conclusions of the first point of the theorem above a *norm preserving extension* of φ_0.

From the second point of the theorem one obtains as corollary a well-known and useful result in normed spaces.

Corollary 2.2.4. *Let (X, p) be an asymmetric seminormed space, $x_0 \in X$ and $X_p^\flat$ its dual. If $p(x_0) > 0$ then*

$$p(x_0) = \sup\{\varphi(x_0) : \varphi \in X_p^\flat,\ \|\varphi|_p \le 1\}\,.$$

Proof. Denote by s the supremum in the right-hand side of the above formula. Since $\varphi(x_0) \le \|\varphi|_p p(x_0) \le p(x_0)$ for every $\varphi \in X_p^\flat$ with $\|\varphi|_p \le 1$, it follows that $s \le p(x_0)$. Choosing $\varphi \in X_p^\flat$ as in Theorem 2.2.2.2, it follows that $p(x_0) = \varphi(x_0) \le s$. □

In the case of asymmetric LCS one obtains the existence of continuous linear extensions.

Proposition 2.2.5. *Let (X, P) be an asymmetric LCS and Y a subspace of X. Then every $\tau(P)$-continuous linear functional on Y has a $\tau(P)$-continuous linear extension to the whole space X.*

Proof. Supposing P directed, then for $\psi \in Y_P^\flat$ there exists $p \in P$ and $\beta \ge 0$, such that

$$\forall y \in Y,\ \ \psi(y) \le \beta p(y)\,.$$

By Theorem 2.2.1 applied to the sublinear functional $q(\cdot) = \beta p(\cdot)$, ψ has a linear extension $\varphi : X \to \mathbb{R}$ such that $\varphi(x) \le \beta p(x)$ for all $x \in X$. It follows that φ is a continuous linear extension of ψ. □

As in the symmetric case, one can prove the existence of some functionals on an asymmetric normed space related to distances to a subspace, see [40, 48, 49]. This result will be applied in Sect. 2.5 to best approximation problems in asymmetric normed spaces.

Let (X, p) be an asymmetric normed space, Y a nonempty subset of X and $x \in X$. Due to the asymmetry of the norm we have to consider two *distances* from x to Y:

$$\begin{aligned} &\text{(i)}\ \ d_p(x, Y) = \inf\{p(y - x) : y \in Y\}, \quad \text{and} \\ &\text{(ii)}\ \ d_p(Y, x) = \inf\{p(x - y) : y \in Y\}\,. \end{aligned} \tag{2.2.3}$$

Observe that $d_p(Y, x) = d_{\bar{p}}(x, Y)$, where $\bar{p}$ is the norm conjugate to p.

Theorem 2.2.6. *Let Y be a subspace of a space with asymmetric norm (X, p) and $x_0 \in X$. If $d := d_p(x_0, Y) > 0$, then there exists a p-bounded linear functional $\varphi : X \to \mathbb{R}$ such that*

$$\text{(i)}\ \varphi|_Y = 0, \quad \text{(ii)}\ \|\varphi|_p = 1, \quad \textit{and} \quad \text{(iii)}\ \varphi(-x_0) = d\,.$$

If $\bar{d} := d_p(Y, x_0) > 0$*, then there exists a p-bounded linear functional* $\psi : X \to \mathbb{R}$ *such that*

$$\text{(j) } \psi|_Y = 0, \quad \text{(jj) } \|\psi|_p = 1, \quad \textit{and} \quad \text{(jjj) } \psi(x_0) = \bar{d}\,.$$

Proof. Suppose first that $\bar{d} = d_p(Y, x_0) > 0$, so that $x_0 \notin Y$. Let $Z := Y \dotplus \mathbb{R}x_0$ ($\dotplus$ stands for the direct sum) and let $\psi_0 : Z \to \mathbb{R}$ be defined by

$$\psi_0(y + tx_0) = t, \ y \in Y, \ t \in \mathbb{R}\,.$$

Then ψ_0 is linear, $\psi_0(y) = 0, \ \forall y \in Y,$ and $\psi_0(x_0) = 1$. For $t > 0$ we have

$$p(y + tx_0) = tp(x_0 + t^{-1}y) \geq t\bar{d} = \bar{d} \cdot \psi_0(y + tx_0)\,,$$

so that

$$\psi_0(y + tx_0) = t \leq \frac{1}{\bar{d}} p(y + tx_0)\,.$$

Since this inequality obviously holds for $t \leq 0$, it follows that $\|\psi_0| \leq 1/\bar{d}$. Let (y_n) be a sequence in Y such that $p(x_0 - y_n) \to \bar{d}$ for $n \to \infty$ and $p(x_0 - y_n) > 0$ for all $n \in \mathbb{N}$. Then

$$\|\psi_0| \geq \psi_0\left(\frac{x_0 - y_n}{p(x_0 - y_n)}\right) = \frac{1}{p(x_0 - y_n)} \to \frac{1}{\bar{d}}\,,$$

implying $\|\psi_0| \geq 1/\bar{d}$. Therefore $\|\psi_0| = 1/\bar{d}$.

If $\psi_1 : X \to \mathbb{R}$ is a linear functional such that

$$\psi_1|_Z = \psi_0 \quad \text{and} \quad \|\psi_1| = \|\psi_0|\,,$$

then the linear functional $\psi = \bar{d} \cdot \varphi_1$ fulfills the conditions (j)–(jjj).

Suppose now that $d = d_p(x_0, Y) > 0$, and let $Z := Y \dotplus \mathbb{R}x_0$. Define $\varphi_0 : Z \to \mathbb{R}$ by

$$\varphi_0(y + tx_0) = -t \iff \varphi_0(y - tx_0) = t \quad \text{for } y \in Y \text{ and } t \in \mathbb{R}\,.$$

Then φ_0 is linear and, for $t > 0$, we have

$$p(y - tx_0) = tp(\frac{1}{t}y - x_0) \geq td = \bar{d} \cdot \varphi_0(y - tx_0)\,,$$

so that

$$\varphi_0(y - tx_0) \leq \frac{1}{d} p(y - tx_0)\,,$$

for $t > 0$. Since this inequality is obviously true if $\varphi_0(y - tx_0) = t \leq 0$, it follows that φ_0 is bounded and $\|\varphi_0| \leq 1/d$. Reasoning as above, one obtains the existence of a functional φ satisfying the conditions (i)–(iii). □

Other extension results can be found in [8, 28, 90, 224]. By studying quasi-uniformities on real vector spaces, Alegre, Ferrer and Gregori [5] were able to prove a Hahn-Banach type extension theorem for pseudo-topological vector spaces. More general and sophisticated versions of the Hahn-Banach theorem are also known, see, for instance, [82], [122], [188], and the book [83]. For an extensive list of references see the survey [30].

2.2.2 The Minkowski gauge functional – definition and properties

A subset Y of a vector space X is called *absorbing* if

$$\forall x \in X,\ \exists t > 0, \quad \text{such that} \quad x \in tY\ .$$

If Y is absorbing, then the *Minkowski functional* (or the *gauge function*) p_Y of the set Y is defined by

$$p_Y(x) = \inf\{t > 0 : x \in tY\}\ .$$

It follows that p_Y is a positive and positively homogeneous functional, and

$$Y \subset \{x \in X : p_Y(x) \le 1\}\ .$$

If Y is convex and absorbing, then p_Y is a positive sublinear functional and

$$\{x \in X : p_Y(x) < 1\} \subset Y \subset \{x \in X : p_Y(x) \le 1\}\ . \tag{2.2.4}$$

Now suppose that (X, P) is an asymmetric LCS and look for conditions on the set Y ensuring the (P, u)-continuity of p_Y.

Proposition 2.2.7. *Let Y be a convex absorbing subset of an asymmetric locally convex space (X, P).*

1. *The Minkowski functional p_Y is (P, u)-continuous if and only if 0 is a τ_P-interior point of Y.*
2. *If p_Y is (P, u)-continuous, then*

$$\tau_P\text{-}\mathrm{int}(Y) = \{x \in X : p_Y(x) < 1\}. \tag{2.2.5}$$

Proof. Suppose the family P to be directed.

1. If 0 is a τ_P-interior point of Y, then there exist $p \in P$ and $r > 0$ such that

$$B_p(0, r) \subset Y \subset \{x \in X : p_Y(x) \le 1\}\ ,$$

i.e.,

$$\forall x \in X, \quad p(x) \le r \Rightarrow p_Y(x) \le 1\ .$$

By Proposition 2.1.1, we have

$$\forall x \in X, \quad p_Y(x) \le \frac{1}{r} p(x)\ ,$$

which, by Proposition 2.1.12, implies the (P, u)-continuity of p_Y.

Conversely, suppose that p_Y is (P, u)-continuous. Since the set $(-\infty; 1)$ is τ_u-open in $\mathbb{R}$, the set $\{x \in X : p_Y(x) < 1\} = p_Y^{-1}((-\infty; 1))$ is τ_P-open, contains 0, and is contained in Y, implying $0 \in \tau_P\text{-}\mathrm{int}(Y)$.

2. Suppose that p_Y is (P,u)-continuous. By the proof of the first point, $\{x \in X : p_Y(x) < 1\} \subset \tau_P\text{-int}(Y)$, so it remains to prove the reverse inclusion.

If $x \in \tau_P\text{-int}(Y)$, then there exist $p_1 \in P$ and $r > 0$ such that $B_{p_1}(x,r) \subset Y$. By Proposition 2.1.12 there exist $p_2 \in P$ and $\beta > 0$ such that $\forall x \in X,\ p_Y(x) \leq \beta p_2(x)$. If $p \in P$ is such that $p \geq p_i,\ i = 1,2,$ then $B_p(x,r) \subset B_{p_1}(x,r) \subset Y$ and

$$\forall x \in X, \quad p_Y(x) \leq \beta p(x) .$$

If $p(x) = 0$, then, by the above inequality, $p_Y(x) = 0 < 1$. If $p(x) > 0$, put $x_\alpha = (1+\alpha)x$ for $\alpha > 0$. If $0 < \alpha < r/p(x)$, then $p(x_\alpha - x) = \alpha p(x) < r$ so that $x_\alpha \in Y$ and $p_Y(x_\alpha) \leq 1$. But then, for any such α we have

$$p_Y(x) = \frac{1}{1+\alpha} p_Y(x_\alpha) \leq \frac{1}{1+\alpha} < 1 . \qquad \square$$

2.2.3 The separation of convex sets

The separation theorems for convex sets are very efficient tools in the treatment of optimization problems in Banach or locally convex spaces. The so far developed machinery allows us to prove the asymmetric analogs of the Eidelheit and Tukey separation theorems (Theorems 2.2.26 and 2.2.28 in [149]). The presentation follows [41].

Theorem 2.2.8. *Let (X,P) be an asymmetric locally convex space and Y_1, Y_2 two disjoint nonempty convex subsets of X with Y_1 τ_P-open.*

Then there exists a τ_P-continuous linear functional $\varphi : X \to \mathbb{R}$ such that

$$\forall y_1 \in Y_1,\ \forall y_2 \in Y_2 \quad \varphi(y_1) < \varphi(y_2) .$$

Proof. Let $y_i^0 \in Y_i$, $i = 1,2,$ and let $x_0 = y_2^0 - y_1^0$. Since the set Y_1 is τ_P-open and the topology τ_P is translation invariant, the set

$$Y := x_0 + Y_1 - Y_2 = \cup\{x_0 - y_2 + Y_1 : y_2 \in Y_2\}$$

is τ_P-open too. It is obvious that Y is also convex.

We have $0 = x_0 + y_1^0 - y_2^0 \in Y$ and $x_0 \notin Y$. Indeed, if $x_0 = x_0 + y_1 - y_2$, for some $y_1 \in Y_1$ and $y_2 \in Y_2$, then the element $y = y_1 = y_2$ would belong to the empty set $Y_1 \cap Y_2$.

By Proposition 2.2.7, the Minkowski functional p_Y of the τ_P-open convex set Y is sublinear, (P,u)-continuous and

$$Y = \{x \in X : p_Y(x) < 1\} . \tag{2.2.6}$$

Since $x_0 \notin Y$, it follows that $p_Y(x_0) \geq 1$. By Proposition 2.2.2.2, there exists a p_Y-bounded linear functional $\psi : X \to \mathbb{R}$ such that $\psi(x_0) = p_Y(x_0)$ and $\psi(x) \leq p_Y(x),\ x \in X$. Taking $\varphi = (1/p_Y(x_0))\psi$ it follows that

$$\varphi(x_0) = 1 \quad \text{and} \quad \forall x \in X,\ \varphi(x) = \frac{1}{p_Y(x_0)}\psi(x) \leq \frac{1}{p_Y(x_0)} p_Y(x) \leq p_Y(x) .$$

By Proposition 2.1.11, the functional φ is (P, u)-continuous. Because Y is τ_P-open and $0 \in Y$, by Proposition 2.2.7 we have $Y = \{x \in X : p_Y(x) < 1\}$. From (2.2.6) and the fact that $\varphi(x_0) = 1$, one obtains

$$\forall y_1 \in Y_1,\ \forall y_2 \in Y_2,\ 1 + \varphi(y_1) - \varphi(y_2) = \varphi(x_0 + y_1 - y_2) \le p_Y(x_0 + y_1 - y_2) < 1\,,$$

implying

$$\forall y_1 \in Y_1,\ \forall y_2 \in Y_2, \quad \varphi(y_1) < \varphi(y_2)\,. \qquad \square$$

We prove now the asymmetric analog of Tukey's separation theorem.

Theorem 2.2.9. *Let (X, P) be an asymmetric locally convex space and Y_1, Y_2 two nonempty disjoint convex subsets of X, with Y_1 τ_P -compact and Y_2 τ_P-closed.*

Then there exists a τ_P-continuous linear functional $\varphi : X \to \mathbb{R}$ such that

$$\sup \varphi(Y_1) < \inf \varphi(Y_2)\,. \tag{2.2.7}$$

Proof. Suppose that P is directed. For $p \in P$ denote by B'_p the open unit p-ball, $B'_p = \{x \in X : p(x) < 1\}$.

Since $Y_1 \cap Y_2 = \emptyset$ and Y_2 is τ_P-closed, for every $y \in Y_1$ there exist $p_y \in P$ and $r_y > 0$ such that

$$(y + 2r_y B'_{p_y}) \cap Y_2 = \emptyset. \tag{2.2.8}$$

The τ_P-open cover $\{y + r_y B'_{p_y} : y \in Y_1\}$ of the τ_P-compact set Y_1, contains a finite subcover $\{y_k + r_k B'_{p_k} : k = 1, 2, \ldots, n\}$, where $p_k = p_{y_k}$ and $r_k = r_{y_k}$ for $k = 1, \ldots, n$. Take $p \in P$ such that $p \ge p_k$, $k = 1, 2, \ldots, n$, put $r := \min\{r_k : k = 1, 2, \ldots, n\}$ and show that

$$(Y_1 + rB'_p) \cap Y_2 = \emptyset. \tag{2.2.9}$$

Indeed, if $y' = y + ru \in Y_2$ for some $y \in Y_1$, $u \in B'_p$, then, choosing $k \in \{1, 2, \ldots, n\}$ such that $y \in y_k + r_k B'_{p_k}$, we have

$$y' = y + ru \in y_k + rB'_p + r_k B'_{p_k} \subset y_k + r_k B'_{p_k} + r_k B'_{p_k} = y_k + 2r_k B'_{p_k}\,,$$

in contradiction to (2.2.8).

The set $Z := Y_1 + rB'_p$ is convex, τ_P-open and disjoint from Y_2. By Theorem 2.2.8, there exists $\varphi \in X^\flat_P$ such that

$$\forall y \in Y_1,\ \forall u \in B'_p,\ \forall y' \in Y_2 \quad \varphi(y) + r\varphi(u) < \varphi(y'). \tag{2.2.10}$$

By Proposition 2.1.11, there exists $q_1 \in P$ and $\beta > 0$ such that $\forall x \in X$, $\varphi(x) \le \beta q_1(x)$. If $q \in P$ is such that $q \ge \max\{p, q_1\}$, then $\varphi(x) \le \beta q(x)$, $x \in X$, and $B'_q \subset B'_p$, so that

$$\forall y \in Y_1,\ \forall u \in B'_q,\ \forall y' \in Y_2 \quad \varphi(y) + r\varphi(u) < \varphi(y'). \tag{2.2.11}$$

By (2.2.10), $\varphi \neq 0$, so that by Propositions 2.1.7 and 2.1.8, $\|\varphi|_q = \sup \varphi(B'_q) > 0$. Passing in (2.2.11) to supremum with respect to $u \in B'_q$, we get

$$\forall y \in Y_1,\ \forall y' \in Y_2 \quad \varphi(y) + r\|\varphi|_q \leq \varphi(y')\ ,$$

implying

$$r\|\varphi|_q + \sup \varphi(Y_1) \leq \inf \varphi(Y_2)\ .$$

It follows that

$$\sup \varphi(Y_1) < \inf \varphi(Y_2)\ . \qquad \square$$

Remark 2.2.10. The inequality in Theorem 2.2.8 can not be reversed, in the sense that, under the same hypotheses on the sets Y_1 and Y_2, we can not find a (P, u)-continuous linear functional ψ on X such that

$$\forall y_2 \in Y_2,\ \forall y_1 \in Y_1 \quad \psi(y_2) < \psi(y_1)\ .$$

This is due, on one hand, to the fact that the functional $-\varphi$ need not be (P, u)-continuous, where φ is the linear functional given by Theorem 2.2.8. On the other hand, analyzing the proof of Theorem 2.2.8, it follows that we should work with the set $Y' := x_0 + Y_2 - Y_1$ which need not be τ_P-open, because the τ_P-openness of Y_1 does not imply the τ_P-openness of $-Y_1$. For instance, if (X, p) is an asymmetric LCS, then the fact that the set Y_1 is τ_p-open implies that $-Y_1$ is $\tau_{\bar{p}}$-open, so that, in this case, there exists a $\bar{p}$-bounded linear functional $\psi : X \to \mathbb{R}$ such that

$$\forall y_2 \in Y_2\ \forall y_1 \in Y_1, \quad \psi(y_2) < \psi(y_1)\ .$$

The same caution must be taken when applying Theorem 2.2.9.

2.2.4 Extreme points and the Krein-Milman theorem

Following the ideas from the symmetric case, one can prove a Krein-Milman type theorem for asymmetric LCS. The proof is based on Tukey's separation theorem and on the fact that the intersection of an arbitrary family of extremal subsets of a convex set Y is an extremal subset of Y, provided it is nonempty. The presentation follows [41]. The asymmetric normed case was treated in [40].

We start by recalling some notions and facts. A point e in a convex subset of a vector space X is called an *extreme point* of Y provided that $(1-t)x + ty = e$ for some $x, y \in Y$ and $0 < t < 1$, implies $x = y = e$. A nonempty convex subset Z of Y is called an *extremal subset* of Y if $(1-t)x + ty \in Z$, for some $x, y \in Y$ and some $0 < t < 1$ implies $x, y \in Z$ (in fact, $[x; y] \subset Z$, by the convexity of Z). An extremal subset is called also a *face* of Y. Obviously, a one-point set $Z = \{e\}$ is an extremal subset of Y if and only if e is an extreme point of Y. Also, if W is an extremal subset of the extremal subset Z of Y then W is an extremal subset of Y too. In particular, if e is an extreme point of an extremal subset Z of Y, then e is an extreme point of Y. The intersection of a family of extremal subsets of Y is an extremal subset of Y provided it is nonempty. We denote by $\operatorname{ext} Y$ the (possibly empty) set of extreme points of the convex set Y.

The following proposition is an immediate consequence of the definitions.

Proposition 2.2.11. *Let Y be a nonempty convex subset of a vector space X and f a linear functional on X.*

If the set $Z = \{z \in Y : f(z) = \sup f(Y)\}$ is nonempty, then it is an extremal subset of Y.

A similar assertion holds for the set $W = \{w \in Y : f(w) = \inf f(Y)\}$.

We can state and prove now the Krein-Milman theorem in the asymmetric case.

Theorem 2.2.12. *Let (X, P) be an asymmetric locally convex space such that the topology τ_P is Hausdorff.*

Then any nonempty convex τ_P-compact subset Y of X coincides with the τ_P-closed convex hull of the set of its extreme points

$$Y = \tau_P\text{-cl-co}(\text{ext Y}) .$$

Proof. All the topological notions will concern the τ_P-topology of X so that we shall omit sometimes "τ_P-" in the following. By Proposition 1.1.63, for every $x \in X$, $x \neq 0$, there exists $p \in P$ such that $p(x) > 0$, so that, by Theorem 2.2.2 (see also Remark 2.2.3.2), there exists $\varphi \in X_P^\flat$ with $\varphi(x) = 1$.

Fact I. *Every nonempty convex compact subset Z of X has an extreme point.*

Let

$$\mathcal{F} := \{F : F \text{ is a closed extremal subset of } Z\} ,$$

and define the order in $\mathcal{F}$ by $F_1 \leq F_2 \iff F_1 \subset F_2$ and show that the set $\mathcal{F}$ is nonempty and downward inductively ordered. Because Y is τ_P-compact and the topology τ_P is Hausdorff, it follows that Y is convex and τ_P-closed, so that $Y \in \mathcal{F}$. Since a totally ordered subfamily $\mathcal{G}$ of $\mathcal{F}$ has the finite intersection property, by the compactness of the set Z the set $G = \cap\mathcal{G}$ is nonempty, closed and extremal. Therefore $G \in \mathcal{F}$ is a lower bound for $\mathcal{G}$. By Zorn's Lemma the ordered set $\mathcal{F}$ has a minimal element F_0. If we show that F_0 is a one-point set, $F_0 = \{x_0\}$, then x_0 will be an extreme point of Z.

Suppose that F_0 contains two distinct points x_1, x_2, and let $p \in P$ be such that $p(x_1 - x_2) > 0$. Let φ be a p-bounded linear functional such that $\varphi(x_1 - x_2) = p(x_1 - x_2) > 0$ (see Theorem 2.2.2.2). It follows that $\varphi \in X^\flat$, so that φ is upper semi-continuous as a mapping from (X, τ_P) to $(\mathbb{R}, |\cdot|)$. By the compactness of the set F_0 the set

$$F_1 = \{x \in F_0 : \varphi(x) = \sup \varphi(F_0)\} = \{x \in F_0 : \varphi(x) \geq \sup \varphi(F_0)\}$$

is nonempty and closed. By Proposition 2.2.11, F_1 is an extremal subset of F_0, and so an extremal subset of Z. Therefore, $F_1 \in \mathcal{F}$, $F_1 \subset F_0$, and $x_2 \in F_0 \setminus F_1$ in contradiction to the minimality of F_0.

Fact II. $Y = \tau_P\text{-cl co}(\operatorname{ext} Y)$.

The inclusion $\operatorname{ext}(Y) \subset Y$ implies $Y_1 := \tau_P\text{-cl co}(\operatorname{ext} Y) \subset Y$. As a closed subset of a compact set, the set Y_1 is convex and compact. Supposing that there exists a point $y_0 \in Y \setminus Y_1$, then, by Theorem 2.2.9, there exists $\varphi \in X^\flat$ such that

$$\sup \varphi(Y_1) < \varphi(y_0). \tag{2.2.12}$$

Using again the upper semi-continuity of φ as a mapping from (X, τ_P) to $(\mathbb{R}, |\cdot|)$, we see that the set

$$F = \{y \in Y : \varphi(y) = \sup \varphi(Y)\} = \{y \in Y : \varphi(y) \geq \sup \varphi(Y)\} ,$$

is nonempty, convex and compact, so that, by Fact I, it has an extreme point e_1. Since F is an extremal subset of Y, it follows that e_1 is an extreme point of Y, implying $e_1 \in Y_1$. Taking into account (2.2.12) we obtain the contradiction

$$\sup \varphi(Y) = \varphi(e_1) \leq \sup \varphi(Y_1) < \varphi(y_0) \leq \sup \varphi(Y) . \qquad \square$$

The following question remains open.

Problem. It is known that in locally convex spaces a kind of converse of the Krein-Milman theorem holds: If Y is convex compact and $Y = \overline{\text{co}}(Z)$ for some subset Z of Y, then $\operatorname{ext}(Y) \subset \bar{Z}$. Is this result true in the asymmetric case too?

Now we shall present the existence of a norm preserving extension that preserves also the extremality of the original functional. In the case of normed spaces the result was obtained by Singer [221] and in the asymmetric case in [40]. The result will be applied in Section 2.5 to the characterization of best approximation elements in asymmetric normed spaces.

Let (X, p) be an asymmetric normed space. In the following theorem, the symbols $B_{Y_p^\flat}$ and $B_{X_p^\flat}$ stand for the closed unit balls of the dual spaces $Y_p^\flat$ and $X_p^\flat$,

$$B_{Y_p^\flat} = \{\varphi \in Y_p^\flat : \|\varphi|_p \leq 1\}$$

and

$$B_{X_p^\flat} = \{\psi \in X_p^\flat : \|\psi|_p \leq 1\} .$$

Theorem 2.2.13. *Let (X, p) be a space with asymmetric norm and Y a subspace of X.*

If φ_0 is an extreme point of the closed ball $\|\varphi_0| \cdot B_{Y_p^\flat}$ then there exists a norm preserving extension φ of φ_0 which is an extreme point of the ball $\|\varphi_0| \cdot B_{X_p^\flat}$ of $X_p^\flat$.

Proof. Denote by $B^\flat$ the closed unit ball of the space $X^\flat$. Because ψ is an extreme point of the ball $B^\flat$ if and only if $r \cdot \psi$ is an extreme point of the ball $rB^\flat$, it is sufficient to prove the theorem for $\|\varphi_0| = 1$. Let φ_0, $\|\varphi_0| = 1$, be an extreme point of the unit ball $B_{Y^\flat}$ of $Y^\flat$ and let

$$E(\varphi_0) = \{\psi \in X^\flat : \psi|_Y = \varphi_0 \text{ and } \|\psi| = 1\} .$$

Clearly, the set $E(\varphi_0)$ convex; by Theorem 2.2.2 it is also nonempty. Further, the set $E(\varphi_0)$ is an extremal subset of $B^\flat$ because $(1-\alpha)\varphi_1 + \alpha\varphi_2 \in E(\varphi_0)$, for some $\varphi_1, \varphi_2 \in B^\flat$ and some $0 < \alpha < 1$, implies $(1-\alpha)\varphi_1|_Y + \alpha\varphi_2|_Y = \varphi_0$, so that, by the extremality of φ_0, we have $\varphi_1|_Y = \varphi_2|_Y = \varphi_0$. Since $\|(1-\alpha)\varphi_1 + \alpha\varphi_2| = 1$ and $\varphi_k \in B^\flat$ it follows that $\|\varphi_k| = 1$, $k = 1, 2$, so that $\varphi_k \in E(\varphi_0)$, $k = 1, 2$.

We show now that the set $E(\varphi_0)$ is a w^*-closed subset of the closed ball B^* of $X^* = (X, p^s)^*$. Let $(\varphi_\gamma : \gamma \in \Gamma)$ be a net in $E(\varphi_0)$ that is w^*-convergent to an element $\varphi \in B^*$, i.e.,

$$\forall x \in X, \quad \varphi_\gamma(x) \to \varphi(x) \quad \text{in } (\mathbb{R}, |\cdot|) .$$

Since for every $x \in X$ and every $\gamma \in \Gamma$, we have $\varphi_\gamma(x) \le p(x)$, it follows that $\varphi(x) \le p(x)$, i.e., $\|\varphi| \le 1$. Also, for every $y \in Y$ and $\gamma \in \Gamma$, $\varphi_\gamma(y) = \varphi_0(y)$, so that $\varphi|_Y = \varphi_0$, and $\|\varphi| \ge \|\varphi_0| = 1$. It follows that $\varphi \in E(\varphi_0)$, showing that $E(\varphi_0)$ is a w^*-closed subset of the w^*-compact set B^*, so it is w^*-compact too.

By the Krein-Milman theorem the convex w^*-compact set $E(\varphi_0)$ agrees with the closed convex hull of its extreme points, so that it has extreme points. Taking an extreme point φ of the extremal subset $E(\varphi_0)$ of $B^\flat$, it follows that φ is an extreme point of the unit ball $B^\flat$ of $X^\flat$ and $\varphi|_Y = \varphi_0$, $\|\varphi| = 1 = \|\varphi_0|$. □

2.3 The fundamental principles

Together with the Hahn-Banach extension theorem, the Open Mapping Theorem and the Closed Graph Theorem are the cornerstones of the whole edifice of classical functional analysis. Although in the asymmetric case they do not hold in full generality, some positive results have been obtained, which will be presented in this section.

2.3.1 The Open Mapping and the Closed Graph Theorems

As it is known, the proofs of two fundamental principles of functional analysis – the Open Mapping Theorem and the Closed Graph Theorem for Banach spaces – rely on Baire's category theorem. Based on Theorem 1.2.44, C. Alegre [2] extended these principles to asymmetric normed spaces.

Theorem 2.3.1 (The Open Mapping Theorem, [2]). *Let (X, p) and (Y, q) be asymmetric normed spaces. Suppose that (X, p) is right-K-complete and Y is Hausdorff* *and a $(q,\bar{q})$-Baire space. If $A : X \to Y$ is linear, surjective and (p,q)-continuous, then for every p-open subset G of X, $A(G)$ is q-open in Y.*

Proof. Let B'_p be the open unit ball of (X,p). Since A is surjective, $A(B'_p)$ is an absorbing convex subset of Y, and so will be the $\bar{q}$-closed convex set $\bar{q}\text{-}\operatorname{cl}(A(B'_p))$, implying $X = \cup_{n=1}^{\infty} n\ \bar{q}\text{-}\operatorname{cl}(A(B'_p))$. Since Y is a $(q,\bar{q})$-Baire space, there exists $n \in \mathbb{N}$ such that $q\text{-}\operatorname{int}\left(n\ \bar{q}\text{-}\operatorname{cl}(A(B'_p))\right) \neq \emptyset$ (see Theorem 1.2.49). It follows

$$q\text{-}\operatorname{int}\left(\bar{q}\text{-}\operatorname{cl}(A(B'_p))\right) \neq \emptyset ,$$

so that, by Proposition 1.1.67.4, there exists $\varepsilon > 0$ such that

$$B_q(0,\varepsilon) \subset \bar{q}\text{-}\operatorname{cl}(A(B'_p)) . \tag{2.3.1}$$

We show that

$$B_q(0,\frac{\varepsilon}{2^2}) \subset A(B'_p) . \tag{2.3.2}$$

Indeed, by (2.3.1), $y \in B_q(0,2^{-2}\varepsilon) \subset \bar{q}\text{-}\operatorname{cl}(A(2^{-2}B'_p)$ implies $B_{\bar{q}}(y,2^{-3}\varepsilon) \cap A(2^{-2}B'_p) \neq \emptyset$, so that there exists $x_1 \in 2^{-2}B'_p$ such that

$$\bar{q}(Ax_1 - y) < \frac{\varepsilon}{2^3} \iff q(y - Ax_1) < \frac{\varepsilon}{2^3} .$$

It follows that

$$y - Ax_1 \in B_q(0,\frac{\varepsilon}{2^3}) \subset \bar{q}\text{-}\operatorname{cl}(A(2^{-3}B'_p) ,$$

so that $B_{\bar{q}}(y - Ax_1, 2^{-4}\varepsilon) \cap A(2^{-3}B'_p) \neq \emptyset$, implying the existence of an element $x_2 \in 2^{-3}B'_p$ such that

$$\bar{q}(Ax_2 + Ax_1 - y) < \frac{\varepsilon}{2^4} \iff q(y - Ax_1 - Ax_2) < \frac{\varepsilon}{2^4} .$$

Continuing in this manner, one obtains the elements $x_k \in 2^{-(k+1)}B'_p$ such that

$$\bar{q}(Ax_n + \cdots + Ax_2 + Ax_1 - y) < \frac{\varepsilon}{2^{n+2}} , \tag{2.3.3}$$

for all $n \in \mathbb{N}$, implying $Ax_1 + \cdots + Ax_n \xrightarrow{\bar{q}} y$ as $n \to \infty$.

Since $p(x_k) < 2^{-k-1}$ the sequence $s_n = x_1 + \cdots + x_n$, $n \in \mathbb{N}$, is p-left-K-Cauchy (see Proposition 1.2.6), or, equivalently, $\bar{p}$-right-K-Cauchy. The right K-completeness of (X,p) implies the right K-completeness of $(X,\bar{p})$, so there exists $x \in X$ such that $\bar{p}(s_n - x) \to 0$ as $n \to \infty$. The (p,q)-continuity of the linear operator A implies its $(\bar{p},\bar{q})$-continuity, so that $Ax_1 + \cdots + Ax_n \xrightarrow{\bar{q}} Ax$ as $n \to \infty$. Since (Y,τ_q) is Hausdorff, $(Y,\tau_{\bar{q}})$ is also Hausdorff, so that $y = Ax$. From

$$\begin{aligned} p(x) \leq & p(x - s_n) + p(s_n) \leq \bar{p}(s_n - x) + p(x_1) + \cdots + p(x_n) \\ < & \bar{p}(s_n - x) + \frac{1}{2^2} + \cdots + \frac{1}{2^{n+1}} \underset{(n\to\infty)}{\longrightarrow} \frac{1}{2} , \end{aligned}$$

it follows that $x \in B'_p$.

Finally, we show that (2.3.2) implies that $A(G)$ is q-open for every p-open subset G of X.

If $y = Ax$ for some $x \in G$, then there exists $r > 0$ such that $x + rB'_p \subset G$. It follows that

$$y + \frac{r\varepsilon}{4} B'_q \subset Ax + rA(B'_p) = A(x + rB'_p) \subset A(G) ,$$

showing that the set $A(G)$ is q-open. □

Remark 2.3.2. Theorem 2.3.1 is proved in [2] under the hypotheses that the asymmetric normed space (X, p) is sequentially right K-complete and (Y, q) is sequentially right K-complete and Hausdorff. The proof is based on Lemma 3 asserting that in an asymmetric normed space (Y, q) which is of second Baire category in itself (and so, by Proposition 1.2.43, a Baire space), q-cl(A) and $\bar{q}$-cl(A) are 0-neighborhoods in (Y, q), for any absorbing and starshaped with respect to 0 subset A of X. While the assertion concerning q-cl(A) follows from the Baire property, it is not founded concerning $\bar{q}$-cl(A). So we reformulated the theorem by asking the space (Y, q) to be $(q, \bar{q})$-Baire.

A consequence of this deep result is the inverse mapping theorem which, in essence, is an equivalent form of the open mapping theorem.

Corollary 2.3.3. *Let (X, p) and (Y, q) be two asymmetric normed spaces. If (X, p) is right K-sequentially and (Y, q) is $(q, \bar{q})$-Baire and Hausdorff, then the inverse of any bijective continuous linear mapping $A : (X, \tau_p) \to (Y, \tau_q)$ is continuous.*

For two asymmetric normed spaces (X, p) and (Y, q) consider $X \times Y$ endowed with the asymmetric norm

$$r(x, y) = p(x) + q(y),\ (x, y) \in X \times Y . \tag{2.3.4}$$

The proof of the results from the following lemma are similar to those in the symmetric case.

Lemma 2.3.4. *Let (X, p), (Y, q) be asymmetric normed spaces.*

1. *The norm r defined by (2.3.4) generates the product topology $\tau_p \times \tau_q$ on $X \times Y$.*
2. *If (X, p) and (Y, q) are right (left) K-sequentially complete, then $(X \times Y, r)$ is right (left) K-sequentially complete.*
3. *A closed subset of a right (left) K-sequentially complete normed space is right (left) K-sequentially complete.*

As in the case of Banach spaces, the closed graph theorem can easily be derived from the open mapping theorem. The graph G_f of a mapping $f : X \to Y$ is the subset of $X \times Y$ given by $G_f = \{(x, y) \in X \times Y : y = f(x)\}$.

Proposition 2.3.5. *If (X, τ), (Y, ν) are topological spaces, with Y Hausdorff, and $f : X \to Y$ is continuous, then the graph G_f of f is closed in $X \times Y$ with respect to the product topology $\tau \times \nu$.*

1. *If the family $\mathcal{A}$ is q-upper pointwise bounded and*

$$\sup_{A\in\mathcal{A}} \sup\{\bar{q}(Ax) : x \in B_p\} = \infty \,, \tag{2.3.9}$$

then the set

$$\{x \in X : \sup_{A\in\mathcal{A}} \bar{q}(Ax) = \infty\} \tag{2.3.10}$$

is p-G_δ and p-dense in X.

2. *Similarly, if the family $\mathcal{A}$ is $\bar{q}$-upper pointwise bounded and*

$$\sup_{A\in\mathcal{A}} \sup\{q(Ax) : x \in B_{\bar{p}}\} = \infty \,, \tag{2.3.11}$$

then the set

$$\{x \in X : \sup_{A\in\mathcal{A}} q(Ax) = \infty\} \tag{2.3.12}$$

is $\bar{p}$-G_δ and $\bar{p}$-dense in X.

This principle has the following curious consequence.

Corollary 2.3.9. *If (X, p) is a right K-complete asymmetric normed space and (Y, q) is an asymmetric normed space, then $L_{p,q}(X, Y) = L_{p,\bar{q}}(X, Y)$.*

Proof. Suppose that there exists $A \in L_{p,q}(X, Y) \setminus L_{p,\bar{q}}(X, Y)$. Then

$$\sup\{\bar{q}(Ax) : x \in B_p\} = \infty \,.$$

Applying Theorem 2.3.8 to the family $\mathcal{A} = \{A\}$, one obtains that $\emptyset = \{x \in X : \bar{q}(Ax) = \infty\}$ is p-dense in X, a contradiction. □

2.3.3 Normed cones

We shall present some results on abstract normed cones as defined in Remark 2.1.4. As we did mention in Remark 2.1.4, the study of the duals of asymmetric normed spaces requires the consideration of normed cones.

A *linear mapping* between two cones X, Y is an additive and positively homogeneous mapping $A : X \to Y$.

An *asymmetric seminorm* on a cone X is a mapping $p : X \to \mathbb{R}^+$ such that

(i) $p(0) = 0$ and $(x, -x \in X \wedge p(x) = p(-x) = 0) \Rightarrow x = 0$
(ii) $p(\alpha x) = \alpha p(x)$;
(iii) $p(x + y) \le p(x) + p(y)$,

for all $x, y \in X$ and $\alpha \ge 0$. If

(iv) $p(x) = 0 \iff x = 0$,

then p is called an *asymmetric norm*.

Starting from an asymmetric seminorm p on a cone X one can define an extended quasi-semimetric e_p on X by the formula

$$e_p(x,y) = \begin{cases} \inf\{p(z) : z \in X,\ y = x + z\} & \text{if} \quad y \in x + X, \\ \infty & \text{otherwise} . \end{cases} \tag{2.3.13}$$

An extended quasi-semimetric $d : X \times X \to [0;\infty]$ on a cone X is called *subinvariant* provided that

$$\begin{aligned} &\text{(i)} \qquad d(x+z, y+z) \le d(x,y), \quad \textit{and} \\ &\text{(ii)} \qquad d(\alpha x, \alpha y) = \alpha d(x,y) , \end{aligned} \tag{2.3.14}$$

for all $x, y, z \in X$ and $\alpha \geq 0$.

For instance, $\mathbb{R}^+$ is a cancellative cone and $u(\alpha) = \alpha$ is an asymmetric norm on $\mathbb{R}^+$. The associated extended quasi-metric, given by $e_u(x,y) = y - x$ if $x \le y$ and $e_u(x,y) = \infty$, otherwise, induces the Sorgenfrey topology on $\mathbb{R}^+$ (see Example 1.1.6).

The topological notions for a normed cone (X,p) will be considered with respect to this extended quasi-semimetric. As before, one associates to e_p the conjugate quasi-semimetric $\bar{e}_p(x,y) = e_p(y,x)$ and the (symmetric) extended semimetric $e_p^s(x,y) = \max\{e_p(x,y), \bar{e}_p(x,y))$.

Some properties of this quasi-semimetric are collected in the following proposition.

Proposition 2.3.10. *Let (X,p) be an asymmetric normed cone.*

1. *The function e_p defined by (2.3.13) is a subinvariant extended quasi-semimetric on X.*
2. *The equality*
$$r \cdot B_{e_p}(x,\varepsilon) = rx + \{y \in X : p(y) \le \varepsilon r\} ,$$
holds for every $x \in X$ and $r, \varepsilon > 0$.
3. *The translations with respect to $+$ and $\cdot$ are $\tau(e_p)$-open, that is, if $Z \subset X$ is $\tau(e_p)$-open, then both $x + Z$ and $r \cdot Z$ are $\tau(e_p)$-open.*

Continuous linear mapping between normed cones have properties similar to those between asymmetric normed spaces, see Proposition 2.1.2.

Proposition 2.3.11 ([234]). *Let $(X,p), (Y,q)$ be asymmetric normed cones and $A : X \to Y$ a linear operator. The following are equivalent.*

1. *The operator A is continuous on X.*
2. *The operator A is continuous at $0 \in X$.*
3. *The operator A is upper bounded on every ball $B_{e_p}[0,r]$.*
4. *There exists $\beta \geq 0$ such that $q(Ax) \le \beta\, p(x)$, for all $x \in X$.*

Proof. The implication $1 \Rightarrow 2$ is trivial and the proofs of the implications $2 \Rightarrow 3$, and $3 \Rightarrow 4$ are similar to those from Proposition 2.1.2.

$4 \Rightarrow 1$. For $x \in X$ and $\varepsilon > 0$ let $r > 0$ be such that $\beta r < \varepsilon$. If $e_p(x, x') < r$, then there exists $z \in X$ such that $x' = x + z$ and $p(z), r$. It follows that $Ax' = Ax + Az$ and $q(Az) \leq \beta p(z) < \beta r < \varepsilon$, proving the continuity of A at x. □

Based on this proposition one can introduce an asymmetric norm on the space $L_{p,q}(X, Y)$ of continuous linear operators between two asymmetric normed cones (X, p) and (Y, q), by

$$\|A|_{p,q} = \sup\{q(Ax) : p(x) \leq 1\} . \tag{2.3.15}$$

It follows that

$$q(Ax) \leq \|A|_{p,q}\, p(x) , \tag{2.3.16}$$

for all $x \in X$ and that $L_{p,q}(X, Y)$ is an asymmetric normed cone with respect to (2.3.15), see [234].

We shall present now some closed graph and open mapping results for normed cones proved by Valero [235]. A uniform boundedness principle for locally convex cones was proved by Roth [211] (see also Roth [212] and the paper [184]).

An asymmetric normed cone is called *bicomplete* if it is complete with respect to the extended metric e_p^s. As it was shown by examples in [235], the Closed Graph and the Open Mapping Theorems do not hold for bicomplete asymmetric normed cones, some supplementary hypotheses being necessary.

A mapping f between two topological spaces (S, ν) and (T, τ) is called *almost continuous* at $s \in S$ if for every open subset V of T such that $f(s) \in V$, the set $\mathrm{cl}_\nu(f^{-1}(V))$ is a ν-neighborhood of s. A subset A of a bitopological space (T, τ_1, τ_2) is called (τ_1, τ_2)-*preopen* if $A \subset \tau_1\text{-}\operatorname{int}(\tau_2\text{-}\operatorname{cl} A)$. A mapping f from a topological space (S, ν) to a bitopological space (T, τ_1, τ_2) is called *almost open* if $f(U)$ is (τ_1, τ_2)-preopen for every ν-open subset U of S.

The closed graph theorem proved by Valero [235] is the following.

Theorem 2.3.12. *Let (X, p) and (Y, q) be two asymmetric normed cones such that the cone Y is right K-sequentially complete with respect to the conjugate extended quasi-metric $\bar{e}_q$. If $A : X \to Y$ is a linear mapping with closed graph in $(X \times Y, \bar{e}_p \times \bar{e}_q)$ which is $(\bar{e}_p, e_q)$-almost continuous at 0, then A is continuous.*

An open mapping theorem holds in similar conditions.

Theorem 2.3.13. *Let (X, p) and (Y, q) be two asymmetric normed cones such that the cone Y is right K-sequentially complete with respect to the conjugate extended quasi-metric $\bar{e}_q$. If $A : X \to Y$ is a linear mapping with closed graph in $(X \times Y, \bar{e}_p \times \bar{e}_q)$ which is almost open as a mapping from $(X, \tau(e_p))$ to $(Y, \tau(e_q), \tau(\bar{e}_q))$, then A is continuous.*

There are also other results on normed cones: the paper [94] discusses the metrizability of the unit ball of the dual of a normed cone, Oltra and Valero [170] study the isometries and bicompletions of normed cones and Valero [234] defines and studies the properties of quotient normed cones (the study of quotient spaces of asymmetric normed spaces is done in [3]). Other properties are investigated in a series of papers by Romaguera, Sánchez Pérez and Valero: in [203] one considers generalized monotone normed cones, quasi-normed monoids are discussed in [201], the dominated extension of functionals, of V-convex functions and duality on cancellative and noncancellative normed cones are treated in [200] and [202], respectively.

2.4 Weak topologies

The aim of this section is to present some basic results on weak topologies on asymmetric normed spaces and on asymmetric LCS. As it is well known, the weak topologies play a crucial role in functional analysis. Their asymmetric counterparts were studied in the fundamental paper [90], where an Alaoglu-Bourbaki type theorem on the weak*-compactness of the closed unit ball of the dual space is proved. The locally convex variant on the weak-compactness of the polar of a 0-neighborhood in an asymmetric LCS was proved in [41]. With an appropriate definition of the bidual, reflexivity is studied and an analogue of Goldstine's theorem on the weak density of a Banach space in its bidual is proved.

2.4.1 The $w^\flat$-topology of the dual space $X_p^\flat$

This is the analog of the w^*-topology on the dual of a normed space, which we shall present following [90].

Let (X, p) be a space with asymmetric norm and $X_p^\flat$ its asymmetric dual. The $w^\flat$-topology on $X_p^\flat$ is the topology admitting as a neighborhood basis of a point $\varphi \in X_p^\flat$ the sets

$$V_{x_1,\ldots,x_n;\varepsilon}(\varphi) = \{\psi \in X_p^\flat : \psi(x_k) - \varphi(x_k) < \varepsilon,\ k = 1, 2, \ldots, n\}\,, \tag{2.4.1}$$

for all $\varepsilon > 0$, and all $n \in \mathbb{N}$ and $x_1, \ldots, x_n \in X$.

The topology $w^\flat$ is derived from a quasi-uniformity $\mathcal{W}_p^\flat$ on $X_p^\flat$ with a basis formed of the sets

$$V_{x_1,\ldots,x_n;\,\varepsilon} = \{(\varphi_1, \varphi_2) \in X_p^\flat \times X_p^\flat : \varphi_2(x_i) - \varphi_1(x_i) \le \varepsilon,\ i = 1, \ldots, n\}\,, \tag{2.4.2}$$

for $n \in \mathbb{N}$, $x_1, \ldots, x_n \in X$ and $\varepsilon > 0$. Note that, for fixed $\varphi_1 = \varphi$, one obtains the neighborhoods from (2.4.1).

By the definition of the topology $w^\flat$, the $w^\flat$-convergence of a net (φ_γ) in $X_p^\flat$ to $\varphi \in X_p^\flat$ is equivalent to

$$\forall x \in X, \quad \varphi_\gamma(x) \to \varphi(x) \quad \text{in} \quad (\mathbb{R}, u)\,.$$

The following proposition shows that, in fact, a stronger result holds.

Proposition 2.4.1. *Let (X,p) be an asymmetric normed space with dual $X_p^\flat$.*

The $w^\flat$-topology is the restriction to $X_p^\flat$ of the w^*-topology of the space $X^* = (X,p^s)^*$. Consequently, it is Hausdorff and the $w^\flat$-convergence of a net (φ_γ) in $X_p^\flat$ to $\varphi \in X_p^\flat$ is equivalent to

$$\forall x \in X, \quad \varphi_\gamma(x) \to \varphi(x) \quad in \quad (\mathbb{R}, |\cdot|)\,.$$

Proof. The first assertion is a direct consequence of the definition of the topology $w^\flat$. The second assertion follows from the remarks that

$$V_{\varepsilon,x} \cap V_{\varepsilon,-x} = \{\psi \in X_p^\flat : |\psi(x) - \varphi(x)| < \varepsilon\}$$

is a $w^\flat$-neighborhood of $0 \in X$ and $X_p^\flat \subset X^*$. □

A deep result in Banach space theory is the Alaoglu-Bourbaki Theorem:

THEOREM. (Alaoglu-Bourbaki) *The closed unit ball B_{X^*} of the dual of a normed space X is w^*-compact.*

The analog of this theorem for asymmetric normed spaces was proved in [90, Theorem 4]. We include a slightly simpler proof of this result.

Proposition 2.4.2. *The closed unit ball $B_p^\flat = B_{X_p^\flat}$ of the space $X_p^\flat$ is a w^*-closed subset of the closed unit ball B^* of the space $X^* = (X,p^s)^*$.*

Proof. Let (φ_γ) be a net in $B_p^\flat$ that is w^*-convergent to an element $\varphi \in X^*$, i.e., for every $x \in X$ the net $(\varphi_\gamma(x))$ converges to $\varphi(x)$ in $(\mathbb{R}, |\cdot|)$. Because $\forall x \in X$, $\varphi_\gamma(x) \le p(x)$, it follows that $\varphi(x) \le p(x)$ for all $x \in X$, showing that $\varphi \in B_p^\flat$. □

Theorem 2.4.3. *The closed unit ball $B_p^\flat$ of the dual $X_p^\flat$ of an asymmetric normed space (X,p) is $w^\flat$-compact.*

Proof. By the Alaoglu-Bourbaki theorem the ball B^* is w^*-compact, so that, as a w^*-closed subset of B^*, the ball $B_p^\flat$ will be w^*-compact too. Since the $w^\flat$-topology is the restriction of w^* to $X_p^\flat$, it follows that the set $B_p^\flat$ is also $w^\flat$-compact. □

Remark 2.4.4. In [41, Proposition 2.11] the Alaoglu-Bourbaki theorem was extended to asymmetric locally convex spaces: the polar of any neighborhood of 0 is a $w^\flat$-compact convex subset of the asymmetric dual cone $X_P^\flat$, see Theorem 2.4.30.

2.4.2 Compact subsets of asymmetric normed spaces

In this subsection we shall present, following [86] and [4], some results on compactness specific to asymmetric normed spaces. The proof will be given in the next subsection within the more general context of asymmetric LCS.

Let (X, p) be an asymmetric normed space. For $x \in X$ put

$$\theta(x) = \{y \in X : p(y - x) = 0\} . \tag{2.4.3}$$

It is clear that $\theta(x) = x + \theta(0)$ and $Y + \theta(0) = \cup\{\theta(y) : y \in Y\}$. Also $\theta(x) = \tau(\bar{p})\text{-cl}(\{x\})$, as can be seen from the equivalences

$$\begin{aligned} y \in \theta(x) &\iff p(y - x) = 0 \iff \bar{p}(x - y) = 0 \\ &\iff \forall \varepsilon > 0,\ \bar{p}(x - y) < \varepsilon \iff \forall \varepsilon > 0,\ x \in B_{\bar{p}}(y, \varepsilon) \\ &\iff y \in \tau(\bar{p})\text{-}\,\mathrm{cl}(\{x\}) . \end{aligned}$$

The following properties hold.

Proposition 2.4.5 ([86]). *Let (X, p) be an asymmetric normed normed space, $x \in X$ and $\varepsilon > 0$. Then $B_p(x, \varepsilon) = \theta(0) + B_p(x, \varepsilon) + \theta(0)$. Also, if Y is a $\tau(p)$-open subset of X, then $Y = Y + \theta(0)$.*

As it is shown by García-Raffi [86] the sets $\theta(x)$ are involved in the study of compactness in asymmetric normed spaces.

Proposition 2.4.6. *Let (X, p) be an asymmetric normed space and $K \subset X$.*

1. *The set K is τ_p-compact if and only if $K + \theta(0)$ is τ_p-compact.*
   If $K + \theta(0)$ is τ_p-compact, $K_0 \subset K + \theta(0)$ and $K_0 + \theta(0) = K + \theta(0)$, then K_0 is τ_p-compact.
2. *A finite sum and a finite union of p-precompact sets is p-precompact.*
3. *The convex hull of a p-precompact set is p-precompact.*
4. *The set K is p-precompact if and only if the $\tau(\bar{p})$-closure of K is p-precompact.*
5. *If $K_0 \subset K \subset K_0 + \theta(0)$ and K_0-is $\tau(p^s)$-compact, then K is p-compact.*

García-Raffi [86] obtained characterizations of finite-dimensional normed spaces similar to those known for normed spaces. In the following proposition all topological notions refer to $\tau(p)$.

Theorem 2.4.7. *Let (X, p) be an asymmetric normed space.*

1. *If X is finite dimensional, of dimension $n \geq 1$, and T_1, then X is topologically isomorphic to the Euclidean space $\mathbb{R}^n$.*
2. *If (X, p) is T_1, then X is finite dimensional if and only if its closed unit ball B_p is $\tau(p)$-compact.*

3. *Suppose that X is finite dimensional. Then X is T_1 if and only if every $\tau(p)$-compact subset of X is $\tau(p)$-closed.*

As it is shown in [4, Example 12], the property 5 from Proposition 2.4.6 does not characterize the p-compactness. This paper contains also some further results on the relations between the $\tau(p^s)$-compactness of K_0 and the p-compactness of K, involving a notion of boundedness called right-boundedness. The unit closed ball B_p of (X, p) is called *right-bounded* if there exists $r > 0$ such that

$$r\, B_p \subset B_{p^s} + \theta(0)\ .$$

Note that the inclusion $B_{p^s} + \theta(0) \subset B_p$ is always true.

Theorem 2.4.8. *Let (X, p) be an asymmetric normed normed space.*

1. ([86]) *If X is finite dimensional and the unit closed ball B_p is right-bounded, then B_p is $\tau(p)$-compact.*
2. ([4]) *Suppose that (X, p) is biBanach with B_p right-bounded with $r = 1$. If $K \subset X$ is p-precompact, then there exists a p^s-compact subset K_0 of K such that $K \subset K_0 + \theta(0)$.*
3. ([4]) *If K_0 is p^s-precompact and $K \subset K_0 + \theta(0)$, then K is outside p-precompact.*

2.4.3 Compact sets in LCS

In this subsection we shall present, following [44], some properties of compact sets in LCS.

A subset Y of a quasi-uniform space $(X, \mathcal{U})$ is called *precompact* if for every $U \in \mathcal{U}$ there exists a finite subset Z of Y such that $Y \subset U[Z]$. The set Y is called *totally bounded* if for every U there exists a finite family $A_1, \dots, A_n$ of subsets of Y such that $A_i \times A_i \subset U$, $i = 1, \dots, n$, and $Y \subset \cup_{i=1}^n A_i$. Note that the total boundedness with respect to $\mathcal{U}$ is equivalent to the total boundedness with respect to the associated uniformity $\mathcal{U}_s$.

If in the above definition of precompactness one asks that the finite set Z be contained in X, then one obtains the notions of *outside precompactness* considered in [4]. Obviously, the precompactness implies the outside precompactness, but the reverse implication is not true, even in asymmetric normed spaces, see Example 1.2.23. In uniform spaces the total boundedness, the precompactness and the outside precompactness agree, and a set is compact if and only if it is totally bounded and complete.

If p is an asymmetric seminorm on a vector space X, we say that a subset Y of X is *p-precompact* if for every $\varepsilon > 0$ there exists a finite subset Z of Y such that

$$Y \subset \bigcup_{z \in Z} B_p(z, \varepsilon). \tag{2.4.4}$$

If for every $\varepsilon > 0$ there exists a finite subset Z of X such that (2.4.4) holds, then the set Y is called *outside p-precompact.* One obtains an equivalent notion if one asks that Y is covered by the family $B_p[z, \varepsilon]$, $z \in Z$, of closed balls. The set Z is called also a (p, ε)-net for Y (in both cases).

A subset of an asymmetric LCS (X, P) is called *precompact* if it is precompact with respect to the quasi-uniformity $\mathcal{U}_P$. The following proposition contains a useful characterization of precompactness in asymmetric LCS in terms of seminorms. The proof follows immediately from the definition of the quasi-uniformity $\mathcal{U}_P$ (the fact that $U_{p,\varepsilon}(x) = B_p(x, \varepsilon)$).

Proposition 2.4.9. *A subset Y of an asymmetric LCS (X, P) is (outside) precompact if and only if it is (outside) p-precompact for every $p \in P$.*

Based on this proposition, the proof of Proposition 1.2.18.3 can be adapted to obtain the following relation between precompactness and outside precompactness.

Proposition 2.4.10. *Let (X, P) be an asymmetric LCS. A subset Y of X is P-precompact if and only if for every $p \in P$ and every $\varepsilon > 0$ there exists a finite subset $\{x_1, \dots, x_n\}$ of X such that $Y \subset \cup_{i=1}^n B_p(x_i, \varepsilon)$ and $Y \cap B_{\bar{p}}(x_i, \varepsilon) \neq \emptyset$ for all $i \in \{1, \dots, n\}$.*

As a consequence of Proposition 1.1.59, one obtains the following relations between various notions of compactness and precompactness. A subset Y of an asymmetric LCS (X, P) is called *P-bounded* provided $\sup\{p(y) : y \in Y\} < \infty$ for every $p \in P$. This is equivalent to the fact that it is absorbed by every $\tau(P)$-neighborhood of 0, that is for every $\tau(P)$-neighborhood V of 0 there exists $\lambda > 0$ such that $\lambda Y \subset V$, or, in other words, Y is topologically bounded.

Proposition 2.4.11. *Let (X, P) be an asymmetric LCS and Y a subset of X.*

1. *If the set Y is P^s-precompact, then it is P-precompact and $\bar{P}$-precompact. The same is true for the outside precompactness.*
2. *If the set Y is $\tau(P^s)$-compact, then it is $\tau(P)$-compact and $\tau(\bar{P})$-compact.*
3. *The outside P-precompact subsets of X are P-bounded. In particular, the P-precompact subsets of X are P-bounded as well.*
4. *A subset of X is P-precompact if and only if its $\tau(\bar{P})$-closure is P-precompact. The same is true for outside P-precompactness.*

Proof. 1. For $\varepsilon > 0$ and $p \in P$ there exists a finite subset $\{y_1, \dots, y_n\}$ of Y such that $Y \subset \cup_{i=1}^n B_{p^s}(y_i, \varepsilon)$. Since $B_{p^s}(y_i, \varepsilon) \subset B_p(y_i, \varepsilon)$, $i = 1, \dots, n$, it follows that $Y \subset \cup_{i=1}^n B_p(y_i, \varepsilon)$, so that Y is P-precompact. Similarly, $B_{p^s}(y_i, \varepsilon) \subset B_{\bar{p}}(y_i, \varepsilon)$, $i = 1, \dots, n$, implies that Y is $\bar{P}$-precompact. The case of outside precompactness can be treated exactly in the same way.

2. This follows from the fact that a compact subset of a topological space remains compact for every coarser topology.

3. For $p \in P$ there exists a finite subset $\{x_1, \ldots, x_n\}$ of X such that $Y \subset \{x_1, \ldots, x_n\} + B_p(0,1)$, implying $p(y) \le \max\{p(x_i) : 1 \le i \le n\} + 1$ for every $y \in Y$.

4. We give a proof different from that in [4]. Suppose first that Y is P-precompact and show that $Z = \tau(\bar{P})\text{-cl}\, Y$ is also P-precompact, which is equivalent to the fact that Z is p-precompact for every $p \in P$.

Let $p \in P$ and $\varepsilon > 0$. Since Y is p-precompact there exists $y_1, \ldots, y_n \in Y$ such that

$$Y \subset \cup_{i=1}^n B_p[y_i, \varepsilon]. \tag{2.4.5}$$

By Proposition 1.1.59.4, the ball $B_p[0,\varepsilon]$ is $\tau(\bar{P})$-closed, so that the set

$$\cup_{i=1}^n B_p[y_i, \varepsilon] = \{y_1, \ldots, y_n\} + B_p[0, \varepsilon]$$

is also $\tau(\bar{P})$-closed. But then, the inclusion (2.4.5) implies

$$\tau(\bar{P})\text{-}\operatorname{cl} Y \subset \cup_{i=1}^n B_p[y_i, \varepsilon]\,.$$

Conversely, suppose that $Z = \tau(\bar{P})\text{-cl}\, Y$ is P-precompact and prove that Y is P-precompact.

For $p \in P$ and $\varepsilon > 0$ there exist $z_1, \ldots, z_n \in Z$ such that

$$Z \subset \cup_{i=1}^n B_p(z_i, \varepsilon). \tag{2.4.6}$$

For every $i \in \{1, \ldots, n\}$ there exists $y_i \in Y \cap B_{\bar{p}}(z_i, \varepsilon)$, that is an $y_i \in Y$ such that $\bar{p}(y_i - z_i) < \varepsilon$, or, equivalently, $p(z_i - y_i) < \varepsilon$.

Let $y \in Y \subset Z$. By (2.4.6) there exists $j \in \{1, \ldots, n\}$ such that $y \in B_p(z_j, \varepsilon)$. But then

$$p(y - y_j) \le p(y - z_j) + p(z_j - y_j) < 2\varepsilon\,,$$

showing that $Y \subset \cup_{i=1}^n B_p(y_i, 2\varepsilon)$.

In the case of outside precompactness, a subset of an outside precompact set is also outside precompact, so the outside precompactness of the $\tau(\bar{P})$-closure of Y implies the outside precompactness of Y. The reverse implication can be proved exactly as in the case of the precompactness. □

Remark 2.4.12. In the case of asymmetric normed spaces, the result from the assertion 4 of the above proposition was proved by García-Raffi [86, Prop. 9].

The following property is a consequence of Propositions 1.1.64.1 and 1.1.67.1, but for the sake of convenience, we give the proof.

Lemma 2.4.13. *Let (X, P) be an asymmetric LCS, $Q \subset P$ and $D \subset \mathbb{R}$ such that the family $\{B_q(0,r) : q \in Q,\, r \in D\}$ is a basis of $\tau(P)$-neighborhoods of 0. Then*

$$\tau(P)\text{-}\operatorname{cl} Y = \bigcap \{Y + B_{\bar{q}}(0,r) : q \in Q,\, r \in D\}, \tag{2.4.7}$$

for every subset Y of X.

Proof of Lemma 2.4.13. Let $x \in \tau(P)\text{-}\operatorname{cl} Y$, $q \in Q$ and $r \in D$. Then $x + B_q(0,r)$ is a $\tau(P)$-neighborhood of x, so that $Y \cap (x + B_q(0,r)) \neq \emptyset$, implying $x + u = y$, for some $u \in B_q(0,r)$ and $y \in Y$. But

$$u \in B_q(0,r) \iff q(u) \leq r \iff \bar{q}(-u) \leq r \iff -u \in B_{\bar{q}}(0,r) ,$$

so that $x = y - u \in Y + B_{\bar{q}}(0,r)$.

Conversely, suppose that x belongs to the intersection from the right-hand side of the equality (2.4.7). For a $\tau(P)$-neighborhood V_0 of 0, let $q \in Q$ and $r \in D$ be such that $B_q(0,r) \subset V_0$. By hypothesis, $x = y + v$ for some $y \in Y$ and $v \in B_{\bar{q}}(0,r)$, which, as above, implies that

$$y = x - v \in x + B_q(0,r) \subset x + V_0 .$$

Consequently, $(x + V_0) \cap Y \neq \emptyset$, showing that $x \in \tau(P)\text{-}\operatorname{cl} Y$. □

Proposition 2.4.14. *Let (X,P) be an asymmetric LCS whose topology $\tau(P)$ is T_1. Then X is finite dimensional if and only if there exists an outside P-precompact $\tau(P)$-neighborhood of 0.*

Proof. Necessity. If $\dim X = m$, then, by Proposition 1.1.68, it is isomorphic, algebraically and topologically, to the Euclidean space $\mathbb{R}^m$. Let $\Phi : \mathbb{R}^m \to X$ be an isomorphism. Since the closed unit ball $B_{\mathbb{R}^m}$ is a compact neighborhood of $0 \in \mathbb{R}^m$, its image by Φ will be a $\tau(P)$-compact neighborhood of $0 \in X$ which will be P-precompact and so outside P-precompact.

Sufficiency. Let $U = B_{p_0}(0,r_0)$ be an outside P-precompact $\tau(P)$-neighborhood of 0. Then there exists a finite subset $\{x_1,\dots,x_n\}$ of X such that

$$U \subset \{x_1,\dots,x_n\} + \frac{1}{2}U$$

implying

$$U \subset Z + \frac{1}{2}U, \tag{2.4.8}$$

where $Z = \operatorname{sp}\{x_1,\dots,x_n\}$ is the linear space generated by $\{x_1,\dots,x_n\}$. By (2.4.8)

$$\frac{1}{2}U \subset \frac{1}{2}Z + \frac{1}{2^2}U = Z + \frac{1}{2^2}U ,$$

so that

$$U \subset Z + \frac{1}{2}U \subset Z + Z + \frac{1}{2^2}U = Z + \frac{1}{2^2}U .$$

Repeating the argument, one obtains

$$U \subset Z + \frac{1}{2^n}U, \tag{2.4.9}$$

for all $n \in \mathbb{N}$.

We show that $\{\frac{1}{2^n}U : n \in \mathbb{N}\}$ is a basis of $\tau(P)$-neighborhoods of 0. For a $\tau(P)$-neighborhood V of 0, there exists $p \in P$ and $r > 0$ such that $B_p(0,r) \subset V$. Since a P-precompact set is topologically bounded (with respect to $\tau(P)$), there exists $\lambda > 0$ such that $\lambda U = \lambda B_{p_0}(0, r_0) \subset B_p(0,r)$. If $n \in \mathbb{N}$ is such that $2^{-n} < \lambda$, then

$$\frac{1}{2^n}U = \frac{1}{2^n}B_{p_0}(0,r_0) \subset \lambda B_{p_0}(0,r_0) \subset B_p(0,r) \subset V .$$

It is easy to check that $\{\frac{1}{2^n}B_{\bar{p}_0}(0,r_0) : n \in \mathbb{N}\}$ is a basis of $\tau(\bar{P})$-neighborhoods of 0, so that, by Lemma 2.4.13 and by (2.4.9),

$$U \subset \bigcap \{Z + \frac{1}{2^n}U : n \in \mathbb{N}\} = \tau(\bar{P})\text{-}\operatorname{cl} Z. \tag{2.4.10}$$

If we show that $\tau(\bar{P})$-cl $Z = Z$, then by (2.4.10), for every $x \in X \setminus \{0\}$ there exists $\lambda > 0$ such that $\lambda x \in U \subset Z$, showing that $X = Z$ is finite dimensional.

Let $x \in \tau(\bar{P})$-cl $Z \setminus Z$. Suppose that $\dim Z = m$ and let $e_1, \dots, e_m$ be an algebraic basis of Z. The space $W = \operatorname{sp}(Z \cup \{x\})$ has dimension $m+1$ and $e_1, \dots, e_m, x$ is an algebraic basis of W. Since $\{\{\frac{1}{2^n}B_{\bar{p}_0}(0,r_0) : n \in \mathbb{N}\}$ is a basis of $\tau(\bar{P})$-neighborhoods of 0, it follows that the topology $\tau(\bar{P})$ is generated by $\bar{p}_0$, so we can work with sequences. Suppose that $z_k = \alpha_{1,k}\, e_1 + \cdots + \alpha_{m,k}\, e_m + 0 \cdot x$, $k \in \mathbb{N}$, is a sequence in Z which converges to $x = 0 \cdot e_1 + \cdots + 0 \cdot e_m + 1 \cdot x$. Since the topology $\tau(P)$ is T_1, Proposition 1.1.63.2, implies that the topology $\tau(\bar{P})$ is also T_1. By Lemma 1.1.70, $\lim_k \alpha_{i,k} = 0$, $i = 1, \dots, m$, and $0 = \lim_k \alpha_{m+1,k} = \alpha_{m+1} = 1$, a contradiction. Consequently, $\tau(\bar{P})$-cl $Z = Z$, and Proposition 2.4.14 is completely proved. □

The following proposition is the analog of a known result in normed spaces. In the case of asymmetric normed spaces it was proved in [4, Proposition 8].

Proposition 2.4.15. *If Y is a precompact subset of an asymmetric LCS (X,P), then the convex hull* $\operatorname{co} Y$ *of Y is also precompact.*

Proof. By Proposition 2.4.9 it is sufficient to show that $\operatorname{co} Y$ is p-precompact for every $p \in P$.

Let $p \in P$ and $\varepsilon > 0$. By the precompactness of Y there exists a finite subset $\{y_1^0, \dots, y_n^0\}$ of Y such that

$$Y \subset \cup_{i=1}^n B_p(y_i^0, \varepsilon). \tag{2.4.11}$$

Let $\Delta_n = \{(\lambda_1, \dots, \lambda_n) \in \mathbb{R}_+^n : \sum_{i=1}^n \lambda_i = 1\}$ be the standard simplex in $\mathbb{R}_+^n$. The mapping $\Phi : \mathbb{R}_+^n \times X^n \to X$ given by $\Phi((\alpha_1, \dots, \alpha_n), (x_1, \dots, x_n)) = \sum_{i=1}^n \alpha_i x_i$ is continuous and $W = \operatorname{co}\{y_1^0, \dots, y_n^0\}$ is the image by this mapping of the compact subset $\Delta_n \times \{y_1^0, \dots, y_n^0\}$ of $\mathbb{R}_+^n \times X^n$, so it is compact and consequently P-precompact.

Therefore, there exists a subset $\{w_1^0, \dots, w_m^0\} \subset W$ such that

$$W \subset \cup_{i=1}^m B_p(w_i^0, \varepsilon). \tag{2.4.12}$$

We show that

$$Y \subset \cup_{i=1}^m B_p(w_i^0, 2\varepsilon). \tag{2.4.13}$$

Let $x \in \operatorname{co} Y$, $x = \sum_{i=1}^l \alpha_i y_i$ for some $\alpha_i \geq 0$, $y_i \in Y$, $i = 1, \ldots, l$, $\sum_{i=1}^l \alpha_i = 1$. By (2.4.11), for every $i \in \{1, \ldots, l\}$ there exists $j(i) \in \{1, \ldots, n\}$ such that $p(y_i - y_{j(i)}^0) \leq \varepsilon$. Putting $w := \sum_{i=1}^l \alpha_i y_{j(i)}^0$, it follows that

$$p(x - w) = p(\sum_{i=1}^l \alpha_i (y_i - y_{j(i)}^0)) \leq \sum_{i=1}^l \alpha_i p(y_i - y_{j(i)}^0) \leq \varepsilon \,.$$

Since $w \in W$, the equality (2.4.12) implies the existence of $i_0 \in \{1, \ldots, m\}$ such that $p(w - w_{i_0}^0) \leq \varepsilon$. But then

$$p(x - w_{i_0}^0) \leq p(x - w) + p(w - w_{i_0}^0) \leq 2\varepsilon \,,$$

showing that (2.4.13) holds. □

Some relations between precompactness and compactness in asymmetric normed spaces were studied in [86] and [4]. These results were extended in [44] to asymmetric LCS.

Let (X, P) be an asymmetric LCS. For $p \in P$ let

$$\theta_{0,p} = \{z \in X : p(z) = 0\}, \quad \text{and}$$
$$\theta_0 = \cap_{p \in P} \theta_{0,p} \,.$$

Let also

$$\theta_{x,p} = \{z \in X : p(z - x) = 0\} = x + \theta_{0,p} \,.$$

It is immediate that θ_x agrees with the $\tau(\bar{P})$-closure of the set $\{x\}$. Indeed

$$\begin{aligned} y \in \tau(\bar{P})\text{-}\operatorname{cl}\{x\} &\iff \forall p \in P,\ \forall \varepsilon > 0,\ \bar{p}(x - y) < \varepsilon \\ &\iff \forall p \in P,\ \bar{p}(x - y) = 0 \\ &\iff \forall p \in P,\ p(y - x) = 0 \\ &\iff y \in \theta_x \,. \end{aligned}$$

As it was shown in [86]

$$B_p(x, \varepsilon) = B_p(x, \varepsilon) + \theta_{0,p} \,.$$

Based on this equality one obtains immediately that

$$Y = Y + \theta_0 \,,$$

for every $\tau(P)$-open subset Y of X. Indeed, $0 \in \theta_0$ implies $Y \subset Y + \theta_0$. Conversely, let $x = y + z$ for some $y \in Y$ and $z \in \theta_0$. Since Y is $\tau(P)$-open there exist $p \in P$ and $\varepsilon > 0$ such that $B_p(y, \varepsilon) \subset Y$, implying $x = y + z \in B_p'(y, \varepsilon) + \theta_0 \subset B_p'(y, \varepsilon) + \theta_{0,p} = B_p'(y, \varepsilon) \subset Y$.

As a consequence of this last equality, one obtains the analog of Proposition 6 from [86].

Proposition 2.4.16. *A subset K of an asymmetric LCS is $\tau(P)$-compact if and only if $K + \theta_0$ is $\tau(P)$-compact.*

Also, if K is $\tau(P)$-compact, then every subset Z of $K + \theta_0$ is $\tau(P)$-compact.

Remark 2.4.17. In the case of an asymmetric normed space (X, p), Alegre et al. [4] give characterizations of τ_p-compact subsets of X. Among other results, they prove, under some supplementary hypotheses, that a subset K of X is τ_p-compact if and only if there exists a τ_{p^s}-compact subset K_0 of X such that $K_0 \subset K \subset K+\theta_0$ ([4, Theorem 20]). It is possible that similar characterizations hold in the locally convex case, a topic for further investigation.

2.4.4 The conjugate operator, precompact operators and a Schauder type theorem

A Schauder type theorem on the compactness of the conjugate of a compact linear operator on an asymmetric normed space was proved in [43]. We shall briefly present this result, referring for the details to the mentioned paper.

For a continuous linear operator $A : (X, p) \to (Y, q)$ between two asymmetric normed spaces, one defines the *conjugate operator* $A^\flat : Y_q^\flat \to X_p^\flat$ by the formula

$$A^\flat \psi = \psi \circ A, \quad \psi \in Y_q^\flat \, . \tag{2.4.14}$$

Concerning the continuity we have.

Proposition 2.4.18. *Let $(X, p), (Y, q)$ be asymmetric normed spaces and $A : X \to Y$ a continuous linear operator.*

1. *The operator $A^\flat : (Y_q^\flat, \|\cdot|_q) \to (X_p^\flat, \|\cdot|_p)$ is additive, positively homogeneous and continuous. So it is also quasi-uniformly continuous with respect to the quasi-uniformities $\mathcal{U}_q^\flat$ and $\mathcal{U}_p^\flat$ on $Y_q^\flat$ and $X_p^\flat$, respectively.*
2. *The operator $A^\flat$ is also quasi-uniformly continuous with respect to the $w^\flat$-quasi-uniformities $\mathcal{W}_q^\flat$ on $Y_q^\flat$ and $\mathcal{W}_p^\flat$ on $X_p^\flat$.*

Proof. 1. It is obvious that $A^\flat$ is properly defined, additive and positively homogeneous.

For every $\psi \in Y_q^\flat$,

$$\|A^\flat \psi|_q = \|\psi \circ A^\flat|_q \le \|\psi|_q \, \|A|_{p,q} \, ,$$

implying the continuity of $A^\flat$, which, in its turn, by the linearity of A, implies the quasi-uniform continuity with respect to the quasi-uniformities $\mathcal{U}_q^\flat$ and $\mathcal{U}_p^\flat$.

2. For $x_1, \ldots, x_n \in X$ and $\varepsilon > 0$ let

$$V = \{(\varphi_1, \varphi_2) \in X_p^\flat \times X_p^\flat : \varphi_2(x_i) - \varphi_1(x_i) \le \varepsilon, \; i = 1, \ldots, n\}$$

be a $w^\flat$-entourage in $X_p^\flat$. Then

$$U = \{(\psi_1, \psi_2) \in Y_q^\flat \times Y_q^\flat : \psi_2(Ax_i) - \psi_1(Ax_i) \le \varepsilon,\ i = 1, \ldots, n\}\,,$$

is a $w^\flat$-entourage in $Y_q^\flat$ and $(A^\flat\psi_1, A^\flat\psi_2) \in V$ for every $(\psi_1, \psi_2) \in U$, proving the quasi-uniform continuity of $A^\flat$ with respect to the $w^\flat$-quasi-uniformities on $Y_q^\flat$ and $X_p^\flat$. □

A linear operator $A : (X, p) \to (Y, q)$ between two asymmetric normed spaces is called (p, q)-*precompact* if the image $A(B_p)$ of the closed unit ball B_p of X by the operator A is a q-precompact subset of Y. We shall denote by $(X, Y)_{p,q}^k$ the set of all (p, q)-precompact operators from X to Y. A subset Y of a quasi-uniform space $(X, \mathcal{U})$ is called $\mathcal{U}$-*precompact* if for every $\varepsilon > 0$ there exists a finite subset Z of Y such that $Y \subset U[Z]$. If there exists a set $Z \subset X$ such that $Y \subset U[Z]$, then Y is called *outside* $\mathcal{U}$-*precompact.* It is clear that a subset Y of an asymmetric normed space (X, p) is (outside) p-precompact if and only if it is (outside) $\mathcal{U}_p$-precompact.

For $\mu \in \{p, \bar{p}, p^s\}$ and $\nu \in \{q, \bar{q}, q_s\}$ denote by $(X, Y)_{\mu,\nu}^\flat$ the cone of all continuous linear operators from (X, μ) to (Y, ν). The space $(X, Y)_s^* := (X, Y)_{p^s,q^s}^\flat$ is the space of all continuous linear operators between the associated normed spaces (X, p^s) and (Y, p^s), which was denoted also by $L(X, Y)$.

On the space $(X, Y)_s^*$ we shall consider several quasi-uniformities. For $\mu \in \{p, \bar{p}, p^s\}$ and $\nu \in \{q, \bar{q}, q^s\}$ let $\mathcal{U}_{\mu,\nu}$ be the quasi-uniformity generated by the basis

$$U_{\mu,\nu;\,\varepsilon} = \{(A, B); A, B \in (X, Y)_s^*,\ \nu(Bx - Ax) \le \varepsilon,\ \forall x \in B_\mu,\},\quad \varepsilon > 0\,, \tag{2.4.15}$$

where $B_\mu = \{x \in X : \mu(x) \le 1\}$ denotes the unit ball of (X, μ). The induced quasi-uniformity on the semilinear subspace $(X, Y)_{\mu,\nu}^\flat$ of $(X, Y)_s^*$ is denoted also by $\mathcal{U}_{\mu,\nu}$ and the corresponding topologies by $\tau(\mu, \nu)$. The uniformity $\mathcal{U}_{p^s,q^s}$ and the topology $\tau(p^s, q^s)$ are those corresponding to the norm (2.1.6) on the space $(X, Y)_s^*$, while the quasi-uniformity $\mathcal{U}_{p,q}$ corresponds to the extended asymmetric norm $\|\cdot|_{p,q}^*$ given by (2.1.23). In the case of the dual space $X_\mu^\flat$ we shall use the notation $\mathcal{U}_\mu^\flat$ for the quasi-uniformity $\mathcal{U}_{\mu,u}$.

Notice that, for $\mu = p^s$ and $\nu = q^s$, the space $(X, Y)_{p^s,q^s}^\flat$ agrees with $(X, Y)_s^*$, the (p^s, q^s)-compact operators are the usual linear compact operators between the normed spaces (X, p^s) and (Y, q^s), so the following proposition extends some well-known results for compact operators on normed spaces. For $\mu \in \{p, \bar{p}, p^s\}$ and $\nu \in \{q, \bar{q}, q_s\}$ one denotes by $(X, Y)_{\mu,\nu}^k$ the set of all (μ, ν)-precompact linear operators from (X, μ) to (Y, ν).

Proposition 2.4.19. *Let $(X, p), (Y, q)$ be asymmetric normed spaces. The following assertions hold.*

1. *$(X, Y)_{\mu,\nu}^k$ is a subcone of the cone $(X, Y)_{\mu,\nu}^\flat$ of all continuous linear operators from X to Y.*
2. *$(X, Y)_{p,q}^k$ is $\tau(p, q^s)$-closed in $(X, Y)_{p,q}^\flat$.*

Proof. 1. We give the proof in the case $\mu = p$ and $\nu = q$. The other cases can be treated similarly.

If $A : X \to Y$ is (p,q)-precompact, then there exists $x_1, \ldots, x_n \in B_p$ such that

$$\forall x \in B_p,\ \exists i \in \{1, \ldots, n\}, \quad q(Ax - Ax_i) \le 1. \tag{2.4.16}$$

If for $x \in B_p$, $i \in \{1, \ldots, n\}$ is chosen according to (2.4.16), then

$$q(Ax) \le q(Ax - Ax_i) + q(Ax_i) \le 1 + \max\{q(Ax_j) : 1 \le j \le n\},$$

showing that the operator A is (p,q)-bounded.

Suppose that $A_1, A_2 : X \to Y$ are (p,q)-precompact and let $\varepsilon > 0$. By the (p,q)-precompactness of the operators A_1, A_2, there exist $x_1, \ldots, x_m$ and $y_1, \ldots, y_n$ in B_p such that

$$\begin{gathered}\forall x \in B_p,\ \exists i \in \{1, \ldots, m\},\ \exists j \in \{1, \ldots, n\},\\ q(A_1x - A_1x_i) \le \varepsilon \ \text{ and } \ q(A_2x - A_2y_j) \le \varepsilon.\end{gathered}$$

It follows that for every $x \in B_p$ there exists a pair (i,j) with $1 \le i \le m$ and $1 \le j \le n$ such that

$$q(A_1x + A_2x - A_1x_i - A_2y_j) \le q(A_1x - A_1x_i) + q(A_2x - A_2y_j) \le 2\varepsilon,$$

showing that $\{A_1x_i + A_2y_j : 1 \le i \le m,\ 1 \le j \le n\}$ is a finite 2ε-net for $(A_1 + A_2)(B_p)$.

The proof of the precompactness of αA, for $\alpha > 0$ and A precompact, is immediate and we omit it.

2. *The $\tau(p,q^s)$-closedness of $(X,Y)^k_{p.q}$.*

Let (A_n) be a sequence in $(X,Y)^k_{p,q}$ which is $\tau(p,q^s)$-convergent to $A \in (X,Y)^\flat_{p,q}$.

For $\varepsilon > 0$ choose $n_0 \in \mathbb{N}$ such that

$$\forall n \ge n_0,\ \forall x \in B_p,\ q(A_nx - Ax) \le \varepsilon \ \text{ and } \ q(Ax - A_nx) \le \varepsilon\,. \tag{2.4.17}$$

Let $x_1, \ldots, x_m \in B_p$ be such that the points $A_{n_0}x_i$, $1 \le i \le m$, form an ε-net for $A_{n_0}(B_p)$. Then for every $x \in B_p$ there exists $i \in \{1, \ldots, m\}$ such that

$$q(A_{n_0}x - A_{n_0}x_i) \le \varepsilon,$$

so that, by (2.4.17),

$$q(Ax - Ax_i) \le q(Ax - A_{n_0}x) + q(A_{n_0}x - A_{n_0}x_i) + q(A_{n_0}x_i - Ax_i) \le 3\varepsilon.$$

Consequently, Ax_i, $1 \le i \le m$, is a 3ε-net for $A(B_p)$, showing that $A \in (X,Y)^k_{p,q}$. □

Now we are prepared to state and prove the analog of the Schauder theorem.

Theorem 2.4.20. *Let $(X,p),(Y,q)$ be asymmetric normed spaces. If the linear operator $A : X \to Y$ is (p,q^s)-precompact, then $A^\flat(B_q^\flat)$ is precompact with respect to the quasi-uniformity $\mathcal{U}_p^\flat$ on $X_p^\flat$.*

Proof. For $\varepsilon > 0$ let

$$U_\varepsilon = \{(\varphi_1,\varphi_2) \in X_p^\flat \times X_p^\flat : \varphi_2(x) - \varphi_1(x) \le \varepsilon,\ \forall x \in B_p\}\ ,$$

be an entourage in $X_p^\flat$ for the quasi-uniformity $\mathcal{U}_p^\flat$.

Since A is (p,q^s)-precompact, there exist $x_1,\ldots,x_n \in B_p$ such that

$$\forall x \in B_p\,,\ \exists j \in \{1,\ldots,n\},\qquad q(Ax - Ax_j) \le \varepsilon \text{ and } q(Ax_j - Ax) \le \varepsilon. \tag{2.4.18}$$

By the Alaoglu-Bourbaki theorem, Theorem 2.4.3, the set $B_q^\flat$ is $w^\flat$-compact, so by the $(w^\flat,w^\flat)$-continuity of the operator $A^\flat$ (Proposition 2.4.18), the set $A^\flat(B_q^\flat)$ is $w^\flat$-compact in $X_p^\flat$. Consequently, the $w^\flat$-open cover of $A^\flat(B_q^\flat)$,

$$V_\psi = \{\varphi \in X_p^\flat : \varphi(x_i) - A^\flat\psi(x_i) < \varepsilon,\ i = 1,\ldots,n\},\ \psi \in B_q^\flat\ ,$$

contains a finite subcover, i.e., there exist $m \in \mathbb{N}$ and $\psi_k \in B_q^\flat$, $1 \le k \le m$, such that

$$A^\flat(B_q^\flat) \subset \bigcup\{V_{\psi_k} : 1 \le k \le m\}\ . \tag{2.4.19}$$

Now let $\psi \in B_q^\flat$. By (2.4.19) there exists $k \in \{1,\ldots,m\}$ such that

$$A^\flat\psi(x_i) - A^\flat\psi_k(x_i) < \varepsilon,\ i = 1,\ldots,n\ .$$

If $x \in B_p$, then, by (2.4.18), there exists $j \in \{1,\ldots,n\}$, such that

$$q(Ax - Ax_j) \le \varepsilon \text{ and } q(Ax_j - Ax) \le \varepsilon\ .$$

It follows that

$$\begin{aligned}
&\psi(Ax) - \psi_k(Ax)\\
&= \psi(Ax - Ax_j) + \psi(Ax_j) - \psi_k(Ax_j) + \psi_k(Ax_j - Ax)\\
&\le q(Ax - Ax_j) + \varepsilon + q(Ax_j - Ax) \le 3\varepsilon\ .
\end{aligned}$$

Consequently,

$$\forall x \in B_p\,,\qquad (A^\flat\psi - A^\flat\psi_k)(x) \le 3\varepsilon\ ,$$

proving that

$$A^\flat(B_q^\flat) \subset U_{3\varepsilon}[\{A^\flat\psi_1,\ldots,A^\flat\psi_m\}]\ . \qquad \square$$

Remark 2.4.21. As a measure of precaution, we have restricted our study to precompact linear operators A on an asymmetric normed space (X,p) with values in another asymmetric normed space (Y,q), meaning that the image $A(B_p)$ of the unit ball of X by A is a q-precompact subset of Y. A compact linear operator should be defined by the condition that $A(B_p)$ is a relatively compact subset of Y, that is the τ_q-closure of $A(B_p)$ is τ_q-compact subset of Y, in concordance to the definition of compact linear operators between normed spaces.

As can be seen from Section 1.2, the relations between precompactness, total boundedness and completeness are considerably more complicated in the asymmetric case than in the symmetric one. The compactness properties of the set $A(B_p)$ need a study of the completeness of the space $(X,Y)^\flat_{\mu,\nu}$ with respect to various quasi-uniformities and various notions of completeness, which could be a topic of further investigation.

2.4.5 The bidual space, reflexivity and Goldstine theorem

The bidual space was introduced in [90], including the definition of the canonical embedding of an asymmetric normed space in its bidual and the definition of a reflexive asymmetric normed space. Further properties were proved in [93].

Let (X,p) be an asymmetric normed space. Denote by $X^\flat_p$ the cone of all continuous linear functionals $f:(X,p)\to(\mathbb{R},u)$ and by X^* the dual space of the associated normed space (X,p^s). On X^* one considers the extended asymmetric norm and the extended norm given by

$$\|f|^* = \sup f(B_p) \quad \text{and} \quad \|f\|^* = \max\{\|f|^*, \|-f|^*\}\,,$$

respectively. The restriction of $\|\cdot|^*$ to $X^\flat_p$ is denoted by $\|\cdot|$. As we have seen, $\operatorname{sp} X^\flat_p = X^\flat_p - X^\flat_p \subset X^*$ (Proposition 2.1.7) and a linear functional $f\in X^*$ belongs to $X^\flat_p$ if and only if $\|f|^*<\infty$ (Proposition 2.1.16).

Let

$$\begin{aligned} X^{**}_e &= \{\varphi : (X^*,\|\cdot\|^*)\to(\mathbb{R},|\cdot|) : \varphi \text{ is linear and continuous}\}\\ &= (X^*,\|\cdot\|^*)^*\,, \end{aligned} \tag{2.4.20}$$

and

$$X^{*\flat}_e = \{\varphi : (X^*,\|\cdot|^*)\to(\mathbb{R},u) : \varphi \text{ is linear and continuous}\} = (X^*,\|\cdot|^*)^\flat\,. \tag{2.4.21}$$

(Here the subscript "e" comes from "extended" and the superscript $*$ means that we consider linear continuous functionals, while the superscript $\flat$ means that the functionals are linear and upper semi-continuous, or, equivalently, continuous to $(\mathbb{R},u)$.)

The set X^{**}_e is a linear space, $X^{*\flat}_e$ is a cone in X^{**}_e, and

$$\operatorname{sp}(X^{*\flat}_e) = X^{*\flat}_e - X^{*\flat}_e \subset X^{**}_e\,.$$

For $\varphi \in X_e^{*\flat}$ let

$$\|\varphi|^{*\flat} = \sup\{\varphi(f) : f \in B_p^\flat\} , \tag{2.4.22}$$

where $B_p^\flat = \{f \in X_p^\flat : \|f| \le 1\}$.

Then $\|\cdot|^{*\flat}$ is an asymmetric norm and the asymmetric normed cone $(X_e^{*\flat}, \|\cdot|^{*\flat})$ is called the *bidual space* of (X, p).

For $x \in X$ define $\Lambda(x) : X^* \to \mathbb{R}$ by

$$\Lambda(x)(f) = f(x), \; f \in X^* . \tag{2.4.23}$$

Proposition 2.4.22. *Let (X, p) be an asymmetric normed space and let Λ be the mapping defined by* (2.4.23). *Then*

$$\Lambda(x) \in X_e^{*\flat} \quad \textit{and} \quad \|\Lambda(x)|^{*\flat} = p(x) ,$$

for every $x \in X$. Moreover, the mapping Λ is linear, so that it defines a linear isometric embedding of (X, p) into the semilinear space $(X_e^{*\flat}, \|\cdot|^{*\flat})$.

Proof. It is obvious that $\Lambda(x)$ is a linear functional on X^*. The inequality

$$\Lambda(x)(f) = f(x) \le \|f|^* p(x) ,$$

valid for every $f \in X^*$, implies $\|\Lambda(x)|^{*\flat} \le p(x)$. Since, by Theorem 2.2.2.2, there exists $f_0 \in B_p^\flat$ such that $f_0(x) = p(x)$, it follows that $\|\Lambda(x)|^{*\flat} = p(x)$.

The linearity of the mapping $\Lambda : X \to X_e^{*\flat}$ is easily verified. □

The space X^* induces a topology on $\mathrm{sp}(X_e^{*\flat})$ denoted by $\sigma(X_e^{*\flat}, X^*)$ having as basis of neighborhoods of 0 the sets

$$V_{f_1,\ldots,f_n;\varepsilon} = \{\varphi \in \mathrm{sp}(X_e^{*\flat}) : |\varphi(f_k)| < \varepsilon, \, k = 1, \ldots, n\} , \tag{2.4.24}$$

for $\varepsilon > 0$, $f_1, \ldots, f_n \in X^*$, $n \in \mathbb{N}$. The neighborhoods of an arbitrary point $\varphi \in \mathrm{sp}(X_e^{*\flat})$ are obtained by translating the neighborhoods of 0.

It is obvious that the topology $\sigma(X_e^{*\flat}, X^*)$ is a locally convex topology on $\mathrm{sp}(X_e^{*\flat})$ generated by the family p_f, $f \in X_e^{*\flat}$, of seminorms, where for $f \in X^*$ the seminorm $p_f : \mathrm{sp}(X_e^{*\flat}) \to \mathbb{R}$ is given by

$$p_f(\varphi) = |\varphi(f)|, \quad \varphi \in \mathrm{sp}(X_e^{*\flat}) . \tag{2.4.25}$$

The following proposition says that, in essence, the spaces X^* and $\mathrm{sp}\, X_e^{*\flat}$ form a dual pair.

Proposition 2.4.23 ([93]). *Let (X, p) be an asymmetric normed space. Then the following hold.*

1. *For each $f \in X^*$ the linear functional $e_f : \mathrm{sp}\, X_e^{*\flat} \to \mathbb{R}$ defined by $e_f(\varphi) = \varphi(f)$, $\varphi \in X_e^{*\flat}$, is continuous from $(\mathrm{sp}\, X_e^{*\flat}, \sigma(X_e^{*\flat}, X^*))$ to $(\mathbb{R}, |\cdot|)$.*
2. *If $\Psi : \mathrm{sp} X_e^{*\flat} \to \mathbb{R}$ is a linear functional, continuous from $(\mathrm{sp} X_e^{*\flat}, \sigma(X_e^{*\flat}, X^*))$ to $(\mathbb{R}, |\cdot|)$, then there exists $f \in X^*$ such that $\Psi = e_f$.*

Proof. The first assertion is almost trivial, because for $f \in X^*$ and $\varepsilon > 0$ the set

$$(e_f)^{-1}((-\varepsilon;\varepsilon)) = \{\varphi \in \mathrm{sp}(X_e^{*\flat}) : |\varphi(f)| < \varepsilon\} = V_{f,\varepsilon}$$

is a $\sigma(X_e^{*\flat}, X^*)$-neighborhood of 0.

To prove the converse assertion 2, let $\Psi : \mathrm{sp}\, X_e^{*\flat} \to \mathbb{R}$ be a linear functional, continuous from $(\mathrm{sp}\, X_e^{*\flat}, \sigma(X_e^{*\flat}, X^*))$ to $(\mathbb{R}, |\cdot|)$. By the characterization of the continuity of linear functionals on locally convex spaces [238, Korollar VIII.2.4], there exist $f_1, \ldots, f_n \in X^*$ and $\beta > 0$ such that

$$\begin{aligned}|\Psi(\varphi)| &\leq \beta \max\{p_{f_k}(\varphi) : 1 \leq k \leq n\}\\ &= \beta \max\{|e_{f_k}(\varphi)| : 1 \leq k \leq n\}\,.\end{aligned}$$

By [238, Lemma VIII.3.3] this implies that Ψ is a linear combination of e_{f_k}, $k = 1, \ldots, n$, $\Psi = \sum_{k=1}^n \alpha_k e_{f_k}$. But then $\Psi = e_f$, where $f = \sum_{k=1}^n \alpha_k f_k$. □

The remarkable theorem of Goldstine on the weak density of B_X in $B_{X^{**}}$ was extended in [93] to asymmetric normed spaces.

Theorem 2.4.24 (Goldstine Theorem). *Let (X, p) be an asymmetric normed space. The image by Λ of the unit ball B_p is $\sigma(X_e^{*\flat}, X^*)$-dense in the unit ball $B_{X_e^{*\flat}}$ of $X_e^{*\flat}$.*

Proof. Denote by C the $\sigma(X_e^{*\flat}, X^*)$-closure of $\Lambda(B_p)$ in $Z = X_e^{*\flat} - X_e^{*\flat}$ and let $\varphi \in C$.

Claim I. $\varphi \in X_e^{*\flat}$.

We have to show that φ is usc from $(X^*, \|\cdot|_p^*)$ to $(\mathbb{R}, |\cdot|)$. Suppose that (f_n) is a sequence in X^* which is $\|\cdot|_p^*$-convergent to $f \in X^*$. Then, given $\varepsilon > 0$, there exists $n_\varepsilon \in \mathbb{N}$ such that

$$\forall n \geq n_\varepsilon, \quad \|f_n - f|_p^* < \varepsilon\,. \tag{2.4.26}$$

Since φ belongs to the $\sigma(X_e^{*\flat}, X^*)$-closure of $\Lambda(B_p)$ in Z, for every $n \in \mathbb{N}$ there exists $x_n \in B_p$ such that

$$|\varphi(f_n - f) - (f_n - f)(x_n)| = |\varphi(f_n - f) - \Lambda(x_n)(f_n - f)| < \varepsilon\,. \tag{2.4.27}$$

But then, by (2.4.27) and (2.4.26),

$$\varphi(f_n) - \varphi(f) < (f_n - f)(x_n) + \varepsilon \leq \|f_n - f|_p^* + \varepsilon < 2\varepsilon\,,$$

for all $n \geq n_\varepsilon$, proving the required upper semi-continuity of the functional φ.

Claim II. $\|\varphi|^{*\flat} \leq 1$.

Let $\varepsilon > 0$ be given. By the definition of C, for every $f \in B_p^\flat$ there exists $x \in B_p$ such that

$$|\varphi(f) - f(x)| = |\varphi(f) - \Lambda(x)(f)| < \varepsilon\,,$$

implying

$$\varphi(f) < f(x) + \varepsilon \le \|f|_p + \varepsilon \le 1 + \varepsilon \,.$$

Since $\varepsilon >$ was arbitrarily taken, it follows that $\varphi(f) \le 1$ for every $f \in B_p^\flat$, so that $\|\varphi|^{*\flat} \le 1$.

By Claim I and Claim II, $C \subset B_{X_e^{*\flat}}$.

Claim III. $B_{X_e^{*\flat}} \subset C$.

Suppose that there exists $\varphi_0 \in C \setminus B_{X_e^{*\flat}}$. Applying Tukey's separation theorem ([149, Theorem 2.28]) in the locally convex space $(Z, \sigma(X_e^{*\flat}, X^*)$ in follows the existence of a $\sigma(X_e^{*\flat}, X^*)$-continuous linear functional $\Psi : Z \to \mathbb{R}$ such that

$$\sup\{\Psi(\varphi) : \varphi \in C\} < \Psi(\varphi_0) \,. \tag{2.4.28}$$

By Proposition 2.4.23.2 there exists $f_0 \in X^*$ such that $\Psi(\varphi) = \varphi(f_0)$, $\varphi \in Z$. It follows that

$$\varphi_0(f_0) = \Psi(\varphi_0) > \sup\{\varphi)(f_0) : \varphi \in C\} \le \sup\{\Lambda(x)(f_0) : x \in B_p\} = \|f_0|_p^* \,,$$

implying $f_0 \in X_p^\flat$. But then

$$\|f_0|_p^* < \varphi_0(f_0) \le \|\varphi_0|^{*\flat} \|f_0|_p^* \le \|f_0|_p^* \,,$$

a contradiction which shows that $B_{X_e^{*\flat}} \subset C$. □

Based on Proposition 2.4.22 one can define the reflexivity of an asymmetric normed space: an asymmetric normed space (X, p) is called *reflexive* if $\Lambda(X) = X_e^{*\flat}$. In spite of the fact that this definition imposes a strong condition on the cone $X_e^{*\flat}$, namely to be a linear space, there are many interesting examples justifying this definition, see [90].

Example 2.4.25. If the normed space (X, p^s) is reflexive, then the asymmetric normed space (X, p) is reflexive in the above sense. In particular any finite-dimensional asymmetric normed space is reflexive.

This follows from the inclusions

$$\Lambda(X) \subset X_e^{*\flat} \subset X_e^{*\flat} - X_e^{*\flat} \subset X^{**} = \Lambda(X) \,.$$

The paper [93] contains a characterization of reflexivity of an asymmetric normed space (X, p) in terms of the completeness of the unit ball B_p with respect to the weak uniformity induced by the space X^* on X that will be presented in what follows.

The locally convex topology $w_e^\flat = \sigma(X_e^{*\flat}, X^*)$ on the space $Z = X_e^{*\flat} - X_e^{*\flat}$ is generated by the uniformity $\mathcal{U}_\flat$ formed of the sets

$$\{(\varphi, \psi) \in X_e^{*\flat} \times X_e^{*\flat} : \psi - \varphi \in V\}, \quad V \in \mathcal{N}(0) \,,$$

where $\mathcal{N}(0)$ denotes the family of all $w_e^\flat$-neighborhoods of $0 \in Z$.

The uniformity $\mathcal{U}_\flat$ is generated by the sets

$$U_{F;\varepsilon} = \{(\varphi,\psi) \in X_e^{*\flat} \times X_e^{*\flat} : \forall f \in F,\ |\varphi(f) - \psi(f)| < \varepsilon\}\,, \tag{2.4.29}$$

for $F \subset X^*$ nonempty finite and $\varepsilon > 0$.

Proposition 2.4.26. *The uniformity $\mathcal{U}_\flat$ given by* (2.4.29) *is complete on $B_{X_e^{*\flat}}$.*

Proof. Consider $Z \subset \mathbb{R}^{X^*}$ and let τ denote the product topology on $(\mathbb{R}, |\cdot|)^{X^*}$. Let $(\varphi_i : i \in I)$ be a net in Z and $\varphi \in Z$. The equivalences

$$\begin{aligned}\varphi_i \xrightarrow{w_e^\flat} \varphi &\iff \forall f \in X^*, \quad \varphi_i(f) \xrightarrow{|\cdot|} \varphi(f)\\ &\iff \varphi_i \xrightarrow{\tau} \varphi\,,\end{aligned}$$

show that $w_e^\flat$ is the restriction of the product topology to Z. The product uniformity of $(\mathbb{R}, |\cdot|)^{X^*}$ is complete since each factor $(\mathbb{R}, |\cdot|)$ is complete. Consequently, the uniformity $\mathcal{U}_\flat$ will be complete on $B_{X_e^{*\flat}}$ provided that $B_{X_e^{*\flat}} = \{\varphi \in Z : \|\varphi|^{*\flat} \le 1\}$ is τ-closed in $(\mathbb{R}, |\cdot|)^{X^*}$.

Let $(\varphi_i : i \in I)$ be a net in $B_{X_e^{*\flat}}$ which is τ-convergent to some $\varphi \in (\mathbb{R}, |\cdot|)^{X^*}$, meaning that

$$\forall f \in X^*, \quad \varphi_i(f) \xrightarrow{|\cdot|} \varphi(f)\,. \tag{2.4.30}$$

The linearity of φ follows from (2.4.30), the linearity of each φ_i, and the fact that the addition and multiplication are continuous operations in $(\mathbb{R}, |\cdot|)$.

Since each φ_i belongs to $B_{X_e^{*\flat}}$,

$$\forall f \in B_p,\ \forall i \in I, \quad \varphi_i(f) \le 1\,.$$

Passing to the limit with respect to $i \in I$, one obtains

$$\forall f \in B_p,\ \varphi(f) \le 1\,,$$

showing that $\|\varphi|^{*\flat} = \sup \varphi(B_p) \le 1$, that is $\varphi \in B_{X_e^{*\flat}}$. □

Denote by w^s the topology on (X,p) induced by the dual $X^* = (X,p^s)^*$. Again this topology is generated by a uniformity $\mathcal{W}_s$ formed of the sets

$$\{(x,y) \in X \times X : x - y \in V\}, \quad V \in \mathcal{V}_s(0)\,,$$

where $\mathcal{V}_s(0)$ denotes the family of all w^s-neighborhoods of $0 \in X$.

The uniformity $\mathcal{W}_s$ is generated by the sets

$$W_{F;\varepsilon} = \{(x,y) \in X \times X : \forall f \in F,\ |f(x) - f(y)| < \varepsilon\}, \tag{2.4.31}$$

for $F \subset X^*$ nonempty finite and $\varepsilon > 0$.

Using this uniformity one can give a characterization of the reflexivity of an asymmetric normed space.

Theorem 2.4.27. *An asymmetric normed space (X,p) is reflexive if and only if the uniformity $\mathcal{W}_s$ is complete on the unit ball B_p of (X,p).*

Proof. Suppose that B_p is complete with respect to the uniformity $\mathcal{W}_s$ and let $\varphi \in B_{X_e^{*\flat}}$. By Goldstine's theorem, Theorem 2.4.24, $\Lambda(B_p)$ is $w_e^\flat$-dense in $B_{X_e^{*\flat}}$, so that for every $F \subset X^*$ nonempty finite and $\varepsilon > 0$ there exists $x_{F,\varepsilon} \in X$ such that

$$\forall f \in F, \quad |f(x_{F,\varepsilon}) - \varphi(f)| = |\Lambda(x_{F,\varepsilon})(f) - \varphi(f)| < \varepsilon\,. \tag{2.4.32}$$

The set $\mathcal{F}(X^*) \times (0;\infty)$ of all these pairs is directed with respect to the order

$$(F_1, \varepsilon_1) \leq (F_2, \varepsilon_2) \iff F_1 \subset F_2 \quad \text{and} \quad \varepsilon_2 \leq \varepsilon_1\,,$$

and (2.4.32) shows that the net $(x_{F,\varepsilon} : (F,\varepsilon) \in \mathcal{F}(X^*) \times (0;\infty))$ is $w_e^\flat$-convergent to φ.

Let us show that the net $(x_{F,\varepsilon})$ is Cauchy with respect to the uniformity $\mathcal{W}_s$. Let $F_0 \subset X^*$ finite nonempty and $\varepsilon_0 > 0$ be given. Then for every $F_1, F_2 \supset F_0$ and $\varepsilon_1, \varepsilon_2 \leq \varepsilon_0$, by the choice of $x_{F,\varepsilon}$ (see (2.4.32)) we have

$$\begin{aligned}|f(x_{F_1,\varepsilon_1}) - f(x_{F_2,\varepsilon_2})| \leq& |f(x_{F_1,\varepsilon_1}) - \varphi(f)| + |f(x_{F_2,\varepsilon_2}) - \varphi(f)| \\ <& \varepsilon_1 + \varepsilon_2 \leq 2\varepsilon_0\,,\end{aligned}$$

for all $f \in F_0$, showing that $x_{F_1,\varepsilon_1} - x_{F_2,\varepsilon_2} \in W_{F;\varepsilon}$. Consequently the net $(x_{F,\varepsilon})$ is $\mathcal{W}_s$-Cauchy, so that, by hypothesis, it is w^s-convergent to some $x \in B_p$. Since the topology w^s is Hausdorff it follows that $\varphi = \Lambda(x)$.

If $\varphi \in X_e^{*\flat} \setminus B_{X_e^{*\flat}}$, then $\psi = \varphi/\|\varphi|^* \in B_{X_e^{*\flat}}$, so that, by the first part of the proof, there exists $y \in B_p$ such that $\psi = \Lambda(y)$, implying $\varphi = \|\varphi|^*\psi$.

We have shown that $\Lambda(X) = X_e^{*\flat}$, i.e., the space $X_e^{*\flat}$ is reflexive.

Conversely, suppose that $\Lambda(X) = X_e^{*\flat}$. Then the uniform spaces $(X, \mathcal{W}_s)$ and $(\Lambda(X), \mathcal{U}_\flat)$ can be identified. By Proposition 2.4.26, $B_{\Lambda(X)}$ is $\mathcal{U}_\flat$-complete, implying that B_X is $\mathcal{W}_s$-complete. □

Remark 2.4.28. It is known that, by the Alaoglu-Bourbaki theorem, the closed unit ball $B_{X^{**}}$ of the bidual X^{**} of a a normed space X is $\sigma(X^{**}, X^*)$-compact, a result that is no longer true in the asymmetric case, as it is shown by the example of the space $(\mathbb{R}, u)$.

Indeed, $\mathrm{id} \in \mathbb{R}_u^\flat = (\mathbb{R}, u)^\flat$, $(\mathbb{R}, |\cdot|)^* = \mathbb{R}^*$, the sequence $t_n = -n$, $n \in \mathbb{N}$, is in the unit ball B_u of $(\mathbb{R}, u)$, but $|\,\mathrm{id}(-n) - \mathrm{id}(-m)| = |n - m| \geq 1$, showing that B_u is not $\sigma(\mathbb{R}_u^\flat, \mathbb{R}^*)$-compact.

2.4.6 Weak topologies on asymmetric LCS

We shall present, following [41], some properties of weak topologies on asymmetric LCS.

The topology $w^\flat$
We consider first the analog of the weak*-topology (w^*-topology) on the dual of a locally convex space. In the case of an asymmetric normed space (X,p) it was considered in [90], see Subsection 2.4.1.

Let (X,P) be an asymmetric locally convex space and $X^\flat = X^\flat_P$ the asymmetric dual cone. A $w^\flat$-neighborhood of an element $\varphi \in X^\flat$ is a subset W of $X^\flat$ for which there exist $x_1,\dots,x_n \in X$ and $\varepsilon > 0$ such that

$$V_{x_1,\dots,x_n;\varepsilon}(\varphi) := \{\psi \in X^\flat : \psi(x_i) - \varphi(x_i) < \varepsilon,\ i = 1,\dots,n\} \subset W\,.$$

The $w^\flat$-convergence of a net $\{\varphi_i, i \in I\}$ to $\varphi \in X^\flat$ is equivalent to the fact that for every $x \in X$ the net $\{(\varphi_i - \varphi)(x), i \in I\}$ converges to 0 in $(\mathbb{R}, u)$, that is

$$\forall x \in X,\ \forall \varepsilon > 0,\ \exists i_0 \in I \quad \text{such that} \quad \forall i \ge i_0, \quad (\varphi_i - \varphi)(x) < \varepsilon\,.$$

Since $X^\flat \subset X^*$ and

$$V_{x;\varepsilon} \cap V_{-x;\varepsilon}(\varphi) = \{\psi \in X^\flat : |(\psi - \varphi)(x)| < \varepsilon\}\,,$$

it follows that the $w^\flat$-topology on $X^\flat$ is induced by the w^*-topology of the space X^*.

Asymmetric polars
Let (X,P) be an asymmetric locally convex space, (X,P^s) the associated locally convex space, $X^\flat$ the asymmetric dual of (X,P) and $X^* = (X,P^s)^*$ the conjugate space of (X,P^s).

The *polar* of a nonempty subset Y of (X,P^s) is defined by

$$Y^\circ = \{x^* \in X^* : \forall y \in Y,\ x^*(y) \le 1\}\,.$$

Define the corresponding set in the case of the asymmetric dual $X^\flat$ by

$$Y^\alpha = Y^\circ \cap X^\flat = \{\varphi \in X^\flat : \forall y \in Y,\ \varphi(y) \le 1\}\,,$$

and call it the *asymmetric polar* of the set Y.

As it is well known, the set Y° is a convex w^*-closed subset of X^* (see, e.g., [238, p. 341]). Since the $w^\flat$-topology on $X^\flat \subset X^*$ is induced by the w^*-topology on X^*, we have the following result.

Proposition 2.4.29. *The asymmetric polar Y^α of a nonempty subset Y of an asymmetric locally convex space (X,P), is a convex $w^\flat$-closed subset of $X^\flat$.*

In the following proposition we prove the asymmetric analog of the Alaoglu-Bourbaki theorem, see, e.g., [238, Satz VIII.3.11].

Theorem 2.4.30. *The asymmetric polar of a neighborhood of the origin of an asymmetric locally convex space* (X, P) *is a convex* $w^\flat$*-compact subset of the asymmetric dual* $X^\flat$.

Proof. Suppose that P is directed. If V is a τ_P-neighborhood of $0 \in X$, then there exist $p \in P$ and $r > 0$ such that $B_p(0, r) \subset V$. Because $p^s(x) \le r$ implies $p(x) \le p^s(x) \le r$, it follows that $B_{p^s}(0, r) \subset B_p(0, r) \subset V$, so that V is a neighborhood of 0 in the locally convex space (X, P). By the Alaoglu-Bourbaki theorem it follows that V° is a convex w^*-compact subset of the dual X^*. Since $w^\flat$-compactness of V^α is equivalent to its w^*-compactness in X^*, it is sufficient to show that the set V^α is w^*-closed in X^*.

Let $\{\varphi_i : i \in I\}$ be a net in V^α that is w^*-convergent to $f \in X^*$. This means that for every $x \in X$ the net $\{\varphi_i(x) : i \in I\}$ converges to $f(x)$ in $(\mathbb{R}, |\cdot|)$. Since for every $v \in V$, $\varphi_i(v) \le 1$, for all $i \in I$, it follows that $f(v) \le 1$ for all $v \in V$. Because f is linear it is sufficient to prove its (P, u)-continuity at $0 \in X$. Consider for some $\varepsilon > 0$ the τ_u-neighborhood $(-\infty\,; \varepsilon)$ of $f(0) = 0 \in \mathbb{R}$. Then $U = \frac{\varepsilon}{2} V$ is a τ_P-neighborhood of $0 \in X$, and for $v \in V$ and $u = \frac{\varepsilon}{2} v \in U$ we have

$$f(u) = \frac{\varepsilon}{2} f(v) \le \frac{\varepsilon}{2} < \varepsilon\,,$$

i.e., $f(U) \subset (-\infty\,; \varepsilon)$, proving the (P, u)-continuity of f at 0.

It follows that $f \in V^\alpha$, so that V^α is w^*-closed in X^*. □

The topology w^α

The weak topology of a locally convex space (X, Q) is defined by the locally convex basis $\mathcal{W}$ formed by the sets of the form

$$V_{x_1^*,\dots,x_n^*;\varepsilon} = \{x \in X : |x_i^*(x)| < \varepsilon,\ 1 \le i \le n\}, \tag{2.4.33}$$

for $\in \mathbb{N}$, $x_1^*, \dots, x_n^* \in X^*$ and $\varepsilon > 0$. Obviously, we can suppose $x_i^* \ne 0$, $i = 1, \dots, n$.

The duality theory for locally convex spaces is based on the following key lemma of algebraic nature.

Lemma 2.4.31. ([238, Lemma VIII.3.3]) *Let* X *be a vector space and* $f, f_1, \dots, f_n : X \to \mathbb{R}$ *linear functionals. The following assertions are equivalent.*

1. $f \in \mathrm{sp}\{f_1, \dots, f_n\}$.
2. *There exists* $L \ge 0$ *such that*

$$\forall x \in X, \quad f(x) \le L \max\{f_1(x), \dots, f_n(x)\}\,.$$

3. $\bigcap_{i=1}^n \ker f_i \subset \ker f$.

In our case this lemma takes the following form.

Lemma 2.4.32. *Let $f, f_1, \ldots, f_n$ be real linear functionals on a vector space X, with $f_1, \ldots, f_n$ linearly independent. Then the following assertions are equivalent.*

1. $\forall x \in X, \quad [f_i(x) \le 0,\ i = 1, \ldots, n \ \Rightarrow\ f(x) \le 0.]$
2. $\exists L \ge 0$ *such that* $\forall x \in X,\ f(x) \le L \max\{f_i(x) : 1 \le i \le n\}$.
3. $\exists a_1, \ldots, a_n \ge 0,$ *such that* $f = \sum_{i=1}^n a_i f_i$.

Proof. Since the implications 2) $\Rightarrow$ 1) and 3) $\Rightarrow$ 2) are obvious, it is sufficient to prove 1) $\Rightarrow$ 3).

If $f_i(x) = 0$ for $i = 1, \ldots, n$, then $f_i(-x) = -f_i(x) = 0,\ i = 1, \ldots, n$, so that $f(x) \le 0$ and $-f(x) = f(-x) \le 0$, implying $f(x) = 0$. Therefore the condition 3) from Lemma 2.4.31 is fulfilled, so that there exist $a_1, \ldots, a_n \in \mathbb{R}$ such that $f = \sum_{i=1}^n a_i f_i$. It remains to show that $a_j \ge 0$ for $j = 1, \ldots, n$. Because $f_1, \ldots, f_n$ are linearly independent, there exist the elements $x_j \in X$ such that $f_i(x_j) = -\delta_{ij} \le 0,\ i, j = 1, 2, \ldots, n,$ where δ_{ij} is the Kronecker symbol. It follows that $f(x_j) \le 0$ and

$$-a_j = \sum_{i=1}^n a_i f_i(x_j) = f(x_j) \le 0\,,$$

for $j = 1, \ldots, n$. □

Let (X, P) be an asymmetric locally convex space and $X^\flat = (X, P)^\flat$ its asymmetric dual cone.

Define the asymmetric weak topology w^α on an asymmetric locally convex space (X, P) as the asymmetric locally convex topology generated by the asymmetric locally convex base $\mathcal{W}_\alpha$ formed of the sets

$$V_{\varphi_1, \ldots, \varphi_n; \varepsilon} = \{x \in X : \varphi_i(x) < \varepsilon,\ 1 \le i \le n\}, \tag{2.4.34}$$

for $n \in \mathbb{N}$, $\varphi_1, \ldots, \varphi_n \in X^\flat$ and $\varepsilon > 0$. The neighborhoods of an arbitrary point $x \in X$ are subsets of X containing a set of the form $x + V_{\varphi_1, \ldots, \varphi_n; \varepsilon} = \{x' \in X : \varphi_i(x' - x) < \varepsilon,\ 1 \le i \le n\}$.

The sets

$$V^-_{\varphi_1, \ldots, \varphi_n; \varepsilon} = \{x \in X : \varphi_i(x) \le \varepsilon,\ 1 \le i \le n\}$$

generate the same topology.

In the following proposition we collect some properties of the topology w^α.

Proposition 2.4.33. *Let (X, P) be an asymmetric locally convex space and $X^\flat = (X, P)^\flat$ its asymmetric dual cone.*

1. *The topology τ_P is finer than w^α.*
2. *For $\varphi \in X^\flat$ and $\varepsilon > 0$ the set $\{x \in X : \varphi(x) < \varepsilon\}$ is w^α-open and $\{x \in X : \varphi(x) \ge \varepsilon\}$ is w^α-closed.*

3. *A net* $\{x_i : i \in I\}$ *in* X *is* w^α*-convergent to* $x \in X$ *if and only if for every* $\varphi \in X^\flat$ *the net* $\{\varphi(x_i\}$ *converges to* $\varphi(x)$ *in* $(\mathbb{R}, u)$*. This means the following:*

$$\forall \varphi \in X^\flat,\ \forall \varepsilon > 0,\ \exists i_0 \text{ such that } \forall i \geq i_0,\ \varphi(x_i - x) < \varepsilon\,.$$

4. *The asymmetric dual* $(X, w^\alpha)^\flat$ *of the asymmetric locally convex space* (X, w^α) *agrees with* $X^\flat$.

Proof. Suppose P is directed.

1. Let $V = V_{\varphi_1,\dots,\varphi_n;\varepsilon}$ be an element of the locally convex basis (2.4.34). Because φ_i are (P, u)-continuous there exist $p_i \in P$ and $L_i \geq 0$ such that

$$\forall x \in X, \quad \varphi_i(x) \leq L_i p_i(x), \text{ for } i = 1, \dots, n\,.$$

The multiball $U = \{x \in X : p_i(x) < \varepsilon/(L+1),\ 1 \leq i \leq n\}$, where $L = \max L_i$, is contained in V, showing that V is a τ_P-neighborhood of $0 \in X$.

2. If $V = \{x \in X : \varphi(x) < \varepsilon\}$ and $x_0 \in V$, then the w^α-neighborhood $\{x \in X : \varphi(x - x_0) < \varepsilon - \varphi(x_0)\}$ of x_0 is contained in V because

$$\varphi(x - x_0) < \varepsilon - \varphi(x_0) \ \Rightarrow\ \varphi(x) = \varphi(x - x_0) + \varphi(x_0) < \varepsilon\,.$$

The assertion 3 follows from definitions.

4. Because τ_P is finer than w^α, the identity map Id: $(X, \tau_P) \to (X, w^\alpha)$ is continuous, implying the (P, u)-continuity of $\varphi\circ\mathrm{Id}$ for any $\varphi \in (X, w^\alpha)^\flat$, i.e., $(X, w^\alpha)^\flat \subset (X, P)^\flat$.

Conversely, if φ is a (P, u)-continuous linear functional, then the set $V = \{x \in X : \varphi(x) < \varepsilon\}$ is a w^α-neighborhood of $0 \in X$ and $\varphi(V) \subset (-\infty; \varepsilon)$ for every $\varepsilon > 0$, proving the (w^α, τ_u)-continuity of φ at 0, and by the linearity of φ, on the whole of X. □

As in the symmetric case the closed convex sets are the same for the topologies τ_P and w^α.

Proposition 2.4.34. *Let* (X, P) *be an asymmetric locally convex space and* Y *a convex subset of* X.

Then Y *is* w^α*-closed if and only if it is* τ_P*-closed.*

Proof. Because τ_P is finer than w^α, it follows that any (not necessarily convex) w^α-closed subset of X is also τ_P-closed.

Suppose now that the convex set Y is τ_P but not w^α-closed. If x_0 is a point in w^α-cl$Y \setminus Y$, then, applying Theorem 2.2.9 to the sets $\{x_0\}$ and Y we get a functional $\varphi \in X^\flat$ such that

$$\varphi(x_0) < \inf \varphi(Y)\,.$$

If $m := \inf \varphi(Y)$, then $V = \{x \in X : \varphi(x - x_0) < 2^{-1}(m - \varphi(x_0))\}$ is a w^α-neighborhood of x_0. Because

$$\varphi(x) = \varphi(x - x_0) + \varphi(x_0) < \frac{m + \varphi(x_0)}{2} < m ,$$

for every $x \in V$, it follows that $V \cap Y = \emptyset$, in contradiction to $x_0 \in w^\alpha\text{-cl}\, Y$. □

The proposition has the following corollary.

Corollary 2.4.35. *Let (X, P) be an asymmetric locally convex space. Then for every subset Z of X the following equality holds:*

$$w^\alpha\text{-cl co}(Y) = \tau_P\text{-cl co}(Y) .$$

Proof. By the definition of the closed convex hull and the preceding proposition we have the equalities

$$\begin{aligned} w^\alpha\text{-cl co}(Y) &= \bigcap\{Y : Y \subset X,\ Y \text{ convex and } w^\alpha\text{-closed}\} \\ &= \bigcap\{Y : Y \subset X,\ Y \text{ convex and } \tau_P\text{-closed}\} \\ &= \tau_P\text{-clco}(Y) . \end{aligned}$$

□

Remark 2.4.36. We can define the asymmetric polar of a subset W of the dual $X^\flat$ of an asymmetric locally convex space (X, P) by

$$W_\alpha = \{x \in X : \forall \varphi \in W,\ \varphi(x) \le 1\} .$$

Since, for $\varphi \in X^\flat$, a set of the form $\{x \in X : \varphi(x) \le 1\}$ is not necessarily τ_P-closed, the set W_α need not be τ_P-closed. Therefore an asymmetric analog of the bipolar theorem (see [238, Satz WIII.3.9]), asserting that

$$(A^\circ)_\circ = \text{cl-co}(A \cup \{0\}) ,$$

for any subset A of a locally convex space (X, Q), does not hold in the asymmetric case.

2.4.7 Asymmetric moduli of rotundity and smoothness

A convex body in a normed space $(X, \|\cdot\|)$ is a bounded closed convex set with nonempty interior. A *rooted convex body* in a normed space X is a pair (K, z) where K is a convex body and z is a fixed point in the interior K, called a *root* for K. If (K, z) is a rooted convex body, then $0 \in \text{int}(K - z) = \text{int}(K) - z$, so that the Minkowski functional p_{K-z} is well defined and it is an asymmetric norm on X satisfying $p_{K-z}(x) > 0$ for $x \neq 0$. The topology defined by p_{K-z} is equivalent to the norm-topology $\tau_{\|\cdot\|}$ of X. This follows from the facts that there exists $0 < r < R$

such that $B_{\|\cdot\|}(0,r) \subset K - z \subset B_{\|\cdot\|}(0,R)$ and $K - z = \{x \in X : p_{K-z}(x) \le 1\}$ is the closed unit ball of the asymmetric normed space (X, p_{K-z}).

Starting from this definition one can introduce geometric properties of convex bodies, as rotundity, local uniform rotundity, smoothness, uniform smoothness, expressed in terms of Minkowski functionals, by analogy with those from Banach space geometry (see, for instance, the book by Megginson [149]). This was done in a series of papers (see [114, 115, 116]) by V. Klee, E. Maluta, C. Zanco and L. Vesely, mainly in connection with the existence of good tilings of normed spaces. A *tiling* of a normed linear space X is a covering $\mathcal{T}$ of X such that each *tile* (member of $\mathcal{T}$) is a convex body and no point of X is interior to more than one tile. In contrast to the genuine theory of tilings in finite-dimensional spaces and an extensive theory for the plane, few significant examples were known and not even a rudimentary theory of tilings in infinite-dimensional spaces was built, see [113]. In the above-mentioned papers substantial progress was made in the study of tilings in infinite-dimensional Banach spaces. Zanco and Zucchi [240] defined and studied some properties of the asymmetric moduli of smoothness, quantitative expressions (in terms of Minkowski functionals) of the rotundity and smoothness properties of convex bodies.

Some uniform geometric properties for families of convex bodies were defined in [115]. We shall restrict these properties to a single convex body, a situation that fits better our needs. A convex body K in a normed space $(X, \|\cdot\|)$ is called *smooth* if for any point $y \in \partial K$ (the boundary of K) there exists exactly one hyperplane $H_y = \{x \in X : x^*(x) = c\}$, for some $x^* \in X^*$ and $c \in \mathbb{R}$, supporting K at y. That is $x^*(y) = c$ and $x^*(x) \le c$ for all $x \in K$. The convex body K is called *rotund* if its boundary ∂K does not contain nontrivial line segments, or, equivalently, if $d(\frac{1}{2}(x+y), \partial K) > 0$ (the distance with respect to $\|\cdot\|$) for every pair of distinct elements $x, y \in \partial K$. The convex body K is called *uniformly rotund* if for every $\varepsilon > 0$ there exists $\delta(\varepsilon) > 0$ such that $d(\frac{1}{2}(x+y), \partial K) \ge \delta(\varepsilon)$ for every $x, y \in K$ with $\|x - y\| \ge \varepsilon$. It is obvious that one obtains the same notion if we require that $x, y \in \partial K$.

A rooted convex body (K, z) is smooth if and only if the Minkowski functional $p = p_{K-z}$ is Gâteaux differentiable on $X \setminus \{z\}$. This means that for any point $x \in X \setminus \{z\}$ there exists a continuous linear functional $p'(x;\cdot) \in X^*$ such that, for every $h \in X$,

$$\lim_{t\to 0} \frac{p(x+th) - p(x)}{t} = p'(x;h) . \tag{2.4.35}$$

With a rooted smooth convex body (K, z) one can associate two *duality mappings* from ∂K to X^* defined for $y \in \partial K$ as follows:

- $J_{(K,z)}(y) =$ the unique functional $x^* \in X^*$ such that $\|x^*\| = 1$ and $H_y - z = \{x \in X : x^*(x) = c\}$, for some $c \in \mathbb{R}$, and
- $J^1_{(K,z)}(y) =$ the unique functional $x^* \in X^*$ such that $H_y - z = \{x \in X : x^*(x) = 1\}$.

If $K = B_X$ (the closed unit ball of $(X, \|\cdot\|)$) and $z = 0$, then $J = J^1$ is the ordinary duality mapping, a very important notion in the geometry of Banach spaces and its applications to nonlinear operator theory, see, for instance, the book by Ciorănescu [36]. Also, it is clear that $J_{K,z_1} = J_{K,z_2}$ for every $z_1, z_2 \in \operatorname{int} K$, so that the duality mapping J_{K,z_1} does not depend on the root z, consequently it can be denoted simply by J_K.

A normed space $(X, \|\cdot\|)$ is called *uniformly smooth* if the norm is uniformly Fréchet differentiable on the unit sphere S_X. This means that, putting $f(x) = \|x\|$, $x \in X$, for every $\varepsilon > 0$ there exists $\delta = \delta(\varepsilon) > 0$ such that for every $x \in S_X$ and every $h \in X$ with $\|h\| \le \delta$

$$|f(x+h) - f(x) - f'(x)h| \le \varepsilon\|h\| ,$$

where $f'(x) \in X^*$ denotes the Fréchet derivative of f at x. The modulus of smoothness of the space X is defined for $\tau \ge 0$ by

$$\rho_X(\tau) = \sup\left\{\frac{1}{2}\left(\|x+y\| + \|x-y\| - 2\right) : x \in S_X,\ \|y\| = \tau\right\} . \tag{2.4.36}$$

The uniform smoothness of X can be characterized by the condition

$$\lim_{\tau \searrow 0} \frac{\rho_X(\tau)}{\tau} = 0 ,$$

see, e.g., [149]. Another characterization can be done in terms of the duality mapping J, namely, the space X is uniformly smooth if and only if the duality mapping $J : S_X \to S_{X^*}$ is norm-to-norm continuous (see [36]).

Starting from this property one can define the *uniform smoothness* of a rooted convex body (K, z) by asking that the duality mapping $J_{K,z} : \partial K \to S_{X^*}$ is norm-to-norm continuous (see [115]). The norm-to-norm continuity of the duality mapping $J_{K,z}$ is equivalent to the norm-to-norm continuity of the duality mapping $J^1_{K,z}$, so that one obtains the same notion of uniform smoothness by working with the duality mapping $J^1_{K,z}$.

A normed space $(X, \|\cdot\|)$ is called *uniformly rotund* (*uniformly convex* by some authors) if for every $\varepsilon > 0$ there exists $\delta = \delta(\varepsilon) > 0$ such that $\|x+y\| \ge 2(1-\delta)$ for every $x, y \in S_X$ with $\|x-y\| \le \varepsilon$.

This property admits also quantitative characterizations in terms of some moduli. We mention two of them.

Clarkson's modulus of uniform rotundity $\delta_X : [0;2] \to [0;1]$ defined by

$$\begin{aligned} \delta_X(\varepsilon) &= \inf\left\{1 - \frac{1}{2}\|x+y\| : x, y \in S_X,\ \|x-y\| \ge \varepsilon\right\} \\ &= \inf\left\{1 - \frac{1}{2}\|x+y\| : x, y \in S_X,\ \|x-y\| = \varepsilon\right\} . \end{aligned} \tag{2.4.37}$$

Gurarii's modulus of uniform rotundity $\gamma_X : [0;2] \to [0;1]$ defined by

$$\gamma_X(\varepsilon) = \inf\{\max_{0\le t\le 1}(1 - \|tx + (1-t)y\|) : x, y \in S_X,\ \|x - y\| = \varepsilon\}\,. \tag{2.4.38}$$

One has $\delta_X \le \gamma_X$, and a 2-dimensional example given in [240] shows that the inequality can be strict. Also it is unknown whether the condition $\|x-y\| = \varepsilon$ in the definition of Gurarii's modulus can be replaced by the condition $\|x - y\| \ge \varepsilon$, as in the definition of Clarkson's modulus. The uniform rotundity of the space X can be characterized in the following way:

$$X \text{ is uniformly rotund} \iff \forall\varepsilon \in (0;2],\ \delta_X(\varepsilon) > 0 \iff \forall\varepsilon \in (0;2],\ \gamma_X(\varepsilon) > 0\,.$$

The *characteristic of convexity* corresponding to the moduli δ_X and γ_X are $\varepsilon_X^0 = \sup\{\varepsilon \in [0;2] : \delta_X(\varepsilon) = 0\} = \{\varepsilon \in [0;2] : \gamma_X(\varepsilon) = 0\}$. In terms of this characteristic of convexity the uniform rotundity of X is characterized by the condition $\varepsilon_X^0 = 0$ and the rotundity by the condition $\delta_X(2) = \gamma_X(2) = 1$. Other geometric properties of the normed space X, as, for instance, the uniform non-squareness can also be expressed in terms of these moduli.

The analogs of these moduli (Gurarii's variant in the case of uniform rotundity) and their relevance for the smoothness and rotundity properties of rooted convex bodies were given in the paper by Zanco and Zucchi [240].

The modulus of smoothness of a rooted convex body (K, z), $\rho_{(K,z)} : [0;\infty) \to [0;1)$ can be defined by replacing the norm in (2.4.36) by the Minkowski functional:

$$\begin{aligned}\rho_{(K,z)}(\tau) = \sup\Big\{\frac{1}{2}(p_{K-z}(x+y) + p_{K-z}(x-y)) - 1 :&\\ x \in \partial(K - z),\ y \in X, p_{K-z}(y) = \tau\Big\}\,.&\end{aligned} \tag{2.4.39}$$

For the sake of simplicity suppose $z = 0$ and put $\rho_K = \rho_{(K,0)}$. The modulus of smoothness is a continuous convex function such that $\rho_K(\tau) \le \lambda_K\tau$, where

$$\lambda_K = \sup\{p_K(-y) : y \in K\} = \sup\{p_{_K\cap K}(y) : y \in K\}\,,$$

could be taken as a possible measure of eccentricity of K with respect to 0. By analogy with the normed case, the asymmetric norm p_K is called K-uniformly Fréchet differentiable if (2.4.35) holds uniformly with respect to $h \in K$ and $x \in \partial K$. One shows ([240, Th. 4.2]) that the following are equivalent:

- the rooted convex body (K, z) is uniformly smooth;
- the Minkowski functional p_{K-z} is K-uniformly Fréchet differentiable on ∂K;
- $\lim_{\tau\searrow 0}\big(\rho_{(K,z)}(\tau)/\tau\big) = 0$ for every $z \in \operatorname{int} K$.

The definition of the modulus of uniform rotundity is more involved and needs to consider the existence of diametral points in K. The *Minkowski diameter* (M-diameter) of a rooted convex body (K, z) is defined by

$$\operatorname{diam}_M(K, z) = \sup\{p_{K-z}(x - y) : x, y \in K\}\,. \tag{2.4.40}$$

A pair of points $x, y \in K$ is called *M-diametral* if

$$\operatorname{diam}_M(K, z) = \max\{p_{K-z}(x - y), p_{K-z}(y - x)\}.$$

The problem of the existence of diametral points of convex bodies with respect to the norm was studied by Garkavi [98] in connection with some minimax and maximin problems. The center of the largest ball (whose radius is denoted by r_K) contained in a convex body K is called an H-center for K, while the center of the smallest ball containing K is called a *Chebyshev center* of K. Both centers could not exist, and if they exist they need not be unique. If x_0 is an H-center for K, then the set $A_K(x_0) = \{y \in K : \|x^0 - y\| = r_K\}$ is called the critical set of the H-center x_0. Garkavi, *loc. cit.*, proved that a Banach space X is reflexive if and only if every convex body in X has an H-center. Also, the space X is finite dimensional if and only if for any convex body K in X and any H-center x_0 of K the set $A_K(x_0)$ of critical points is nonempty.

By a compactness argument it follows that if X is finite dimensional, then every rooted convex body admits M-diametral points. In the case of M-diametral points the authors show in [240] that, for every $p \in [1; \infty)$, the space ℓ_p contains a rooted convex body $(K, 0)$ whose M-diameter is not attained. The general problem of the validity of Garkavi's results for M-diameters remains open, with the conjecture that the existence of M-diametral points for every rooted convex body is equivalent to the finite dimensionality of X. For a rooted convex body (K, z) in a normed space let

$$\Delta_{(K,z)}(\varepsilon) = \inf\left\{ \max_{t\in[0;1]} (1 - p_{K-z}(tx + (1-t)y)) : x, y \in K - z,\, p_{K-z}(x - y) \geq \varepsilon \right\}. \tag{2.4.41}$$

If (K, z) has diametral points, then $\Delta_{(K,z)}$ is defined on $[0; \operatorname{diam}_M(K, z)]$ with values in $[0; 1]$ and, if (K, z) does not have diametral points, then $\Delta_{(K,z)}$ is defined on $[0; \operatorname{diam}_M(K, z))$ with values in $[0; 1)$.

Based on this notion, one defines the modulus of uniform rotundity $\gamma_{(K,z)}$ of a rooted convex body (K, z) by the conditions:

- $\gamma_{(K,z)}(\varepsilon) = \Delta_{(K,z)}(\varepsilon), \quad$ for $0 \leq \varepsilon < \operatorname{diam}_M(K, z)$,
- $\gamma_{(K,z)}(\operatorname{diam}_M(K, z)) = \Delta_{(K,z)}(\operatorname{diam}_M(K, z)), \quad$ if $\dim X = 2$,

and

- $\gamma_{(K,z)}(\operatorname{diam}_M(K, z)) = \inf\{\gamma_{(Y\cap(K-z),0)}(\operatorname{diam}_M(Y \cap (K - z), 0)) :$ Y is a 2-dimensional subspace of $X\}$, in general.

The last formula is justified by the fact that a similar result holds for both moduli of uniform rotundity δ_X and γ_X.

One shows that the function $\gamma_{(K,z)}$ is continuous on some interval $[0; \beta)$, where the number $\beta > 0$ depends on the geometric properties of the convex body K, expressed in terms of the so-called directional M-diameters of K. Also the convex body K is rotund if and only if $\inf\{\gamma_{(K,z)}(\operatorname{diam}_M(K, z)) : z \in \operatorname{int} K\} = 1\}$, and is uniformly rotund if and only if $\varepsilon^0_{(K,z)} = 0$ for every $z \in \operatorname{int} K$, where $\varepsilon^0_{(K,z)} = \sup\{\varepsilon \in [0; \operatorname{diam}_M(K,z)) : \gamma_{(K,z)}(\varepsilon) = 0\}$ is the characteristic of convexity corresponding to the asymmetric modulus $\gamma_{(K,z)}$.

If a Banach space X contains a rooted convex body $(K, 0)$ with $\varepsilon^0_{(K,0)} < 2$, then X is superreflexive. Recall that a Banach space X is called superreflexive if it admits an equivalent uniformly rotund renorming (it is known that any uniformly rotund Banach space is reflexive).

If $0 \in \operatorname{int} K$, then one defines the polar set of K by $K^\pi = \{x^* \in X^* : \forall y \in K,\ x^*(y) \le 1\}$. If B_X is the unit ball of a normed space $(X, \|\cdot\|)$, then B_X^π is the unit ball of the dual space. Based on Corollary 2.2.4 from the next section, it follows that this relation is also true in the asymmetric case: $B_{X^\flat} = B_p^\pi$. The well-known duality relation between uniform smoothness (US) and uniform rotundity (UR) holds in this case too:

- the convex body K is UR $\iff$ the polar set K^π is US.

Remark 2.4.37. The results presented above stand in a normed space. It would be of interest to study these properties in an asymmetric normed space, and to see their significance for the properties of the corresponding asymmetric normed space. For instance, is there any connection between the asymmetric uniform rotundity and the reflexivity of the asymmetric normed space (as defined in Subsection 2.4.5), like in the case of Banach spaces?

The approximation properties of subsets of a Banach space heavily depend on the geometric properties of the underlying space, see, for instance, the survey [42]. It would be interesting to see to what extent can these properties be extended to asymmetric normed spaces?

2.5 Applications to best approximation

The aim of this section is to study best approximation in asymmetric normed spaces. Due to the asymmetry of the norm, two kind of distances from a point to a set have to be considered, exemplified on Ascoli's formula for the distance to a closed hyperplane. Some characterization and duality results for best approximation by elements in closed convex sets and by elements in sets with convex bounded complement are proved.

As it is known, the natural framework for treating the problem of best approximation is that of normed spaces, see the books by Singer [222, 223], so that it is very natural to consider the corresponding problem in asymmetric normed spaces. Some problems of best approximation with respect to an asymmetric norm, including approximation in spaces of continuous or integrable functions, were considered by Duffin and Karlovitz [70]. Dunham [71] treated the problem of best approximation by elements of a finite-dimensional subspace of an asymmetric normed normed space and proved existence results, uniqueness results (guaranteed by the rotundity of the asymmetric norm), and found some conditions ensuring the continuity of the metric projection. Pfankuche-Winkler [176] considered the best approximation problem in some asymmetric normed spaces of Orlicz type. De Blasi and Myjak [59] proved some generic existence results for the problem of best approximation with respect to an asymmetric norm in a Banach space. Similar problems were considered by Li and Ni [146] and Ni [167].

As it is well known, any closed convex subset of a Hilbert space is Chebyshev. A famous problem in best approximation theory is that of the convexity of Chebyshev sets: must any Chebyshev subset of a Hilbert space be convex? There are a lot of results in this direction as presented, for instance, in the survey paper by Balaganskii and Vlasov [20], but the general problem is still unsolved. In some of the papers dealing with this problem one works with asymmetric norms as, for instance, in Alimov [12].

2.5.1 Characterizations of nearest points in convex sets and duality

As in the normed case, linear functionals are useful in characterizing the nearest points and for the duality results in best approximation in asymmetric normed spaces. In the following we shall present some results obtained in the papers [40, 48, 49]. Let (X,p) be an asymmetric normed space, Y a nonempty subset of X and $x \in X$. Recall that, due to the asymmetry of the norm, we have to consider two *distances* from x to Y:

$$\text{(i)}\ d_p(x,Y) = \inf\{p(y-x) : y \in Y\} \quad \text{and} \quad \text{(ii)}\ d_p(Y,x) = \inf\{p(x-y) : y \in Y\}\,. \tag{2.5.1}$$

Observe that $d_p(Y,x) = d_{\bar{p}}(x,Y)$, where $\bar{p}$ is the norm conjugate to p. Let also

$$P_Y(x) = \{y \in Y : p(y-x) = d_p(x,Y)\}$$

and

$$\bar{P}_Y(x) = \{y \in Y : p(x-y) = d_p(Y,x)\}\,,$$

denote the metric projections on Y. An element y in $P_Y(x)$ is called a *p-nearest point* to x in Y, while an element $\bar{y}$ in $\bar{P}_Y(x)$ is called a *$\bar{p}$-nearest point* to x in Y.

The set Y is called:

- *p-proximinal* if $P_Y(x) \neq \emptyset$ for every $x \in X$,
- *p-semi-Chebyshev* if $\#P_Y(x) \leq 1$ for every $x \in X$ (i.e., every $x \in X$ has at most one p-nearest point in Y),
- *p-Chebyshev* if $\#P_Y(x) = 1$ for every $x \in X$ (i.e., every $x \in X$ has exactly one p-nearest point in Y).

The corresponding notions for the conjugate norm $\bar{p}$ are defined similarly.

A consequence of Theorem 2.2.6 is the following characterization of nearest points.

Theorem 2.5.1. *Let (X,p) be a space with asymmetric norm, Y a subspace of X and x_0 a point in X.*

1. *Let $d = d_p(x_0, Y) > 0$. An element $y_0 \in Y$ is a p-nearest point to x_0 in Y if and only if there exists a p-bounded linear functional $\varphi : X \to \mathbb{R}$ such that*
$$\text{(i) } \varphi|_Y = 0, \quad \text{(ii) } \|\varphi|_p = 1, \quad \text{(iii) } \varphi(-x_0) = p(y_0 - x_0)\,.$$
2. *Let $\bar{d} = d_{\bar{p}}(x_0, Y) > 0$. An element $y_1 \in Y$ is a $\bar{p}$-nearest point to x_0 in Y if and only if there exists a p-bounded linear functional $\psi : X \to \mathbb{R}$ such that*
$$\text{(j) } \psi|_Y = 0, \quad \text{(jj) } \|\psi|_p = 1, \quad \text{(jjj) } \psi(x_0) = p(x_0 - y_1)\,.$$

Proof. 1. Suppose that $y_0 \in Y$ is such that $p(y_0 - x_0) = d = d_p(x_0, Y) > 0$. By Theorem 2.2.6, there exists $\varphi \in X_p^\flat$, $\|\varphi|_p = 1$, such that $\varphi|_Y = 0$ and $\varphi(-x_0) = d = p(y_0 - x_0)$.

Conversely, if for $y_0 \in Y$ there exists $\varphi \in X^\flat$ satisfying the conditions (i)–(iii), then for every $y \in Y$,
$$p(y - x_0) \geq \varphi(y - x_0) = \varphi(y_0 - x_0) = p(y_0 - x_0)\,,$$
implying $p(y_0 - x_0) = d_p(x_0, Y)$.

The second assertion can be proved in a similar way. □

For a nonempty subset Y of an asymmetric normed space (X,p) put
$$Y^\perp = Y_p^\perp = \{\varphi \in X_p^\flat : \varphi|_Y = 0\}\,.$$

A consequence of Eidelheit's Separation Theorem (Theorem 2.2.8) is the following duality result for best approximation by elements of convex sets in asymmetric normed spaces. These extend results obtained in the case of normed spaces by Nikolski [169], Garkavi [96, 97], Singer [221] (see also Singer's book [222, Appendix I] and [99]). Some duality results in the asymmetric case were proved also by Babenko [19]. The case of so-called p-convex sets was considered in [47] and [37, 38].

Theorem 2.5.2. *For a nonempty convex subset Y of a space (X,p) with asymmetric norm and $x_0 \in X$, the following duality relations hold:*
$$d_p(x_0, Y) = \sup_{\|\varphi|_p \leq 1} \inf_{y \in Y} \varphi(y - x_0) \tag{2.5.2}$$
and
$$d_p(Y, x_0) = \sup_{\|\varphi|_p \leq 1} \inf_{y \in Y} \varphi(x_0 - y). \tag{2.5.3}$$

If $d_p(x_0, Y) > 0$, then there exists $\varphi_0 \in X_p^\flat$, $\|\varphi_0|_p = 1$, such that $d_p(x_0, Y) = \inf\{\varphi_0(y - x_0) : y \in Y\}$, i.e., the supremum in the right-hand side of the relation (2.5.2) is attained.

A similar result holds for the second duality relation.

Proof. We shall prove first the relation (2.5.2) and obtain (2.5.3) as an immediate consequence. Let $i = d_p(x_0, Y)$ and denote by s the quantity in the right-hand side of the relation (2.5.2).

For any $\varphi \in X_p^\flat$ with $\|\varphi|_p \leq 1$ we have

$$\forall y \in Y,\ \varphi(y - x_0) \leq p(y - x_0)\ ,$$

implying

$$\forall \varphi \in X_p^\flat \text{ with } \|\varphi|_p \leq 1, \quad \inf\{\varphi(y - x_0) : y \in Y\} \leq i\ ,$$

so that $s \leq i$. Taking $\varphi = 0$ in the definition of s it follows that $s \geq 0$, so that $s = i = 0$ if $i = 0$.

Suppose now $i > 0$ and let

$$Z := \{x \in X : p(x - x_0) < i\}\ .$$

It follows that Z is nonempty, convex, τ_p-open and $Z \cap Y = \emptyset$. By the first separation theorem (Theorem 2.2.8), there exists $\psi \in X_p^\flat$ such that

$$\forall z \in Z\ \forall y \in Y, \quad \psi(z) < \psi(y)\ .$$

Putting $\varphi = (1/\|\psi|_p)\psi$ we have $\|\varphi|_p = 1$ and

$$\forall z \in Z\ \forall y \in Y, \quad \varphi(z - x_0) < \varphi(y - x_0). \tag{2.5.4}$$

Since

$$\sup\{\varphi(z - x_0) : z \in Z\} = \sup\{\varphi(w) : p(w) < i\} = i\|\varphi|_p = i\ ,$$

the inequality (2.5.4) yields

$$i = \sup\{\varphi(z - x_0) : z \in Z\} \leq \inf\{\varphi(y - x_0) : y \in Y\} \leq s\ ,$$

so that $s = i$.

Let us show now that the relation (2.5.3) follows from (2.5.2). Since the relation (2.5.3) holds for $\bar{p}$ too, by Proposition 2.1.7.5 we can write

$$\begin{aligned}
\inf_{y \in Y} p(x_0 - y) &= \sup\{\inf \psi(Y - x_0) : \psi \in X_{\bar{p}}^\flat,\ \|\psi|_{\bar{p}} \leq 1\} \\
&= \sup\{\inf(-\psi)(x_0 - Y) : -\psi \in X_p^\flat,\ \| - \psi|_p \leq 1\} \\
&= \sup\{\inf \varphi(x_0 - Y) : \varphi \in X_p^\flat,\ \|\varphi|_p \leq 1\}\ ,
\end{aligned}$$

showing that the relation (2.5.3) holds too.

Finally, suppose $i > 0$ and let (φ_n) be a sequence in the unit ball $B_{X_p^\flat}$ of $X_p^\flat$ such that $\lim_n \inf\{\varphi_n(y - x_0) : y \in Y\} = s$ or, equivalently,

$$\lim_n (\inf \varphi_n(Y) - \varphi_n(x_0)) = s. \tag{2.5.5}$$

Since the ball $B_{X_p^\flat}$ is a w^*-compact subset of B_{X^*} (see Proposition 2.4.2), it follows that the sequence (φ_n) contains a subnet $(\psi_j : j \in J)$ which is w^*-convergent to an element $\varphi_0 \in B_{X_p^\flat}$. It follows that

$$\forall z \in X, \quad \lim_j \psi_j(z) = \varphi_0(z) .$$

Then, by (2.5.5), $\lim_j \inf \psi_j(Y) = \varphi_0(x_0) + s$. Since, for every $y \in Y$, $\psi_j(y) \geq \inf \psi_j(Y)$, by passing to the limit for $j \in J$ we get

$$\forall y \in Y, \quad \varphi_0(y) \geq \varphi_0(x_0) + s ,$$

so that

$$s \leq \inf \varphi_0(Y) - \varphi_0(x_0) = \inf \varphi_0(Y - x_0) .$$

The definition of s implies $s = \inf \varphi_0(Y - x_0)$.

It remains to show that $\|\varphi_0|_p = 1$. If $\|\varphi_0|_p < 1$, then $\lambda = 1/\|\varphi_0|_p > 1$ and the functional $\psi_0 = \lambda\varphi_0$ satisfies $\|\psi_0|_p = 1$ and

$$s \geq \inf \psi_0(Y - x_0) = \lambda \inf \varphi_0(Y - x_0) = \lambda s ,$$

a contradiction, since $s = i > 0$. Observe that $s > 0$ implies $\varphi_0 \neq 0$, so that $\|\varphi_0|_p > 0$ and λ is properly defined. □

Based on this duality result one obtains the following characterization of nearest points.

Theorem 2.5.3. *Let (X, p) be a space with asymmetric norm, Y a nonempty subset of X, $x \in X$, and $y_0 \in Y$.*

If there exists a functional $\varphi_0 \in X_p^\flat$ such that

$$\text{(i) } \|\varphi_0|_p = 1, \quad \text{(ii) } \varphi_0(y_0 - x) = p(y_0 - x), \quad \text{(iii) } \varphi_0(y_0) = \inf \varphi_0(Y) , \tag{2.5.6}$$

then y_0 is a p-nearest point to x in Y.

Similarly, if for $z_0 \in Y$, there exists a functional $\psi_0 \in X_p^\flat$ such that

$$\text{(j) } \|\psi_0|_p = 1, \quad \text{(jj) } \psi_0(x - z_0) = p(x - z_0), \quad \text{(jjj) } \psi_0(z_0) = \sup \psi_0(Y) , \tag{2.5.7}$$

then z_0 is a $\bar{p}$-nearest point to x in Y.

Conversely, if Y is convex, $d_p(x, Y) > 0$, an y_0 is a p-nearest point to x in Y, then there exists a functional $\varphi_0 \in X_p^\flat$ satisfying the conditions (i)–(iii) *from above.*

Similarly, if $z_0 \in Y$ is a $\bar{p}$-nearest point to x in Y with $p(x - z_0) > 0$, then there exists a functional $\psi_0 \in X_p^\flat$ satisfying the conditions (j)–(jjj).

Proof. Suppose that $\varphi_0 \in X_p^\flat$ satisfies the conditions (i)–(iii). Then for every $y \in Y$,

$$p(y_0-x) = \varphi_0(y_0-x) = \varphi_0(y_0)-\varphi_0(x) = \inf \varphi_0(Y)-\varphi_0(x) \le \varphi_0(y-x) \le p(y-x) ,$$

showing that

$$p(y_0 - x) = \inf\{p(y - x) : y \in Y\} = d_p(x, Y) .$$

If $z_0 \in Y$ and $\psi_0 \in X_p^\flat$ are such that the conditions (j)–(jjj) hold, then

$$p(x-z_0) = \psi_0(x-z_0) = \psi(x)-\sup \psi_0(Y) = \inf \psi_0(x-Y) \le \psi_0(x-y) \le p(x-y) ,$$

for all $y \in Y$.

Suppose now that Y is convex and y_0 is a p-nearest point to x in Y, $p(y_0-x) = d_p(x, Y) > 0$. Let $\varphi_0 \in X_p^\flat$ be the functional whose existence is stated in the second part of Theorem 2.5.3, i.e., $\|\varphi_0|_p = 1$ and $\inf \varphi_0(Y - x) = d_p(x, Y) = p(y_0 - x)$.

Then

$$p(y_0 - x) = d_p(x, Y) = \inf \varphi_0(Y - x) \le \varphi(y_0 - x) \le p(y_0 - x) ,$$

implying $\varphi_0(y_0) = \inf \varphi_0(Y)$ and $\varphi_0(y_0 - x) = p(y_0 - x)$.

If $z_0 \in Y$ is such that $p(x - z_0) = d_p(Y, x) > 0$, then, by the second part of Theorem 2.5.3, there exists $\psi_0 \in X_p^\flat$ such that $\|\psi_0|_p = 1$, and $\inf \psi_0(x - Y) = \sup\{\inf \psi(x - Y) : \psi \in X_p^\flat, \|\psi|_p \le 1\} = d_p(Y, x)$. Then, by the duality relation (2.5.7), we have

$$p(x_0 - z_0) = d_p(Y, x) = \inf \psi_0(x - Y) \le \psi_0(x - z_0) \le p(x - z_0)$$

implying $\psi_0(x - z_0) = p(x - z_0)$ and $\psi_0(x) - \psi(z_0) = \inf \psi_0(x - Y) = \psi_0(x) - \sup \psi_0(Y)$, so that $\psi_0(z_0) = \sup \psi_0(Y)$. □

When Y is a subspace of a space with asymmetric norm (X, p), one obtains the following characterization of nearest points. Denote by $Y^\perp = \{\varphi \in X_p^\flat : \varphi|_Y = 0\}$ *the annihilator* of Y in $X_p^\flat$.

Corollary 2.5.4. *Let Y be a subspace of (X, p), $x \in X$ and $y_0 \in Y$. If there exists $\varphi_0 \in Y^\perp$ such that*

$$(\text{i}') \ \|\varphi_0|_p = 1 \quad \textit{and} \quad (\text{ii}') \ \varphi_0(y_0 - x) = p(y_0 - x) ,$$

then $p(y_0 - x) = d_p(x, Y)$, i.e., y_0 is a p-nearest point to x in Y.

Conversely, if $y_0 \in Y$ is such that $p(y_0 - x) = d_p(x, Y) > 0$, then there exists a functional $\varphi_0 \in Y^\perp$ which satisfies the conditions (i′) *and* (ii′).

Similarly, in order that $z_0 \in Y$ be a $\bar{p}$-nearest point to $x \in X$ it is sufficient and, if $d_p(Y, x) > 0$ also necessary, to exist a functional $\psi_0 \in Y^\perp$ such that

$$(j') \ \|\psi_0|_p = 1 \quad \textit{and} \quad (jj') \ \psi_0(x - z_0) = p(x - z_0) .$$

Proof. If $\varphi_0 \in Y^\perp$ satisfies the conditions (i′) and (ii′), then $\varphi_0(y_0) = 0 = \inf \varphi_0(Y)$, so that, by Theorem 2.5.3, it is a p-nearest point to x.

Conversely, if $p(y_0 - x) = d_p(x, Y) > 0$, then, by the necessity part of the same theorem, there exists $\varphi_0 \in X_p^\flat$ satisfying the conditions (i)–(iii). By (iii) we have

$$\forall y' \in Y,\ \varphi_0(y' - y_0) \le 0 \iff \forall y \in Y,\ \varphi_0(y) \le 0 \iff \forall y \in Y,\ \varphi_0(y) = 0\,,$$

showing that $\varphi_0 \in Y^\perp$.

The case of a $\bar{p}$-nearest point z_0 is treated similarly. □

Based on Theorem 2.2.13 one can prove the following characterization theorem in terms of the extreme points of the unit ball of the dual space $X_p^\flat$. In the case of a normed space X the result was obtained by Singer [221] when Y is a subspace of X and by Garkavi [97] for convex sets. The asymmetric case was treated in [40].

Theorem 2.5.5. *Let (X, p) be a space with asymmetric norm, Y a nonempty subset of X, $x \in X$ and $y_0 \in Y$.*

If for every $y \in Y$ there is a functional $\varphi = \varphi_y$ in the unit ball $B_p^\flat$ of $X_p^\flat$ such that

$$\text{(i) } \varphi(y_0 - x) = p(y_0 - x) \quad \text{and} \quad \text{(ii) } \varphi(y_0 - y) \le 0\,,$$

then y_0 is a p-nearest point to x in Y.

Conversely, if Y is convex and $y_0 \in Y$ is such that

$$p(y_0 - x) = d_p(x, Y) > 0\,,$$

then for every $y \in Y$ there exists an extreme point $\varphi = \varphi_y$ of the unit ball $B_p^\flat$ of $X_p^\flat$, satisfying the conditions (i) *and* (ii) *from above.*

Similarly, if $z_0 \in Y$ is such that for every $y \in Y$ there exists a functional $\psi = \psi_y \in B_p^\flat$ such that

$$\text{(j) } \psi(x - z_0) = p(x - z_0) \quad \text{and} \quad \text{(jj) } \psi(y - z_0) \le 0\,,$$

then z_0 is a $\bar{p}$-nearest point to x in Y.

Conversely, if $z_0 \in Y$ is such that $p(x - z_0) = d_p(Y, x) > 0$, then for every $y \in Y$ there exists an extreme point $\psi = \psi_y$ of the unit ball $B_p^\flat$ of $X_p^\flat$ satisfying the conditions (j) *and* (jj).

Proof. Suppose that $y_0 \in Y$ is such that for every $y \in Y$ there is a functional $\varphi = \varphi_y$ in $X_p^\flat$ satisfying the conditions (i) and (ii). Then, for every $y \in Y$,

$$p(y_0 - x) = \varphi(y_0 - x) = \varphi(y_0 - y) + \varphi(y - x) \le \varphi(y - x) \le p(y - x)\,,$$

showing that $p(y_0 - x) = d_p(x, Y)$.

Suppose now that Y is convex and that $y_0 \in Y$ is a p-nearest point to x in Y such that $d_p(x,Y) = p(y_0 - x) > 0$. The equalities

$$p(y_0 - x) = \inf\{p(y-x) : y \in Y\} = \inf\{p(w) : w \in Y - x\}$$

show that $y_0 - x$ is a p-nearest point to 0 in $Y - x$. For $y \in Y \setminus \{y_0\}$ let $Z := \mathrm{sp}\{y_0 - x, y - x\}$ – the space generated by $y_0 - x$ and $y - x$, and let $W := Z \cap (Y - x)$. Since $y_0 - x$ is a p-nearest point to 0 in W, by Theorem 2.5.3, there exists $\psi_0 \in Z_p^\flat$, $\|\psi_0|_p = 1$, such that

$$\psi_0(y_0 - x) = p(y_0 - x) \quad \text{and} \quad \psi_0(y_0 - x) = \inf \psi_0(W)\,.$$

It follows that

$$\psi_0(y_0 - x) \le \psi_0(y - x) \iff \psi_0(y_0 - y) \le 0\,.$$

The set $B_{Z_p^\flat}$ is a w^*-compact convex subset of the two-dimensional space Z^*, so that, by the Carathéodory and Krein-Milman theorems,

$$\psi_0 = \sum_{i=1}^{r} \alpha_i \psi_i \tag{2.5.8}$$

where $1 \le r \le 3$, $\alpha_i > 0$, $\sum_i \alpha_i = 1$, and ψ_i are extreme points of the set $B_{Z_p^\flat}$. The equality $\psi_0(y_0 - x) = p(y_0 - x)$ and (2.5.8) imply $\psi_i(y_0 - x) = p(y_0 - x)$, $i = 1, \dots, r$. Also, since $\psi_0(y_0 - y) \le 0$, at least one of the ψ_i, say ψ_1, must satisfy $\psi_1(y_0 - y) \le 0$.

By Theorem 2.2.13, ψ_1 has a norm preserving extension $\varphi \in X_p^\flat$ that is an extreme point of the unit ball $B_p^\flat$. The functional φ satisfies all the requirements of the theorem.

The case of $\bar{p}$-nearest points can be reduced to that of p-nearest points, by working in the space $(X, \bar{p})$, taking into account the equality $d_p(Y, x) = d_{\bar{p}}(x, Y)$, the fact that $\psi \in X_{\bar{p}}^\flat$ if and only if $-\psi \in X_p^\flat$ and that $\|\psi|_{\bar{p}} = \| -\psi|_p$ (see Proposition 2.1.7.5). Also, as it is easily seen, ψ is an extreme point of the unit ball $B_{\bar{p}}^\flat$ of $X_{\bar{p}}^\flat$ if and only if $-\psi$ is an extreme point of the ball $B_p^\flat$. □

2.5.2 The distance to a hyperplane

As it was shown in [48], the well-known formula for the distance to a closed hyperplane in a normed space (the so-called Arzelá formula) has an analog in spaces with asymmetric norm. Remark that in this case we have to work with both of the distances d_p and $d_{\bar{p}}$ given by (2.2.3).

Proposition 2.5.6. *Let (X, p) be a space with asymmetric norm, $\varphi \in X_p^\flat$, $\varphi \neq 0$, $c \in \mathbb{R}$,*

$$H = \{x \in X : \varphi(x) = c\}$$

the hyperplane corresponding to φ *and* c, *and*

$$H^{<} = \{x \in X : \varphi(x) < c\} \quad \text{and} \quad H^{>} = \{x \in X : \varphi(x) > c\} ,$$

the open half-spaces determined by H.

1. *We have*

$$d_{\bar{p}}(x_0, H) = \frac{\varphi(x_0) - c}{\|\varphi|_p} \tag{2.5.9}$$

   for every $x_0 \in H^{>}$, *and*

$$d_p(x_0, H) = \frac{c - \varphi(x_0)}{\|\varphi|_p} \tag{2.5.10}$$

   for every $x_0 \in H^{<}$.

2. *If there exists an element* $z_0 \in X$ *with* $p(z_0) = 1$ *such that* $\varphi(z_0) = \|\varphi|_p$ *then every element in* $H^{>}$ *has a* $\bar{p}$*-nearest point in* H, *and every element in* $H^{<}$ *has a* p*-nearest point in* H.
3. *If there is an element* $x_0 \in H^{>}$ *having a* $\bar{p}$*-nearest point in* H, *or there is an element* $x_0' \in H^{<}$ *having a* p*-nearest point in* H, *then there exists an element* $z_0 \in X$, $p(z_0) = 1$, *such that* $\varphi(z_0) = \|\varphi|_p$. *It follows that, in this case, every element in* $H^{>}$ *has a* $\bar{p}$*-nearest point in* H, *and every element in* $H^{<}$ *has a* p*-nearest point in* H.

Proof. 1. Let $x_0 \in H^{>}$. Then, for every $h \in H$, $\varphi(h) = c$, so that

$$\varphi(x_0) - c = \varphi(x_0 - h) \le \|\varphi|p(x_0 - h) ,$$

implying

$$d_{\bar{p}}(x_0, H) \ge \frac{\varphi(x_0) - c}{\|\varphi|} .$$

By Proposition 2.1.8.1, there exists a sequence (z_n) in X with $p(z_n) = 1$, such that $\varphi(z_n) \to \|\varphi|$ and $\varphi(z_n) > 0$ for all $n \in \mathbb{N}$. Then

$$h_n := x_0 - \frac{\varphi(x_0) - c}{\varphi(z_n)} z_n$$

belongs to H and

$$d_{\bar{p}}(x_0, H) \le p(x_0 - h_n) = \frac{\varphi(x_0) - c}{\varphi(z_n)} \to \frac{\varphi(x_0) - c}{\|\varphi|} .$$

It follows that $d_{\bar{p}}(x_0, H) \ge (\varphi(x_0) - c)/\|\varphi|$, so that formula (2.5.9) holds.

To prove (2.5.10), observe that for $h \in H$,

$$c - \varphi(x_0') = \varphi(h - x_0') \le \|\varphi|p(h - x_0') ,$$

implying

$$d_p(x_0', H) \geq \frac{c - \varphi(x_0')}{\|\varphi|} .$$

If the sequence (z_n) is as above then

$$h_n' := \frac{c - \varphi(x_0')}{\varphi(z_n)} z_n + x_0'$$

belongs to H and

$$d_p(x_0', H) \leq p(h_n' - x_0') = \frac{c - \varphi(x_0')}{\varphi(z_n)} \to \frac{c - \varphi(x_0')}{\|\varphi|} ,$$

so that $d_p(x_0', H) \geq (c - \varphi(x_0'))/\|\varphi|$, and formula (2.5.10) holds too.

2. Let $z_0 \in X$ be such that $p(z_0) = 1$ and $\varphi(z_0) = \|\varphi|$. Then, for $x_0 \in H^>$ and $x_0' \in H^<$, the elements

$$h_0 := x_0 - \frac{\varphi(x_0) - c}{\varphi(z_0)} z_0 \quad \text{and} \quad h_0' := \frac{c - \varphi(x_0')}{\varphi(z_0)} z_0 + x_0'$$

belong to H,

$$p(x_0 - h_0) = \frac{\varphi(x_0) - c}{\|\varphi|} = d_{\bar{p}}(x_0, H) \quad \text{and} \quad p(h_n' - x_0') = \frac{c - \varphi(x_0')}{\|\varphi|} = d_p(x_0', H) .$$

If an element $x_0 \in H^>$ has a $\bar{p}$-nearest point $h_0 \in H$ then

$$p(x_0 - h_0) = d_{\bar{p}}(x_0, H) = \frac{\varphi(x_0) - c}{\|\varphi|} = \frac{\varphi(x_0 - h_0)}{\|\varphi|} .$$

It follows that $z_0 = (x_0 - h_0)/p(x_0 - h_0)$ satisfies the conditions $p(z_0) = 1$ and $\varphi(z_0) = \|\varphi|$. If an element $x_0' \in H^<$ has a p-nearest point h_0' in H, then $z_0' = (h_0' - x_0')/p(h_0' - x_0')$ satisfies $p(z_0') = 1$ and $\varphi(z_0') = \|\varphi|$. □

Remark that, according to the assertions 2 and 3 of the above proposition, the hyperplane H generated by a functional $\varphi \in X_p^\flat$ has some proximinality properties if and only if the functional φ attains its norm on the unit ball of X, a situation similar to that in normed spaces.

2.5.3 Best approximation by elements of sets with convex complement

Best approximation by elements of sets with convex complement was considered by Klee [112], in connection with the still unsolved problem of convexity of Chebyshev sets in Hilbert space (see the survey [20]). Klee conjectured that if a Hilbert space contains a non convex Chebyshev set, then it contains a Chebyshev set whose complement is convex and bounded. The conjecture was solved affirmatively by Asplund [16] who proposed the term *Klee cavern* to designate a set whose complement is convex and bounded. This term was used by Franchetti and Singer [81] who proved duality and characterization results for best approximation by elements of caverns as well as some existence results. In [47] some of these results were extended to sets with p-convex complement. In the paper [40] it was shown that the duality result proved by Franchetti and Singer, *loc cit.*, holds in spaces with asymmetric norm too. The proof is based on the formula for the distance to a hyperplane, Proposition 2.5.6.

We call a subset Y of (X,p) *upper p-bounded* if there exists $r>0$ and $x\in X$ such that $Y\subset B_p[x,r]$ or, equivalently, if $\sup p(Y)<\infty$.

The duality result is the following.

Theorem 2.5.7. *Let (X,p) be a space with asymmetric norm, Z a τ_p-open, upper p-bounded convex subset of X and $Y=X\setminus Z$.*

Then for every $x\in Z$ the following duality relation holds:

$$d_p(x,Y)=\inf\{\sup\varphi(Y)-\varphi(x):\varphi\in X_p^\flat,\ \|\varphi|_p=1\}. \tag{2.5.11}$$

Proof. Let $d=d_p(x,Y)$ and denote by λ the quantity in the right-hand side of the relation (2.5.11).

For $\varphi\in X_p^\flat$, $\|\varphi|_p=1$, let $c=\sup\varphi(Z)$. Because Z is upper p-bounded and $\varphi\le p$, it follows that c is finite. Also $\varphi(z)<1$ for every $z\in Z$. Indeed, as the set Z is τ_p-open, for every $z\in Z$ there exists $r>0$ such that $B_p(z,r)\subset Z$. Choosing $u\in X$ such that $\varphi(u)=1$, it follows that $p(u)\ge\varphi(u)=1$ and $z+(r/p(u))u\in B_p(z,r)\subset Z$, so that

$$c\ge\varphi\left(z+\frac{r}{p(u)}u\right)=\varphi(z)+\frac{r}{p(u)}>\varphi(z)\,.$$

Therefore, $\varphi(y)=c$ implies $y\notin Z \iff y\in Y$, i.e., the hyperplane $H=\{x'\in X:\varphi(x')=c\}$ is contained in Y.

Taking into account this fact and the formula (2.5.10) we get

$$d=d_p(x,Y)\le d_p(x,H)=\frac{c-\varphi(x)}{\|\varphi|_p}=c-\varphi(x)=\sup\varphi(Y)-\varphi(x)\,.$$

Since $\varphi\in X_p^\flat$, $\|\varphi|_p=1$, was arbitrarily chosen it follows that $d\le\lambda$.

To prove the reverse inequality, observe that $y\in Y$ implies $Z\cap\{y\}=\emptyset$, so that, by the first separation theorem (Theorem 2.2.8), there exists $\varphi\in X_p^\flat$ such that

$$\forall z\in Z,\quad \varphi(z)<\varphi(y)\,.$$

Dividing (if necessary) this inequality by $\|\varphi|_p$, we can suppose $\|\varphi|_p=1$, so that

$$\lambda\le\sup\varphi(Z)-\varphi(x)\le\varphi(y-x)\le p(y-x)\,,$$

for any $y\in Y$, implying $\lambda\le d$. □

2.5.4 Optimal points

García Raffi and Sánchez Pérez [95] propose a finer approach to the best approximation problem in asymmetric normed spaces.

Let (X, p) be an asymmetric normed space and $p^s(x) = \max\{p(x), p(-x)\}$ the associated norm on X. A norm p_0^s on X that is equivalent to p^s is called a *p-associated norm* on X. For $Y \subset X$ nonempty and $x \in X$ a point $y_0 \in P_Y(x)$ is called p_0^s*-optimal distance point* provided that

$$p_0^s(y_0 - x) \le p_0^s(y - x) , \tag{2.5.12}$$

for all $y \in P_Y(x)$. A p^s-optimal distance point is called simply an *optimal distance point*. The set of all p_0^s-optimal distance points to x in $P_Y(x)$ is denoted by $O_{Y,p_0^s}(x)$ and that of optimal distance points by $O_Y(x)$.

As it is remarked in [95], the size of the set $P_Y(x)$ of nearest points to x in Y depends on the set $\theta(x)$ given by (2.4.3).

Proposition 2.5.8. *Let (X, p) be an asymmetric normed space, $Y \subset X$, $x, y \in X$. If $P_Y(x) \neq \emptyset$, then*

1. $y \in P_Y(x) \;\Rightarrow\; \theta(y) \cap Y \subset P_Y(x)$.
2. $\theta(y) \cap Y \neq \emptyset \;\Rightarrow\; p(y - x) \ge d_p(x, Y)$.
3. $P_Y(x) = \bigcup\{\theta(y) \cap Y : y \in B_p[x, d]\}$, *where* $d = d(x, Y)$.

The paper contains also the following existence results for optimal distance points.

Theorem 2.5.9. *Let (X, p) be an asymmetric normed space, p_0^s a p-associated norm on X, Y a nonempty convex subset of X and $x \in X$ such that $P_Y(x) \neq \emptyset$.*

1. *If the normed space (X, p_0^s) is strictly convex, then there is at most one p_0^s-optimal distance point in $P_Y(x)$.*
2. *In any of the following cases there is at least one p_0^s-optimal distance point in $P_Y(x)$:*

   (i) *the set $P_Y(x)$ is locally compact (in particular, if Y is contained in a finite-dimensional subspace of X), or*

   (ii) *(X, p^s) is a reflexive Banach space and Y is p^s-closed.*

2.5.5 Sign-sensitive approximation in spaces of continuous or integrable functions

As we did mention, asymmetric normed spaces and the notation $\|\cdot|$ for an asymmetric norm were introduced and studied by Krein and Nudelman [129, Ch. IX, §5] in their book on the moment problem. As an example they considered some spaces of continuous functions with an asymmetric norm given by a weight function. Consider a pair $\varphi = (\varphi_+, \varphi_-)$ of continuous strictly positive functions on an interval $[a;b]$ and denote by $B(\varphi)$ the space of all continuous functions on $[a;b]$ equipped with the norm

$$\|f| = \max_{a\le t\le b}\left\{\frac{f^+(t)}{\varphi_+(t)} + \frac{f^-(t)}{\varphi_-(t)}\right\}, \tag{2.5.13}$$

for $f \in C[a;b]$, where $f^+(t) = \max\{f(t),0\}$ and $f^-(t) = \max\{-f(t),0\}$. (It follows that $f = f^+ - f^-$ and $|f| = f^+ + f^-$.)

The asymmetric norm (2.5.13) is topologically equivalent to the usual sup-norm on $C[a;b]$ (and coincides with it for $\varphi_+ = \varphi_- \equiv 1$), so that the dual of $B(\varphi)$ agrees with the dual of $C[a;b]$, that is with the space of all functions with bounded variation on $[a;b]$ with the total variation norm. The authors give in [129] the expression of the asymmetric norm of continuous linear functionals in terms of (φ_+,φ_-) and study some extremal problems in this space.

Later Dolzhenko and Sevastyanov [67, 69] (see also the survey [68]) considered some best approximation problems in spaces of continuous functions on an interval $\Delta = [a;b]$ and studied the existence and uniqueness (finite-dimensional Chebyshev subspaces) and gave characterizations of the best approximation (alternance and Kolmogorov type criteria). They consider a pair $w = (w_+,w_-)$ of nonnegative functions and the asymmetric norm

$$\|f|_w = \sup\{|w_+(t)f^+(t) - w_-(t)f^-(t)| : t\in\Delta\}. \tag{2.5.14}$$

Asymmetric norms on spaces of integrable functions are defined analogously:

$$\|f|_{p,w}^p = \int_a^b |w_+(t)f^+(t) - w_-(t)f^-(t)|^p dt, \tag{2.5.15}$$

for $1\le p<\infty$.

The study of sign-sensitive approximation is considerably more complicated than the usual approximation (in the sup-norm or in L^p-norms) and requires a fine analysis, based on the properties of weight functions. This analysis is done in several papers, mainly by Russian and Ukrainian authors. Among these we mention Dolzhenko and Sevastyanov with the papers quoted above, Sevastyanov [217], Babenko [17, 18], Kozko [125, 126, 127, 128], Ramazanov [180, 181, 182, 183], Simonov [219], Simonov and Simonova [220], and the references quoted in these papers.

2.6 Spaces of semi-Lipschitz functions

One of the most important classes of asymmetric normed spaces is that of semi-Lipschitz functions on a quasi-metric space. This section is concerned with their study, with emphasis on various completeness results and applications to best approximation in quasi-metric spaces.

The properties of the spaces of semi-Lipschitz functions were studied by Romaguera and Sanchis [204, 206] and Romaguera, Sánchez-Álvarez and Sanchis [198]. The paper by Mustăţa [160] is concerned with the behavior of the extreme points of the unit ball in spaces of semi-Lipschitz functions.

A good presentation of properties of spaces of Lipschitz functions on metric spaces is given in the book by Weaver [236].

2.6.1 Semi-Lipschitz functions – definition and the extension property

Let (X, ρ) be a metric space and $(Y, \|\cdot\|)$ a normed space. A function $f : X \to Y$ is called *Lipschitz* if there exists $L \geq 0$ such that

$$\|f(x) - f(y)\| \leq L\rho(x, y) \tag{2.6.1}$$

for all $x, y \in X$. A number $L \geq 0$ satisfying (2.6.1) is called a *Lipschitz constant* for f. The space of all Lipschitz functions from X to Y is denoted by $\mathrm{Lip}_{\rho,\|\cdot\|}(X, Y)$, respectively by $\mathrm{Lip}_\rho(X)$ when $Y = (\mathbb{R}, |\cdot|)$.

The formula

$$\|f\|_{\rho,\|\cdot\|} = \sup\left\{\frac{\|f(x) - f(y)\|}{\rho(x, y)} : x, y \in X,\ \rho(x, y) > 0\right\} \tag{2.6.2}$$

defines a norm on the space $\mathrm{Lip}_{\rho,\|\cdot\|}(X, Y)$, that is $\left(\mathrm{Lip}_{\rho,\|\cdot\|}(X, Y), \|\cdot\|_{\rho,\|\cdot\|}\right)$ is a normed space which is complete, provided Y is a Banach space, see [236]. The number $\|f\|_{\rho,\|\cdot\|}$ is the smallest Lipschitz constant for f.

Suppose now that (X, ρ) is a quasi-metric space, (Y, q) an asymmetric normed space. A function $f : X \to \mathbb{R}$ is called *semi-Lipschitz* provided there exists a number $L \geq 0$ such that

$$q(f(x) - f(y)) \leq L\,\rho(x, y)\ , \tag{2.6.3}$$

for all $x, y \in X$. A number $L \geq 0$ for which (2.6.3) holds is called a *semi-Lipschitz constant* for f and we say that f is L-semi-Lipschitz. We denote by $\mathrm{SLip}(X, Y)$ ($\mathrm{SLip}_{\rho,d}(X, Y)$ if more precision is needed) the set of all semi-Lipschitz functions from X to Y.

In particular, if Y is the space $(\mathbb{R}, u)$, $u(\alpha) = \alpha^+$, (see Example 1.1.3), the condition (2.6.3)is equivalent to

$$f(x) - f(y) \leq L\,\rho(x, y)\ , \tag{2.6.4}$$

for all $x, y \in X$. In this case one uses the notation $\mathrm{SLip}_\rho(X) = \mathrm{SLip}_\rho(X, \mathbb{R})$.

A function $f : X \to Y$ is called $\le_{\rho,q}$-*monotone* if $q(f(x)-f(y)) = 0$ whenever $\rho(x,y) = 0$. In particular, a function $f : X \to \mathbb{R}$ is $\le_{\rho,u}$-monotone, called $\le_{\rho}$-*monotone*, if and only if $f(x) \le f(y)$ whenever $\rho(x,y) = 0$.

Obviously, a semi-Lipschitz function is $\le_{\rho,q}$-monotone. Since the topology τ_ρ is T_1 if and only if $\rho(x,y) = 0 \iff x = y$, (Proposition 1.1.8) it follows that any function on a T_1 quasi-metric space is $\le_{\rho,q}$-monotone.

Remark 2.6.1. It is clear that for $\alpha, \beta \in \mathbb{R}$,

$$\alpha \le \beta \iff \alpha - \beta \le 0 \iff u(\alpha - \beta) = (\alpha - \beta)^+ = 0.$$

If p is an asymmetric seminorm on a vector space X, then

$$x \le_p y \iff p(x-y) = 0$$

defines an order relation on X. Similarly, in a quasi-semimetric space (X, ρ)

$$x \le_\rho y \iff \rho(x,y) = 0$$

also defines an order relation.

Taking into account these order relations, the $\le_{\rho,q}$-monotonicity can be expressed by the condition

$$x \le_\rho y \implies f(x) \le_q f(y),$$

justifying the term monotonicity.

Suppose now that (X, ρ) is a quasi-metric space and (Y, q) an asymmetric normed space. For an arbitrary function $f : X \to Y$ put

$$\|f|_{\rho,q} = \sup\left\{\frac{q(f(x)-f(y))}{\rho(x,y)} : x, y \in X,\ \rho(x,y) > 0\right\}, \tag{2.6.5}$$

and $\|f|_\rho = \|f|_{\rho,u}$ when Y is $(\mathbb{R}, u)$.

Proposition 2.6.2. *Let (X, ρ) be a quasi-metric space and (Y, q) an asymmetric normed space.*

1. *The set $\mathrm{SLip}_{\rho,q}(X,Y)$ is a cone in the linear space $\mathrm{Lip}_{\rho^s,q^s}(X,Y)$ of all Lipschitz functions from the metric space (X, ρ^s) to the normed space (Y, q^s) and $\|f\|_{\rho^s,q^s} \le \|f|_{\rho,q}$ for all $f \in \mathrm{SLip}_{\rho,q}(X,Y)$.*
2. *If f is semi-Lipschitz, then $\|f|_{\rho,q}$ is the smallest semi-Lipschitz constant for f.*
3. *A function $f : X \to Y$ is semi-Lipschitz if and only if f is $\le_{\rho,q}$-monotone and $\|f|_{\rho,q} < \infty$.*

Proof. 1. It is clear that $f+g,\ \alpha f \in \mathrm{SLip}_{\rho,q}(X,Y)$ for all $f,g \in \mathrm{SLip}_{\rho,q}(X,Y)$ and $\alpha \geq 0$.

If $f \in \mathrm{SLip}_{\rho,q}(X,Y)$, then

$$q(f(x)-f(y)) \leq L\rho(x,y) \leq L\rho^s(x,y),\ \ x,y \in X\ ,$$

implying

$$q^s(f(x)-f(y)) \leq L\rho^s(x,y)\ ,$$

for all $x,y \in X$, so that $f \in \mathrm{Lip}_{\rho^s,q^s}(X,Y)$ and $\|f\|_{\rho^s,q^s} \leq \|f|_{\rho,q}$.

2. The inequality $q(f(x)-f(y))/\rho(x,y) \leq \|f|_{\rho,q}$ implies $f(x)-f(y) \leq \|f|_\rho\,\rho(x,y)$ for all $x,y \in X$ with $\rho(x,y) > 0$. Since a semi-Lipschitz function is $\leq_{\rho,q}$-monotone, $\rho(x,y) = 0$ implies $q(f(x)-f(y)) = 0 = \|f|_{\rho,q}\,\rho(x,y)$. Consequently, $\|f|_{\rho,q}$ is a semi-Lipschitz constant for f.

Suppose that L is a semi-Lipschitz constant for f. Then

$$q(f(x)-f(y))/\rho(x,y) \leq L,$$

whenever $\rho(x,y) > 0$, so that $\|f|_{\rho,q} \leq L$, showing that $\|f|_{\rho,q}$ is the smallest semi-Lipschitz constant for f.

The above reasonings show also the validity of 3. □

The following example shows that the inequality from Proposition 2.6.2.1 can be strict.

Example 2.6.3. On a three point set set $X = \{x_1,x_2,x_3\}$ consider the quasi-metric $\rho(x_1,x_2) = 1$, $\rho(x_2,x_1) = 2, \rho(x_1,x_3) = \rho(x_3,x_1) = 2$, $\rho(x_2,x_3) = \rho(x_3,x_2) = 2$, and the function $f : X \to \mathbb{R}$ given by $f(x_1) = 1$, $f(x_2) = f(x_3) = 2$. Then $\rho^s(x_i,x_j) = 2$ for $i \neq j$ and

$$\begin{aligned}\|f|_{\rho,|\cdot|} &= \max\left\{\frac{|f(x_i)-f(x_j)|}{\rho(x_i,x_j)} : 1 \leq i,j \leq 3,\ i \neq j\right\} = 1\\ &> \frac{1}{2} = \max\left\{\frac{|f(x_i)-f(x_j)|}{\rho^s(x_i,x_j)} : 1 \leq i,j \leq 3,\ i \neq j\right\} = \|f|_{\rho^s,|\cdot|}.\end{aligned}$$

The following proposition puts in evidence some useful semi-Lipschitz functions.

Proposition 2.6.4. *Let (X,ρ) be a quasi-metric space, $y \in X$ and $Y \subset X$ nonempty.*

1. *The functions $\rho(\cdot,y) : X \to \mathbb{R}$ and $d(\cdot,Y) : X \to \mathbb{R}$ are semi-Lipschitz with semi-Lipschitz constant* 1.
2. *For fixed $a \in X$, the functions $f(x) = \rho(a,x_0) - \rho(a,x)$ and $g(x) = \rho(x,a) - \rho(x_0,a)$ belong to $\mathrm{SLip}_{\rho,0}(X)$ and $\|f|_\rho$, $\|g|_\rho \leq 1$.*

Proof. 1. The inequality

$$\rho(x,y) \le \rho(x,x') + \rho(x',y) , \tag{2.6.6}$$

valid for $x, x' \in X$, shows that the function $\rho(\cdot, y)$ is semi-Lipschitz. Since the inequality (2.6.6) holds for all $y \in Y$ and fixed x, x', passing to infimum with respect to $y \in Y$ one obtains $d(x,Y) \le \rho(x,x') + d(x',Y)$, which means that the function $d(\cdot, Y)$ is semi-Lipschitz, too.

The assertions from 2 follow from 1. □

An important result in the study of Lipschitz functions on metric spaces is the extension of Lipschitz functions, usually known as Kirszbraun's extension theorem, see, for instance, the book [237].

In the case of semi-Lipschitz functions a similar result was proved by Mustăţa [158] (see also [157, 164]). The extension problem for semi-Lipschitz functions on quasi-metric spaces was considered also by Matoušková [148]. The paper [87] discusses the existence of an extension of an asymmetric norm defined on a cone K to an asymmetric norm defined on the generated linear space $X = K - K$.

Proposition 2.6.5. *Let (X, ρ) be a quasi-metric space, Y a nonempty subset of X and $f : Y \to \mathbb{R}$ an L-semi-Lipschitz function.*

1. *The functions F, G defined for $x \in X$ by*

$$F(x) = \inf\{f(y) + L\rho(x,y) : y \in Y\} \tag{2.6.7}$$

and

$$G(x) = \sup\{f(y') - L\rho(y',x) : y' \in Y\} \tag{2.6.8}$$

are L-semi-Lipschitz extensions of f.

2. *Any other L-semi-Lipschitz extension H of f satisfies the inequalities*

$$G \le H \le F . \tag{2.6.9}$$

Proof. 1. Let $y, y' \in Y$ and $x \in X$. The inequalities $f(y') - f(y) \le L\,\rho(y',y) \le L\,\rho(y',x) + L\,\rho(x,y)$ imply

$$f(y') - L\,\rho(y',x) \le f(y) + L\,\rho(x,y) .$$

Passing to supremum with respect to $y' \in Y$ and to infimum with respect to $y \in Y$, it follows that G, F are well defined and $G \le F$.

Let $x \in Y$. Then $f(x) \le f(y) + L\,\rho(x,y)$ for every $y \in Y$, implies $f(x) \le F(x)$. Similarly, $f(y') - L\,\rho(y',x) \le f(x)$ implies $G(x) \le f(x)$. Taking $y = x$ in (2.6.7) and $y' = x$ in (2.6.8), it follows that $F(x) \le f(x)$ and $G(x) \ge f(x)$, so that $G(x) = f(x) = F(x)$ for every $x \in Y$.

To conclude, we have to show that the functions F, G are semi-Lipschitz. Let $x, x' \in X$. The inequalities

$$F(x) \le f(y) + L\,\rho(x,y) \le f(y) + L\,\rho(x,x') + L\,\rho(x',y)\,,$$

valid for all $y \in Y$, yield $F(x) \le F(x') + L\,\rho(x,x')$, showing that F is semi-Lipschitz.

Similar reasonings show that G is semi-Lipschitz too.

2. Let H be an L-semi-Lipschitz extension of f and $x \in X$. Since

$$H(x) \le H(y) + L\,\rho(x,y) = f(y) + L\,\rho(x,y)\,,$$

for every $y \in Y$, passing to infimum with respect to $y \in Y$ one obtains $H(x) \le F(x)$.

Similarly $f(y') - H(x) = H(y') - H(x) \le L\,\rho(y',x,)$ implies

$$f(y') - L\,\rho(y',x) \le H(x)\,,$$

for every $y' \in Y$. Passing to supremum with respect to $y' \in Y$ one obtains $G(x) \le H(x)$. □

The following corollary shows the existence of norm-preserving extensions of semi-Lipschitz functions.

Corollary 2.6.6. *Let X, Y and f be as in Proposition 2.6.5 and*

$$\|f|_\rho = \sup\{u(f(y) - f(y'))/\rho(y,y') : y, y' \in Y,\ \rho(y,y') > 0\}\,,$$

and let F, G be given by (2.6.7) and (2.6.8) for $L = \|f|_\rho$. Then

1. *The functions F and G are semi-Lipschitz norm-preserving extensions of f, that is*
$$(i)\ F|_Y = G|_Y = f \quad \text{and} \quad (ii)\ \|F|_\rho = \|G|_\rho = \|f|_\rho\,.$$
2. *Any other semi-Lipschitz norm-preserving extension H of f satisfies the inequalities*
$$G \le H \le F\,.$$

2.6.2 Properties of the cone of semi-Lipschitz functions – linearity

In the case of real-valued Lipschitz functions on a metric space (X, ρ) the function $\|\cdot\| : \mathrm{Lip}(X,\mathbb{R}) \to \mathbb{R}$ given by

$$\|f\|_\rho = \sup\left\{\frac{|f(x) - f(y)|}{\rho(x,y)} : x, y \in X,\ \rho(x,y) > 0\right\},$$

is only a seminorm on $\mathrm{Lip}(X,\mathbb{R})$ because $\|f\|_\rho = 0$ for any constant function. To obtain a norm one can proceed in two ways:

- one takes the quotient space of $\mathrm{Lip}(X,\mathbb{R})$ with respect to the subspace of constant functions, or
- one fixes a point $x_0 \in X$ and one considers the subspace

$$\mathrm{Lip}_0(X,\mathbb{R}) = \{f \in \mathrm{Lip}(X,\mathbb{R}) : f(x_0) = 0\}$$

with the induced norm.

We shall follow the second way. A metric space with an a priori fixed point x_0 is called a *pointed metric space.* If, in addition, X is a vector space then one takes usually $x_0 = 0$, the null element of X.

For a pointed quasi-metric space (X,ρ,x_0) and an asymmetric normed space (Y,q) let

$$\mathrm{SLip}^0_{\rho,q}(X,Y) = \left\{f \in \mathrm{SLip}_{\rho,q}(X,Y) : f(x_0) = 0\right\}. \tag{2.6.10}$$

In the case $Y = (\mathbb{R},u)$ one uses the notation $\mathrm{SLip}^0_\rho(X)$.

The following proposition contains some simple remarks concerning the relations between ρ- and $\bar{\rho}$-semi-Lipschitz functions, obtained in [215] and in [206] in the case $Y = \mathbb{R}$.

Proposition 2.6.7. *Let (X,ρ) be a quasi-metric space, (Y,q) an asymmetric normed space and $f : X \to Y$ a function.*

1. *The function f belongs to* $\mathrm{SLip}_{\rho,q}(X,Y)$ *(resp. to* $\mathrm{SLip}^0_{\rho,q}(X,Y)$*) if and only if $-f$ belongs to* $\mathrm{SLip}_{\bar{\rho},q}(X,Y)$ *(resp. to* $\mathrm{SLip}^0_{\bar{\rho},q}(X,Y)$*). Also the following equality holds,*

$$\|f|_{\rho,q} = \|-f|_{\bar{\rho},q}\,.$$

The correspondence $f \mapsto -f$ is an isometric isomorphism between the normed cones

$$\left(\mathrm{SLip}_{\rho,q}(X,Y), \|\cdot|_{\rho,q}\right) \quad \textit{and} \quad \left(\mathrm{SLip}_{\bar{\rho},q}(X,Y), \|\cdot|_{\bar{\rho},q}\right),$$

respectively between

$$\left(\mathrm{SLip}^0_{\rho,q}(X,Y), \|\cdot|_{\rho,q}\right) \quad \textit{and} \quad \left(\mathrm{SLip}^0_{\bar{\rho},q}(X,Y), \|\cdot|_{\bar{\rho},q}\right).$$

2. *The sets* $\mathrm{SLip}_{\rho,q}(X,Y) \cap \mathrm{SLip}_{\bar{\rho},q}(X,Y)$ *and* $\mathrm{SLip}^0_{\rho,q}(X,Y) \cap \mathrm{SLip}^0_{\bar{\rho},q}(X,Y)$ *are linear spaces.*

A problem discussed in [206] and [215] is: under what conditions is the cone $\mathrm{SLip}_{\rho,0}(X)$ a linear space?

We mention also the following result from [215] and [206] (the case $Y = \mathbb{R}$).

Theorem 2.6.8. *Let (X,ρ) be a quasi-metric space and (Y,q) an asymmetric normed space. The following are equivalent.*

1. $\mathrm{SLip}^0_{\rho,q}(X,Y) = \mathrm{SLip}^0_{\bar{\rho},q}(X,Y)$.
2. $\mathrm{SLip}^0_{\rho,q}(X,Y)$ *is a linear space.*

3. $\mathrm{SLip}^0_{\bar\rho,q}(X,Y)$ *is a linear space.*
4. $\mathrm{SLip}^0_{\bar\rho,q}(X,Y) \subset \mathrm{SLip}^0_{\rho,q}(X,Y)$.
5. $\mathrm{SLip}^0_{\rho,q}(X,Y) \subset \mathrm{SLip}^0_{\bar\rho,q}(X,Y)$.

Similar equivalence results hold for the spaces $\mathrm{SLip}_{\rho,q}(X,Y)$ *and* $\mathrm{SLip}_{\bar\rho,q}(X,Y)$.

Proof. Observe that $\mathrm{SLip}^0_{\rho,q}(X,Y)$ is a linear space if and only if $(\mathrm{SLip}^0_{\rho,q}(X,Y),+)$ is a group and similarly for $\mathrm{SLip}^0_{\bar\rho,q}(X,Y)$.

$1 \Rightarrow 2$. By Proposition 2.6.2, $\mathrm{SLip}^0_{\rho,q}(X,Y) = \mathrm{SLip}^0_{\rho,q}(X,Y) \cap \mathrm{SLip}^0_{\bar\rho,q}(X,Y)$ is a linear space.

$2 \Rightarrow 3$. Since $f \in \mathrm{SLip}^0_{\bar\rho,q}(X,Y) \iff -f\,\mathrm{SLip}^0_{\rho,q}(X,Y)$ and $\mathrm{SLip}^0_{\rho,q}(X,Y)$ is a linear space, it follows that $f = -(-f) \in \mathrm{SLip}^0_{\rho,q}(X,Y) \iff -f\,\mathrm{SLip}^0_{\bar\rho,q}(X,Y)$.

$3 \Rightarrow 4$. Follows from $f \in \mathrm{SLip}^0_{\bar\rho,q}(X,Y) \Rightarrow -f \in \mathrm{SLip}^0_{\bar\rho,q}(X,Y)$ and $-f \in \mathrm{SLip}^0_{\bar\rho,q}(X,Y) \iff f \in \mathrm{SLip}^0_{\rho,q}(X,Y)$.

The proof of $4 \Rightarrow 5$ is similar to that of the above implication, and $5 \Rightarrow 4$ follows from the equality $\bar{\bar\rho} = \rho$.

The implication $5 \Rightarrow 1$ follows from the equivalence $4 \iff 5$. □

Remark 2.6.9. The equality $\mathrm{SLip}^{x_0}_{\rho,q}(X,Y) = \mathrm{SLip}^{x_0}_{\bar\rho,q}(X,Y)$ does not depend on the point x_0, in the sense that if $\mathrm{SLip}^{x_0}_{\rho,q}(X,Y) = \mathrm{SLip}^{x_0}_{\bar\rho,q}(X,Y)$ for some $x_0 \in X$, then $\mathrm{SLip}^{x_1}_{\rho,q}(X,Y) = \mathrm{SLip}^{x_1}_{\bar\rho,q}(X,Y)$ for any other point $x_1 \in X$.

Indeed, the correspondence $f \mapsto f - f(x_0)$ applies $\mathrm{SLip}^{x_1}_{\rho,q}(X,Y)$ onto $\mathrm{SLip}^{x_0}_{\bar\rho,q}(X,Y)$ and preserves the Lipschitz constants.

In the case $Y = \mathbb{R}$ one obtains the following condition on the metric ρ.

Proposition 2.6.10 ([206]). *Let (X,ρ,x_0) be a pointed quasi-metric space. If*

$$\mathrm{SLip}^0_\rho(X) = \mathrm{SLip}^0_{\bar\rho}(X),$$

then the topology τ_ρ is T_1 and there exist $\beta,\beta' > 0$ such that

$$\rho(x,x_0) \le \beta\rho(x_0,x) \quad \textit{and} \quad \rho(x_0,x) \le \beta'\rho(x,x_0)\,, \tag{2.6.11}$$

for all $x \in X$.

Proof. Let $a,b \in X$ be such that $\rho(a,b) = 0$. By Proposition 2.6.4, the function $f(x) = \rho(x,a) - \rho(x_0,a)$, $x \in X$, is ρ-semi-Lipschitz, and so belongs to $\mathrm{SLip}^0_{\rho,q}(X) = \mathrm{SLip}^0_{\bar\rho}(X)$, so that there exists $\lambda > 0$ such that $\rho(x,a) - \rho(y,a) = f(x) - f(y) \le \lambda\rho(y,x)$, for all $x,y \in X$. Taking $x = b$ and $y = a$ one obtains $\rho(b,a) \le \lambda\rho(a,b) = 0$. Since ρ is a quasi-metric, $\rho(b,a) = \rho(a,b) = 0$ implies $a = b$. By Proposition 1.1.8.3 the topology τ_ρ is T_1 (and $\tau_{\bar\rho}$ as well).

The function $g(x) = \rho(x,x_0)$, $x \in X$, belongs to $\mathrm{SLip}^0_\rho(X) = \mathrm{SLip}^0_{\bar\rho}(X)$, so there exists $\beta > 0$ such that $\rho(x,x_0) = f(x) - f(x_0) \le \beta\rho(x_0,x)$. The existence of $\beta' > 0$ satisfying the second inequality is proved similarly. □

The above results have the following consequences.

Corollary 2.6.11. *Let* (X, ρ, x_0) *be a pointed quasi-metric space.*

1. *If* $\mathrm{SLip}^0_\rho(X) = \mathrm{SLip}^0_{\bar\rho}(X)$, *then* $\tau_\rho = \tau_{\bar\rho}$ *and hence the topology* τ_ρ *is metrizable.*
2. $(\mathrm{SLip}^0_\rho(X), +, \|\cdot|_\rho)$ *is a topological group if and only if* $\mathrm{SLip}^0_\rho(X) = \mathrm{SLip}^0_{\bar\rho}(X)$ *and* $\tau_\rho = \tau_{\bar\rho}$.

We mention also the following result.

Proposition 2.6.12. *Let* (X, ρ, x_0) *be a pointed quasi-metric space. The following assertions are equivalent.*

1. $\mathrm{SLip}^0_\rho(X)$ *is a vector space and* $\|\cdot|_\rho$ *is a complete norm on it, that is* $(\mathrm{SLip}^0_\rho(X), \|\cdot|_\rho)$ *is a Banach space.*
2. $\mathrm{SLip}^0_\rho(X) = \mathrm{SLip}^0_{\bar\rho}(X)$ *and* $\|\cdot|_\rho = \|\cdot|_{\bar\rho}$.
3. (X, ρ) *is a metric space.*

Proof. $1 \Rightarrow 2$. By Proposition 2.6.8, $\mathrm{SLip}^0_\rho(X) = \mathrm{SLip}^0_{\bar\rho}(X)$. Since $\|\cdot|_\rho$ is a norm on $\mathrm{SLip}^0_\rho(X)$ it follows $\|f|_{\bar\rho} = \|-f|_\rho = \|f|_\rho$.

$2 \Rightarrow 3$. If $\mathrm{SLip}^0_\rho(X) = \mathrm{SLip}^0_{\bar\rho}(X)$, then, by Proposition 2.6.10, the topology τ_ρ is T_1. Consequently, if ρ is not a metric, then there exists a pair $a, b \in X$ such that $\rho(a,b) > \rho(b,a) > 0$. The function $f(x) = \rho(x,b) - \rho(x_0,b)$, $x \in X$, is in $\mathrm{SLip}^0_\rho(X)$ and $\|f|_\rho \le 1$ (see Proposition 2.6.4). By hypothesis $f \in \mathrm{SLip}^0_{\bar\rho}(X)$. But

$$\|f|_{\bar\rho} = \sup\{u(f(x) - f(y))/\rho(y,x) : \rho(y,x) > 0\} \ge \frac{u(f(a) - f(b))}{\rho(b,a)} = \frac{\rho(a,b)}{\rho(b,a)} > 1\,,$$

in contradiction to the hypothesis $\|f|_{\bar\rho} = \|f|_\rho$.

$3 \Rightarrow 1$. If (X, ρ) is a metric space, then $\mathrm{SLip}^0_\rho(X) = \mathrm{SLip}^0_{\bar\rho}(X) = \mathrm{Lip}_0(X)$ – the space of Lipschitz functions vanishing at x_0. It is well known that $\mathrm{Lip}_0(X)$ is a Banach space with respect to the Lipschitz norm $\|\cdot\|_\rho = \|\cdot|_\rho = \|\cdot|_{\bar\rho}$ (see [236]). □

Remark 2.6.13. Concerning the validity of Proposition 2.6.12 in the case of spaces of semi-Lipschitz functions with values in an asymmetric normed space (Y, q), Sánchez-Álvarez [215] has shown that $2 \iff 3$ and $3 \Rightarrow 1$, but the implication $1 \Rightarrow 3$ does not hold in general.

2.6.3 Completeness properties of the spaces of semi-Lipschitz functions

In order to treat some completeness questions for spaces of semi-Lipschitz functions, one defines an extended quasi-metric on $\mathrm{SLip}_\rho(X)$ by the formula

$$\delta_\rho(f,g) = \sup\left\{\frac{u((g-f)(x) - (g-f)(y))}{\rho(x,y)} : x, y \in X,\ \rho(x,y) > 0\right\}. \qquad (2.6.12)$$

For $f, g \in \mathrm{SLip}_\rho(X)$ put also

$$\bar{\delta}_\rho(f,g) = \delta_\rho(g,f) \quad \text{and} \quad \delta^s_\rho(f,g) = \delta_\rho(f,g) \vee \bar{\delta}_\rho(f,g) .$$

Because $[\varphi(x) \vee 0] \vee [(-\varphi(x)) \vee 0] = |\varphi(x)|$,

$$\begin{aligned} \delta^s_\rho(f,g) &= \delta_\rho(f,g) \vee \delta_\rho(g,f) \\ &= \sup\left\{ \frac{|(g-f)(x) - (g-f)(y)|}{\rho(x,y)} : x, y \in X,\ \rho(x,y) > 0 \right\} . \end{aligned}$$

In fact δ_ρ can be considered as an extended quasi-metric on the linear space $\mathrm{SLip}_\rho(X) - \mathrm{SLip}_\rho(X)$ generated by $\mathrm{SLip}_\rho(X)$ in $\mathrm{Lip}_\rho(X)$.

Remark 2.6.14. If the topology τ_ρ is T_1 (equivalently, $\rho(x,y) > 0$ whenever $x \neq y$), then

$$\bar{\delta}_\rho(f,g) = \delta_{\bar{\rho}}(f,g) .$$

Indeed

$$\begin{aligned} \delta_{\bar{\rho}}(f,g) &= \sup\left\{ \frac{u\left((g-f)(x) - (g-f)(y)\right)}{\rho(y,x)} : x, y \in X,\ x \neq y \right\} \\ &= \sup\left\{ \frac{u\left((f-g)(x) - (f-g)(y)\right)}{\rho(x,y)} : x, y \in X,\ x \neq y \right\} \\ &= \delta_\rho(g,f) = \bar{\delta}_\rho(f,g) . \end{aligned}$$

The following example, given by Romaguera and Sanchis [204], shows that δ_ρ could be effectively an extended quasi-metric.

Example 2.6.15. For $x, y \in \mathbb{R}$ let $\rho(x,y) = x - y$ if $x \geq y$ and $\rho(x,y) = 1$ if $x < y$, i.e., $(\mathbb{R}, \rho)$ is the Sorgenfrey line (see Example 1.1.6). The identity mapping $\mathrm{id} : \mathbb{R} \to \mathbb{R}$ is semi-Lipschitz with $\|\,\mathrm{id}\,|_\rho = 1$, so that $\delta_\rho(0, \mathrm{id}) = 1$, but $\delta_\rho(\mathrm{id}, 0) = \infty$ because $\sup\{((y-x) \vee 0)/\rho(x,y) : x \neq y\} = \infty$.

Theorem 2.6.16. *Let (X, ρ) be a quasi-metric space.*

1. ([204]) *The space $\mathrm{SLip}^0_\rho(X)$ is bicomplete with respect to the extended quasi-metric δ_ρ, that is complete with respect to the extended metric $\delta^s_\rho = \delta_\rho \vee \bar{\delta}_\rho$.*
2. ([206]) *The extended quasi-metric space $(\mathrm{SLip}^0_\rho(X), \bar{\delta}_\rho)$ is right K-complete.*

Proof. 1. Let (f_n) be a δ^s_ρ-Cauchy sequence in $\mathrm{SLip}^0_\rho(X)$. Then

$$\begin{aligned} \delta^s_\rho(f_n, f_{n+k}) &= \delta_\rho(f_n, f_{n+k}) \vee \delta_\rho(f_{n+k}, f_n) \\ &= \sup\left\{ \frac{|(f_{n+k} - f_n)(u) - (f_{n+k} - f_n)(v)|}{\rho(u,v)} : u, v \in X,\ \rho(u,v) > 0 \right\} , \end{aligned}$$

so that for every $\varepsilon > 0$ there exists $n_\varepsilon \in \mathbb{N}$ such that

$$\frac{|(f_{n+k} - f_n)(u) - (f_{n+k} - f_n)(v)|}{\rho(u,v)} \leq \varepsilon \,, \tag{2.6.13}$$

for all $n \geq n_\varepsilon$, $k \in \mathbb{N}$ and all $u, v \in X$ with $\rho(u,v) > 0$.

Claim I. For every $x \in X$, $(f_n(x))_{n\in\mathbb{N}}$ is a Cauchy sequence in $(\mathbb{R}, |\cdot|)$.

Let $x \neq x_0$ be fixed and $\varepsilon' > 0$. If $\rho(x, x_0) > 0$ let $n_0 \in \mathbb{N}$ be such that (2.6.13) holds for $\varepsilon := \varepsilon'/\rho(x, x_0)$. Taking $u = x$ and $v = x_0$ in (2.6.13) it follows that

$$|(f_{n+k} - f_n)(x)| \leq \varepsilon\rho(x, x_0) = \varepsilon' \,,$$

for all $n \geq n_\varepsilon$ and all $k \in \mathbb{N}$. If $\rho(x_0, x) > 0$, then use (2.6.13) with $\varepsilon := \varepsilon'/\rho(x_0, x)$, $u = x_0$ and $v = x$ to obtain

$$|(f_{n+k} - f_n)(x)| \leq \varepsilon\rho(x_0, x) = \varepsilon' \,,$$

for all $n \geq n_\varepsilon$ and all $k \in \mathbb{N}$.

Consequently $(f_n(x))_{n\in\mathbb{N}}$ is a Cauchy sequence in $(\mathbb{R}, |\cdot|)$ for every $x \in X$, so that we can define a function $f : X \to \mathbb{R}$ by $f(x) = \lim_n f_n(x)$, $x \in X$.

To end the proof we have to show that $f \in \mathrm{SLip}^0_\rho(X)$ and that $f_n \xrightarrow{\delta^s_\rho} f$.

Claim II. $f \in \mathrm{SLip}^0_\rho(X)$.

Let $m \in \mathbb{N}$ be such that (2.6.13) holds for $\varepsilon = 1$. Then for every $x, y \in X$ with $\rho(x, y) > 0$ and $k \in \mathbb{N}$,

$$|(f_{m+k} - f_m)(x) - (f_{m+k} - f_m(y)| \leq \rho(x, y) \,,$$

yielding for $k \to \infty$,

$$|(f - f_m)(x) - (f - f_m)(y)| \leq \rho(x, y) \,.$$

But then

$$\begin{aligned} f(x) - f(y) &= (f - f_m)(x) - (f - f_m)(y) + f_m(x) - f_m(y) \\ &\leq (1 + \|f_m|_\rho)\, \rho(x, y) \,, \end{aligned}$$

for all $x, y \in X$ with $\rho(x, y) > 0$. Since the pointwise limit of a sequence of $\leq_\rho$-monotone functions is $\leq_\rho$-monotone, it follows that $f \in \mathrm{SLip}^0_\rho(X)$.

Claim III. $f_n \xrightarrow{\delta^s_\rho} f$.

For $\varepsilon > 0$ let $n_\varepsilon \in \mathbb{N}$ be chosen according to (2.6.13). Then for every $x, y \in X$ with $\rho(x, y) > 0$

$$\forall n \geq n_\varepsilon, \;\; \forall k \in \mathbb{N}, \quad |(f_{n+k} - f_n)(x) - (f_{n+k} - f_n)(y)| \leq \rho(x, y) \,,$$

which yields for $k \to \infty$,

$$\forall n \geq n_\varepsilon, \quad |(f - f_n)(x) - (f - f_n)(y)| \leq \varepsilon\rho(x, y) .$$

It follows that

$$\forall n \geq n_\varepsilon, \quad \delta_\rho^s(f, f_n) \leq \varepsilon ,$$

proving Claim III.

2. Let (f_n) be a right $\bar{\delta}_\rho$-K-Cauchy sequence in $\mathrm{SLip}_\rho^0(X)$. Then for every $\varepsilon > 0$ there exists $n_\varepsilon \in \mathbb{N}$ such that

$$\forall n \geq n_\varepsilon, \ \forall k \in \mathbb{N}, \quad \bar{\delta}_\rho(f_{n+k}, f_n) \leq \varepsilon ,$$

that is

$$\frac{u\left((f_{n+k} - f_n)(u) - (f_{n+k} - f_n)(v)\right)}{\rho(u, v)} \leq \varepsilon , \tag{2.6.14}$$

for all $n \geq n_\varepsilon$, all $k \in \mathbb{N}$ and all $u, v \in X$ with $\rho(u, v) > 0$.

Claim I. For every $x \in X$, $(f_n(x))_{n\in\mathbb{N}}$ is a Cauchy sequence in $(\mathbb{R}, |\cdot|)$.

Let $x \in X \setminus \{x_0\}$ and $\varepsilon' > 0$.

Case 1. $\rho(x, x_0) > 0$ and $\rho(x_0, x) > 0$. Take n_0 such that (2.6.14) holds for $\varepsilon := \varepsilon'/\rho^s(x, x_0)$. Taking first $u = x, \ v = x_0$ and then $u = x_0, \ v = x$, one obtains

$$(f_{n+k} - f_n)(x) \leq \varepsilon\rho(x, x_0) \leq \varepsilon\rho^s(x, x_0) = \varepsilon', \quad \text{respectively}$$
$$(f_{n+k} - f_n)(x) \leq \varepsilon\rho(x_0, x) \leq \varepsilon\rho^s(x, x_0) = \varepsilon' .$$

Consequently

$$\forall n \geq n_0, \quad |f_{n+k}(x) - f_n)(x)| \leq \varepsilon' ,$$

showing that the sequence $(f_n(x)$ is $|\cdot|$-Cauchy.

Case 2. $\rho(x, x_0) > 0$ and $\rho(x_0, x) = 0$. Reasoning like above, given $\varepsilon > 0$ there exists $m_0 \in \mathbb{N}$ such that

$$\forall n \geq m_0, \ \forall k \in \mathbb{N}, \quad f_{n+k}(x) - f_n(x) \leq \varepsilon . \tag{2.6.15}$$

Since the functions f_n are $\leq_\rho$-monotone, $0 = f_n(x_0) \leq f_n(x)$ for all $n \in \mathbb{N}$.

Applying (2.6.14) for $\varepsilon = 1/\rho(x, x_0)$ and $u = x, \ v = x_0$, it follows that there exists $m_1 \in \mathbb{N}$ such that $f_{m_1+k}(x) - f_{m_1}(x) \leq 1$, implying $0 \leq f_{m_1+k}(x) \leq f_{m_1}(x) + 1$ for all $k \in \mathbb{N}$. Consequently, the sequence $(f_n(x))$ is bounded, so it contains a subsequence $f_{n_i}(x))_{i\in\mathbb{N}}$ converging to some $f(x) \in \mathbb{R}$.

Let $i_0 \in \mathbb{N}$ such that $n_{i_0} > m_0$ and

$$\forall i \geq i_0, \quad |f_{n_i}(x) - f(x)| \leq \varepsilon . \tag{2.6.16}$$

For $n > n_{i_0}$ and $k \in \mathbb{N}$ let $n_j > n+k$. Then, combining (2.6.15) and (2.6.16), one obtains

$$\begin{aligned} & f_n(x) - f_{n+k}(x) \\ & \quad = f_n(x) - f_{n_{i_0}}(x) + f_{n_{i_0}}(x) - f(x) + f(x) - f_{n_j}(x) + f_{n_j}(x) - f_{n+k}(x) \\ & \quad \le 4\varepsilon \,. \end{aligned} \tag{2.6.17}$$

The inequalities (2.6.15) and (2.6.17) show that $(f_n(x))$ is a $|\cdot|$-Cauchy sequence.

Case 3. $\rho(x, x_0) = 0$ and $\rho(x_0, x) > 0$. In this case $f_n(x) \le f_n(x_0) = 0$ and a reasoning analogous to that made in Case 2 shows that the sequence $(f_n(x))$ is Cauchy in this case too.

Consequently we can define a function $f : X \to \mathbb{R}$ by $f(x) = \lim_n f_n(x)$, $x \in X$.

To end the proof we have to show that $f \in \mathrm{SLip}_\rho^0(X)$ and that $f_n \xrightarrow{\bar{\delta}_\rho} f$.

Claim II. $f \in \mathrm{SLip}_\rho^0(X)$.

Applying (2.6.14) for $\varepsilon = 1$, it follows that there exists $m_1 \in \mathbb{N}$ such that for all $x, y \in X$ with $\rho(x, y) > 0$,

$$\forall k \in \mathbb{N}, \quad \frac{(f_{m_1+k} - f_{m_1})(x) - (f_{m_1+k} - f_{m_1})(y)}{\rho(x, y)} \le 1 \,,$$

yielding, for $k \to \infty$ and after some calculation,

$$\frac{f(x) - f(y)}{\rho(x, y)} \le \frac{f_{m_1})(x) - f_{m_1})(y)}{\rho(x, y)} + 1 \le \| f_{m_1} |_\rho + 1 \,.$$

As a pointwise limit of a sequence of $\le_\rho$-monotone functions, the function f is $\le_\rho$-monotone, so it belongs to $\mathrm{SLip}_\rho^0(X)$.

Claim III. $f_n \xrightarrow{\bar{\delta}_\rho} f$.

For $\varepsilon > 0$ let $n_\varepsilon \in \mathbb{N}$ such that (2.6.14) holds. Considering $n \ge n_\varepsilon$ fixed and letting $k \to \infty$, one obtains

$$\frac{u\left((f - f_n)(u) - (f - f_n)(v)\right)}{\rho(u, v)} \le \varepsilon \,,$$

for all $u, v \in X$ with $\rho(u, v) > 0$ and all $n \ge n_\varepsilon$. It follows that

$$\bar{\delta}(f, f_n) \le \varepsilon \,,$$

for all $n \ge n_\varepsilon$, proving Claim III. □

Remark 2.6.17. Sánchez-Álvarez [215] studied the completeness of the space $\mathrm{SLip}^0_{\rho,q}(X,Y)$, for (X,ρ) a quasi-metric space and (Y,q) an asymmetric normed space, with respect to the extended quasi-metric

$$\delta_{\rho,q}(f,g) = \sup\left\{\frac{q((g-f)(x)-(g-f)(y))}{\rho(x,y)} : x,y\in X,\ \rho(x,y)>0\right\}. \tag{2.6.18}$$

He proved that if the asymmetric normed space (Y,q) is bicomplete, then $\mathrm{SLip}^0_{\rho,q}(X,Y)$ is complete with respect to the extended metric

$$\delta^s_{\rho,q}(f,g) = \delta_{\rho,q}(f,g)\vee\delta_{\rho,q}(g,f)$$

([215, Theorem 3.5]).

Assertion 2 from Theorem 2.6.16 was extended to the case when (Y,q) is a finite-dimensional bicomplete asymmetric normed space (Theorem 4.1 in the same paper).

We mention also the following result from [215].

Proposition 2.6.18. *Let (X,ρ) be a T_1 quasi-metric space and (Y,q) a bicomplete asymmetric normed space. Then the linear space $\mathrm{SLip}^0_{\rho,q}(X,Y)\cap\mathrm{SLip}^0_{\bar\rho,q}(X,Y)$ is complete with respect to the norm*

$$\|f\|_{\rho,q} = \sup\left\{\frac{q^s(f(x)-f(y))}{\rho(x,y)} : x,y\in X,\ \rho(x,y)>0\right\}.$$

Proof. Observe that

$$\begin{aligned}\|f\|_{\rho,q} &= \sup\left\{\frac{q(f(x)-f(y))\vee q(f(y)-f(x))}{\rho(x,y)} : x,y\in X,\ \rho(x,y)>0\right\}\\ &= \sup\left\{\frac{q(f(x)-f(y))}{\rho(x,y)} : x,y\in X,\ \rho(x,y)>0\right\}\\ &\qquad\vee\sup\left\{\frac{q(f(y)-f(x))}{\rho(x,y)} : x,y\in X,\ \rho(x,y)>0\right\}\\ &= \|f|_{\rho,q}\vee\|-f|_{\rho,q} = \|f|^s_{\rho,q},\end{aligned}$$

so that $\|\cdot|^s_{\rho,q}$ is the norm associated to the asymmetric norm $\|\cdot|_{\rho,q}$ given by (2.6.5).

The proof of the completeness follows the line of the proof of the assertion 1 in Theorem 2.6.16. □

Remark 2.6.19. If the metric space (X,ρ) is T_1, then

$$\mathrm{SLip}^0_{\rho,q}(X,Y)\cap\mathrm{SLip}^0_{\bar\rho,q}(X,Y) = \mathrm{Lip}^0\left((X,\rho^s),(Y,q^s)\right),$$

and the norm $\|\cdot\|_{\rho,q}$ agrees with the Lipschitz norm $\|\cdot\|_{\rho^s,q^s}$ given by (2.6.2).

Indeed,

$$\|f\|_{\rho,q} = \sup\left\{\frac{q^s(f(x)-f(y))}{\rho(x,y)} : x,y \in X,\ x \neq y\right\}$$
$$= \sup\left\{\frac{q^s(f(x)-f(y))}{\rho^s(x,y)} : x,y \in X,\ x \neq y\right\} = \|f\|_{\rho^s,q^s} .$$

Doitchinov [62, 63, 64] defined and studied a notion of completeness for quasi-metric spaces with the aim to obtain a satisfactory theory of completion (see [62] for the quasi-metric case and [66] for quasi-uniform spaces). A sequence (x_n) in a quasi-metric space (X,ρ) is called *D-Cauchy* if there exists another sequence (y_n) such that $\lim_{m,n} \rho(y_m, x_n) = 0$. The quasi-metric space (X,ρ) is called D-complete if every D-Cauchy sequence converges. A quasi-metric space (X,ρ) is called *balanced* if for all sequences $(x_n),(y_n)$ such that $\lim_{m,n} \rho(y_m, x_n) = 0$ and for every $x,y \in X$ and $r_1, r_2 \geq 0$, $\rho(x,x_n) \leq r_1$ and $\rho(y_n,y) \leq r_2$ for all $n \in \mathbb{N}$, implies $\rho(x,y) \leq r_1 + r_2$. The concept of balancedness, meaning a kind of symmetry of a quasi-metric space, was also introduced by Doitchinov in [63], to develop a satisfactory theory of completion. He proved that a balanced quasi-metric generates a Hausdorff and completely regular topology, see Proposition 1.1.11 and Subsection 1.2.6.

The following completeness result was proved in [198].

Theorem 2.6.20. *Let (X,ρ) be a T_1 quasi-metric space and (Y,q) an asymmetric normed space. If the asymmetric normed space (Y,q) is biBanach, then the space $\mathrm{SLip}^0_{\rho,q}(X,Y)$ is balanced and D-complete with respect to the extended quasi-metric $\delta_{\rho,q}$ defined by* (2.6.18).

Proof. The metric $\boldsymbol{\delta_{\rho,q}}$ is balanced on $\mathrm{SLip}^0_{\rho,q}(X,Y)$.

Let $(f_m),(g_n)$ be two sequences in $\mathrm{SLip}^0_{\rho,q}(X,Y)$ and $f,g \in \mathrm{SLip}^0_{\rho,q}(X,Y)$ such that

$$\begin{aligned}
&\text{(i)} && \lim_{m,n\to\infty} \delta_{\rho,q}(f_m, g_n) = 0,\\
&\text{(ii)} && \forall m \in \mathbb{N},\ \ \delta_{\rho,q}(f_m, f) \leq r_1,\\
&\text{(iii)} && \forall n \in \mathbb{N},\ \ \delta_{\rho,q}(g, g_n) \leq r_2\ ,
\end{aligned} \tag{2.6.19}$$

for some some numbers $r_1, r_2 > 0$.

We have to show that $\delta_{\rho,q}(g,f) \leq r_1 + r_2$.

Since the metric ρ is T_1, $\rho(x,y) > 0 \iff x \neq y$, so that the extended quasi-metric (2.6.18) is given by

$$\delta_{\rho,q}(h_1,h_2) = \sup_{x\neq y} \frac{q((h_2-h_1)(x)-(h_2-h_1)(y))}{\rho(x,y)} ,$$

for $h_1, h_2 \in \mathrm{SLip}^0_{\rho,q}(X,Y)$.

Let $x \neq y$ be a fixed pair of distinct elements in X. Then, by condition (i) in (2.6.19), for every $\varepsilon > 0$ there exists $n_\varepsilon \in \mathbb{N}$ such that

$$\frac{q((g_n - f_m)(u) - (g_n - f_m)(v))}{\rho(u, v)} \leq \varepsilon \,, \tag{2.6.20}$$

for all $m, n \geq n_\varepsilon$ and all $u, v \in X$ with $u \neq v$.

It follows that

$$\begin{aligned} & q((f-g)(x) - (f-g)(y)) \\ & \quad \leq q((f - f_{n_\varepsilon})(x) - (f - f_{n_\varepsilon})(y)) + q((f_{n_\varepsilon} - g_{n_\varepsilon})(x) - (f_{n_\varepsilon} - g_{n_\varepsilon})(y)) \\ & \qquad + q((g_{n_\varepsilon} - g)(x) - (g_{n_\varepsilon} - g)(y) \leq r_1\rho(x, y) + \varepsilon\rho(y, x) + r_2\rho(x, y) \,. \end{aligned} \tag{2.6.21}$$

The inequality for the middle term in the second row follows from (2.6.20) and the equality

$$q((f_{n_\varepsilon} - g_{n_\varepsilon})(x) - (f_{n_\varepsilon} - g_{n_\varepsilon})(y)) = q((g_{n_\varepsilon} - f_{n_\varepsilon})(y) - (g_{n_\varepsilon} - f_{n_\varepsilon})(x)) \,.$$

Since $\varepsilon > 0$ is arbitrary in (2.6.21), it follows that

$$q((f-g)(x) - (f-g)(y)) \leq q((f - f_{n_\varepsilon})(x) - (f - f_{n_\varepsilon})(y)) \leq r_1\rho(x, y) + r_2\rho(x, y) \,,$$

for all $x, y \in X$ with $x \neq y$, that is $\delta_{\rho,q}(g, f) \leq r_1 + r_2$.

The metric $\delta_{\rho,q}$ is D-complete on $\mathrm{SLip}^0_{\rho,q}(X, Y)$.

Suppose that (f_n) is a D-Cauchy sequence in $\mathrm{SLip}^0_{\rho,q}(X, Y)$ and let (g_m) be a sequence in $\mathrm{SLip}^0_{\rho,q}(X, Y)$ such that

$$\lim_{m,n\to\infty} \delta_{\rho,q}(g_m, f_n) = 0 \,, \tag{2.6.22}$$

that is (g_n) is a cosequence for (f_n).

Claim I. *For every $x \in X$, $(f_n(x))$ is a Cauchy sequence in (Y, q^s).*

Let $x \in X$, $x \neq x_0$. For $\varepsilon > 0$ let $n_\varepsilon \in \mathbb{N}$ be chosen according to (2.6.20). Taking in (2.6.20) first $u = x$, $v = x_0$ and then $u = x_0$, $v = x$, one obtains

$$\begin{aligned} & q((f_n - g_m)(x) \leq \varepsilon\rho(x, x_0) \leq \varepsilon\rho^s(x, x_0), \quad \text{respectively} \\ & q((g_m - f_n)(x) \leq \varepsilon\rho(x_0, x) \leq \varepsilon\rho^s(x, x_0) \,, \end{aligned}$$

implying

$$q^s((f_n - g_m)(x) \leq \varepsilon\rho^s(x, x_0) \,, \tag{2.6.23}$$

for all $m, n \geq n_\varepsilon$. But then,

$$q^s((f_n - f_m)(x) \leq q^s((f_n - g_{n_\varepsilon})(x) + q^s((g_{n_\varepsilon} - f_m)(x) \leq 2\varepsilon\rho^s(x_0, x) \,,$$

for all $m, n \geq n_\varepsilon$, proving that $(f_n(x))$ is a Cauchy sequence in the Banach space (Y, q^s), so it converges to some $y \in Y$.

It follows that we can define a function $f : X \to Y$ by $f(x) = \lim_n f_n(x)$, $x \in X$. To end the proof we have to show that $f \in \mathrm{SLip}^0_{\rho,q}(X, Y)$ and $f_n \to f$ with respect to the extended quasi-metric $\delta_{\rho,q}$.

Observe that, by (2.6.23), the sequence $(g_m(x))$ also converges to $f(x)$ in the norm q^s.

Claim II. $f \in \mathrm{SLip}^0_{\rho,q}(X, Y)$.

Let $n_1 \in \mathbb{N}$ be such that (2.6.20) holds for $\varepsilon = 1$. Then, taking into account (2.6.19).(ii), one gets

$$\begin{aligned} q(f(u) - f(v)) &\leq q((f - f_{n_1})(u) - (f - f_{n_1})(v)) \\ &\quad + q((f_{n_1} - g_{n_1})(u) - (f_{n_1} - g_{n_1})(v)) + q((g_{n_1})(u) - (g_{n_1})(v)) \\ &\leq (r_1 + 1 + \|g_{n_1}|_{\rho,q})\rho(u, v) , \end{aligned}$$

for all $u, v \in X$, proving that $f \in \mathrm{SLip}^0_{\rho,q}(X, Y)$.

Claim III. *$f_n \to f$ with respect to the extended quasi-metric $\delta_{\rho,q}$.*

Given $\varepsilon > 0$ let n_ε be chosen according to (2.6.20) and let $x \neq y$ in X, arbitrary, but fixed for the moment.

Let $n \geq n_\varepsilon$. Since $\lim_m q^s(g_m(u) - f(u)) = 0$ for every $u \in X$, there exists $m \geq n_\varepsilon$ such that

$$q^s(g_m(x)f(x)) \leq \varepsilon\rho(x, y) \quad \text{and} \quad q^s(g_m(y) - f(y)) \leq \varepsilon\rho(x, y) . \tag{2.6.24}$$

But then

$$\begin{aligned} &q((f_n - f)(x) - (f_n - f)(y)) \\ &\quad \leq q((f_n - g_m)(y) - (f_n - g_m)(y)) + q((g_m - f)(x) - (g_m - f)(y)) \\ &\quad \leq 3\varepsilon\rho(x, y) , \end{aligned}$$

because, by (2.6.24),

$$\begin{aligned} q((g_m - f)(x) - (g_m - f)(y)) &\leq q^s((g_m - f)(x) - (g_m - f)(y)) \\ &\leq q^s((g_m - f)(x))(x)) + q^s((g_m - f)(x))(y)) \leq 2\varepsilon\rho(x, y) . \end{aligned}$$

Since the points $x \neq y$ were arbitrarily chosen in X, it follows that $\delta_{\rho,q}(f, f_n) \leq 3\varepsilon$ for all $n \geq n_\varepsilon$, proving the convergence of the sequence (f_n) to f with respect to the extended quasi-metric $\delta_{\rho,q}$. □

Let (X, p) be a T_1 asymmetric normed space and $X^\flat_p$ its dual. Then $X^\flat_p$ is contained in the cone $\mathrm{SLip}^0_{\bar{p}}(X)$, where

$$\bar{\rho}(x, y) = p(x - y) = \bar{p}(y - x) = \rho_{\bar{p}}(x, y), \;\; x, y \in X .$$

Indeed

$$\varphi \in X_\rho^\flat \ \Rightarrow\ \forall x, y \in X,\ \ \varphi(x-y) \le \|\varphi|_p\, p(x-y) = \|\varphi|_p\, \bar{\rho}(x,y)\ \Rightarrow \varphi \in \mathrm{SLip}_{\bar{\rho}}^0(X)\,.$$

It follows also that $\|\varphi|_{\bar{\rho}} \le \|\varphi|_p$. In fact we have equality:

$$\|\varphi|_{\bar{\rho}} = \sup\{\frac{u(\varphi(x-y))}{p(x-y)} : x \neq y\} = \sup\{u(\varphi(z)) : p(z) \le 1\} = \|\varphi|_p\,.$$

It follows that the restriction $\delta_p^\flat$ of the extended quasi-metric $\delta_{\bar{\rho}}$ to $X_p^\flat$ is given by

$$\begin{aligned}\delta_p^\flat(\varphi,\psi) &= \sup\{\frac{u((\psi-\varphi)(x-y))}{p(x-y)} : x \neq y\} \\ &= \sup\{u((\psi-\varphi)(z)) : X,\, p(z) \le 1\} = \|\psi - \varphi|_p\,,\end{aligned}$$

that is it agrees with the extended quasi-metric associated to the extended asymmetric norm $\|\cdot|_{p,u}$, see (2.1.23) and Proposition 2.1.3.

Theorem 2.6.21 ([198]). *If (X,p) is a T_1 asymmetric normed space, then the dual space $X_p^\flat$ is balanced and D-complete with respect to the extended quasi-metric $\delta_p^\flat$.*

Proof. Since $X_p^\flat \subset \mathrm{SLip}_{\bar{\rho}}^0(X)$ and the balancedness is a hereditary property it follows that $X_p^\flat$ is balanced with respect to the extended metric $\delta_p^\flat$.

If (φ_n) is a D-Cauchy sequence in $(X_p^\flat, \delta_p^\flat)$, then it is D-Cauchy in $(\mathrm{SLip}_\rho^0(X), \delta_{\bar{\rho}})$, so that, by Theorem 2.6.20, it converges to some $\varphi \in \mathrm{SLip}_{\bar{\rho}}^0(X)$. By the proof of the same theorem, for every $x \in X$ the sequence $(\varphi_n(x))$ converges in $(\mathbb{R}, |\cdot|)$ to $\varphi(x)$, implying the linearity of the limit φ. Since φ is semi-Lipschitz, there exists $\beta > 0$ such that

$$\varphi(x) = \varphi(x) - \varphi(0) \le u\,(\varphi(x) - \varphi(0)) \le \beta\, \bar{\rho}(x,0) = \beta\, p(x)\,,$$

showing that $\varphi \in X_p^\flat$. □

2.6.4 Applications to best approximation in quasi-metric spaces

Semi-Lipschitz functions can be used to study some best approximation problems in quasi-metric spaces. The notions can transposed from asymmetric normed spaces (see (2.2.3)) to quasi-metric spaces. Let (X,ρ) be a quasi-metric space. For $Y \subset X$ and $x \in X$ put

$$\begin{aligned} d(x,Y) &= \inf\{\rho(x,y) : y \in Y\} \quad \text{and} \\ d(Y,x) &= \inf\{\rho(y,x) : y \in Y\}\,, \end{aligned}$$

see (1.1.22). Consider also the set

$$Y^\perp = \{f \in \mathrm{SLip}_\rho^0(X) : f|_Y = 0\}\,.$$

The following result is the semi-Lipschitz analog of Theorem 2.5.1.

Proposition 2.6.22. *Let (X, ρ) be a quasi-metric space, $Y \subset X$ nonempty, $x_0 \in X$ such that $d(x_0, Y) > 0$ and $y_0 \in Y$. Then y_0 is a nearest point to x_0 in Y if and only if there exists $f \in Y^{\perp}$ such that*

$$\text{(i)} \quad \|f|_{\rho} = 1 \quad \textit{and} \quad \text{(ii)} \quad \rho(x_0, y_0) = f(x_0) - f(y_0) .$$

In the case of metric spaces and Lipschitz functions similar results were obtained by Mustăţa [154, 155], who proved also many results on the characterization of the approximation properties in a quasi-metric space in terms of the semi-Lipschitz functions defined on it. Other results on best approximation and extensions were obtained in [160, 163].

Let $(X, \|\cdot\|)$ be a normed space and X^* its dual. For a subspace Y of X put

$$Y^{\perp} = \{x^* \in X^* : x^*|_Y = 0\}$$

and denote by $E_Y(y^*)$ the set of all norm-preserving extensions of a continuous linear functional y^* on Y, that is,

$$E_Y(y^*) = \{x^* \in X^* : x^*|_Y = y^* \text{ and } \|x^*\| = \|y^*\|\} .$$

Phelps [177] proved the following remarkable result relating the approximation properties of the space $Y^{\perp}$ and the extension properties of the space Y.

Theorem 2.6.23 (Phelps [177]). *Let Y be a closed subspace of a normed space X. Then $Y^{\perp}$ is a proximinal subspace of X^* and for every $x^* \in X^*$ the following equality holds:*

$$P_{Y^{\perp}}(x^*) = E_Y(x^*|_Y) .$$

Consequently, $Y^{\perp}$ is a Chebyshev subspace of X^* if and only if every $y^* \in Y^*$ has a unique norm-preserving extension to the whole of X.

Extensions of this result to spaces of Lipschitz functions on metric spaces and to spaces of semi-Lipschitz functions on quasi-metric spaces were given by Mustăţa [156, 157, 159, 161, 162, 165] (see also the paper [39] containing a survey of various situations where a Phelps type result can occur). An iterative approximation method to find the global minimum of a semi-Lipschitz function is proposed in [166]. Romaguera and Sanchis give in [205] characterizations of preferences on separable quasi-metric spaces admitting semi-Lipschitz utility functions, with applications to theoretical computer science.

Bibliography

[1] G.E. Albert, *A note on quasi-metric spaces*, Bull. Amer. Math. Soc. **47** (1941), 479–482.

[2] C. Alegre, *Continuous operators on asymmetric normed spaces*, Acta Math. Hungar. **122** (2009), no. 4, 357–372.

[3] C. Alegre and I. Ferrando, *Quotient subspaces of asymmetric normed linear spaces*, Bol. Soc. Mat. Mexicana (3) **13** (2007), no. 2, 357–365.

[4] C. Alegre, I. Ferrando, L.M. García-Raffi, and E.A. Sánchez Pérez, *Compactness in asymmetric normed spaces*, Topology Appl. **155** (2008), no. 6, 527–539.

[5] C. Alegre, J. Ferrer, and V. Gregori, *Quasi-uniformities on real vector spaces*, Indian J. Pure Appl. Math. **28** (1997), no. 7, 929–937.

[6] ———, *On a class of real normed lattices*, Czechoslovak Math. J. **48(123)** (1998), no. 4, 785–792.

[7] ———, *On pairwise Baire bitopological spaces*, Publ. Math. Debrecen **55** (1999), no. 1–2, 3–15.

[8] ———, *On the Hahn-Banach theorem in certain linear quasi-uniform structures*, Acta Math. Hungar. **82** (1999), no. 4, 325–330.

[9] E. Alemany and S. Romaguera, *On half-completion and bicompletion of quasi-metric spaces*, Comment. Math. Univ. Carolin. **37** (1996), no. 4, 749–756.

[10] ———, *On right K-sequentially complete quasi-metric spaces*, Acta Math. Hungar. **75** (1997), no. 3, 267–278.

[11] A.R. Alimov, *The Banach-Mazur theorem for spaces with nonsymmetric distance*, Uspekhi Mat. Nauk **58** (2003), no. 2, 159–160.

[12] ———, *Convexity of Chebyshev sets in a subspace*, (Russian) Mat. Zametki **78** (2005), no. 1, 3–15; translation in Math. Notes **78** (2005), no. 1–2, 3–13.

[13] A. Andrikopoulos, *Completeness in quasi-uniform spaces*, Acta Math. Hungar. **105** (2004), no. 1–2, 151–173.

[14] ———, *A larger class than the Deák one for the coincidence of some notions of quasi-uniform completeness using pairs of filters*, Studia Sci. Math. Hungar. **41** (2004), no. 4, 431–436.

[15] J.A. Antonino and S. Romaguera, *Equinormal metrics and upper semi-continuity*, Math. Jap. **36** (1991), No.1, 147–151.

[16] E. Asplund, *Čebyšev sets in Hilbert space*, Trans. Amer. Math. Soc. **144** (1969), 235–240.

[17] V.F. Babenko, *Nonsymmetric approximations in spaces of summable functions*, Ukrain. Mat. Zh. **34** (1982), no. 4, 409–416.

[18] ———, *Nonsymmetric extremal problems of approximation theory*, Dokl. Akad. Nauk SSSR **269** (1983), no. 3, 521–524.

[19] ———, *Duality theorems for certain problems of the theory of approximation*, Current problems in real and complex analysis, Akad. Nauk Ukrain. SSR Inst. Mat., Kiev, 1984, pp. 3–13.

[20] V.S. Balaganskii and L.P. Vlasov, *The problem of the convexity of Chebyshev sets*, Uspekhi Mat. Nauk **51** (1996), no. 6(312), 125–188.

[21] H.L Bentley, H. Herrlich, and W.N. Hunsaker, *A Baire category theorem for quasi-metric spaces*, Indian J. Math. **37** (1995), no. 1, 27–30

[22] T. Bîrsan, *Sur les espaces bitopologiques connexes*, An. Şti. Univ. "Al. I. Cuza" Iaşi Secţ. I a Mat. (N.S.) **14** (1968), 293–296.

[23] ———, *Compacité dans les espaces bitopologiques*, An. Şti. Univ. "Al. I. Cuza" Iaşi Secţ. I a Mat. (N.S.) **15** (1969), 317–328.

[24] ———, *Sur les espaces bitopologiques complètement réguliers*, An. Şti. Univ. "Al. I. Cuza" Iaşi Secţ. I a Mat. (N.S.) **16** (1970), 29–34.

[25] ———, *Contribution à l'étude des groupes bitopologiques*, An. Şti. Univ. "Al. I. Cuza" Iaşi Secţ. I a Mat. (N.S.) **19** (1973), no. 2, 297–310.

[26] ———, *Transitive quasi-uniformities and zero-dimensional bitopological spaces*, An. Şti. Univ. "Al. I. Cuza" Iaşi Secţ. I a Mat. (N.S.) **20** (1974), no. 2, 317–322.

[27] S. Bodjanová, *Some basic notions of mathematical analysis in oriented metric spaces*, Math. Slovaca **31** (1981), no. 3, 277–289.

[28] P.A. Borodin, *The Banach-Mazur theorem for spaces with an asymmetric norm and its applications in convex analysis*, Mat. Zametki **69** (2001), no. 3, 329–337.

[29] M.K. Bose, A.R. Choudhury, and A. Mukharjee, *On bitopological paracompactness*, Mat. Vesnik **60** (2008), no. 4, 255–259.

[30] G. Buskes, *The Hahn-Banach theorem surveyed,* Dissertationes Math. (Rozprawy Mat.) **327** (1993), 49 pp.

[31] J.W. Carlson and T.L. Hicks, *On completeness in quasi-uniform spaces*, J. Math. Anal. Appl. **34** (1971), 618–627.

[32] S.B. Chen, G. Tan and Z. Mao, *On convergence in quasi-metric spaces*, J. Wuhan Univ. Sci. and Techn. (Natural Science Ed.) **28** (2005), no. 4.

[33] S.A. Chen, W. Li, S.P. Tian and Z.Y. Mao, *On optimization problems in quasi-metric spaces*, Proc. 5th International Conf. on Machine Learning and Cybernetics, Dalian, 13–16 Aug. 2006, p. 865–870.

[34] S. Chen, S. Tian and Z. Mao, *On Caristi's fixed point theorem in quasi-metric spaces*, Dyn. Contin. Discrete Impuls. Syst. Ser. A Math. Anal. **13A** (2006), Part 3, suppl., 1150–1157.

[35] S.A. Chen, W. Li, D. Zou and S.B. Chen, *Fixed point theorems in quasi-metric spaces*, Proc. 6th International Conf. on Machine Learning and Cybernetics, Hong Kong, 19–22 Aug. 2007, pp. 2499–2504.

[36] I. Ciorănescu, *Geometry of Banach spaces, Duality mappings and nonlinear problems*, Kluwer Academic Publishers, Dordrecht, 1990.

[37] S. Cobzaş, *On a theorem of V.N. Nikolski on characterization of best approximation for convex sets*, Anal. Numér. Théor. Approx. **19** (1990), no. 1, 7–13.

[38] ______, *Some remarks on the characterization of nearest points*, Studia Univ. Babeş-Bolyai, Math. **35** (1990), no. 2, 54–56.

[39] ______, *Phelps type duality results in best approximation*, Rev. Anal. Numér. Théor. Approx. **31** (2002), no. 1, 29–43.

[40] ______, *Separation of convex sets and best approximation in spaces with asymmetric norm*, Quaest. Math. **27** (2004), no. 3, 275–296.

[41] ______, *Asymmetric locally convex spaces*, Int. J. Math. Math. Sci. (2005), no. 16, 2585–2608.

[42] ______, *Geometric properties of Banach spaces and the existence of nearest and farthest points*, Abstr. Appl. Anal. (2005), no. 3, 259–285.

[43] ______, *Compact operators on spaces with asymmetric norm*, Stud. Univ. Babeş-Bolyai Math. **51** (2006), no. 4, 69–87.

[44] ______, *Compact and precompact sets in asymmetric locally convex spaces*, Topology Appl. **156** (2009), no. 9, 1620–1629.

[45] ______, *Functional analysis in asymmetric normed spaces*, arXiv:1006.1175v1 (2010).

[46] ______, *Completeness in quasi-metric spaces and Ekeland variational principle*, Topology Appl. **158** (2011), no. 8, 1073–1084.

[47] S. Cobzaş and I. Muntean, *Duality relations and characterizations of best approximation for p-convex sets*, Anal. Numér. Théor. Approx. **16** (1987), no. 2, 95–108.

[48] S. Cobzaş and C. Mustăţa, *Extension of bounded linear functionals and best approximation in spaces with asymmetric norm*, Rev. Anal. Numér. Théor. Approx. **33** (2004), no. 1, 39–50.

[49] ______, *Best approximation in spaces with asymmetric norm*, Rev. Anal. Numér. Théor. Approx. **35** (2006), no. 1, 17–31.

[50] J. Collins and J. Zimmer, *An asymmetric Arzelà-Ascoli theorem*, Topology Appl. **154** (2007), no. 11, 2312–2322.

[51] J.J. Conradie and M.D. Mabula, *Convergence and left-K-sequential completeness in asymmetrically normed lattices,* Acta Math. Hungar. (accepted).

[52] I.E. Cooke and I.L. Reilly, *On bitopological compactness*, J. London Math. Soc. (2) **9** (1974/75), 518–522.

[53] M.C. Datta, *Projective bitopological spaces*, J. Austral. Math. Soc. **13** (1972), 327–334.

[54] ———, *Projective bitopological spaces. II*, J. Austral. Math. Soc. **14** (1972), 119–128.

[55] ———, *Paracompactness in bitopological spaces and an application to quasi-metric spaces*, Indian J. Pure Appl. Math. **8** (1977), no. 6, 685–690.

[56] J. Deák, *On the coincidence of some notions of quasi-uniform completeness defined by filter pairs*, Studia Sci. Math. Hungar. **26** (1991), no. 4, 411–413.

[57] ———, *A bitopological view of quasi-uniform completeness. I, II*, Studia Sci. Math. Hungar. **30** (1995), no. 3–4, 389–409, 411–431.

[58] ———, *A bitopological view of quasi-uniform completeness. III*, Studia Sci. Math. Hungar. **31** (1996), no. 4, 385–404.

[59] F.S. De Blasi and J. Myjak, *On a generalized best approximation problem*, J. Approx. Theory **94** (1998), no. 1, 54–72.

[60] A. Di Concilio, *Spazi quasimetrici e topologie ad essi associate* (Italian), Rend. Accad. Sci. Fis. Mat., IV. Ser., Napoli **38** (1971), 113–130 (1972).

[61] D. Doitchinov, *On completeness of quasi-uniform spaces*, C. R. Acad. Bulgare Sci. **41** (1988), no. 7, 5–8.

[62] ———, *Completeness and completions of quasi-metric spaces*, Rend. Circ. Mat. Palermo (2) Suppl. (1988), no. 18, 41–50, Third National Conference on Topology (Trieste, 1986).

[63] ———, *On completeness in quasi-metric spaces*, Topology Appl. **30** (1988), no. 2, 127–148.

[64] ———, *Cauchy sequences and completeness in quasi-metric spaces*, Pliska Stud. Math. Bulgar. **11** (1991), 27–34.

[65] ———, *A concept of completeness of quasi-uniform spaces*, Topology Appl. **38** (1991), no. 3, 205–217.

[66] ———, *Completeness and completion of quasi-uniform spaces*, Trudy Mat. Inst. Steklov. **193** (1992), 103–107.

[67] E.P. Dolzhenko and E.A. Sevastyanov, *Approximations with a sign-sensitive weight* (*existence and uniqueness theorems*), Izv. Ross. Akad. Nauk Ser. Mat. **62** (1998), no. 6, 59–102.

[68] ———, *Sign-sensitive approximations*, J. Math. Sci. (New York) **91** (1998), no. 5, 3205–3257, Analysis, 10.

[69] ———, *Approximation with a sign-sensitive weight (stability, applications to snake theory and Hausdorff approximations)*, Izv. Ross. Akad. Nauk Ser. Mat. **63** (1999), no. 3, 77–118.

[70] R.J. Duffin and L.A. Karlovitz, *Formulation of linear programs in analysis. I. Approximation theory*, SIAM J. Appl. Math. **16** (1968), 662–675.

[71] Ch.B. Dunham, *Asymmetric norms and linear approximation*, Congr. Numer. **69** (1989), 113–120, Eighteenth Manitoba Conference on Numerical Mathematics and Computing (Winnipeg, MB, 1988).

[72] B.P. Dvalishvili, *Bitopological Spaces: Theory, Relations with Generalized Algebraic Structures, and Applications*, North-Holland Mathematics Studies, 199, Elsevier Science B.V., Amsterdam, 2005.

[73] R. Engelking, *General topology*, Monografie Matematyczne, Tom 60, PWN–Polish Scientific Publishers, Warsaw, 1985 (2nd edition).

[74] M. Fabian, P. Habala, P. Hájek, V. Montesinos Santalucía, J. Pelant, and V. Zizler, *Functional Analysis and Infinite-Dimensional Geometry*, CMS Books in Mathematics/Ouvrages de Mathématiques de la SMC, 8, Springer-Verlag, New York, 2001.

[75] J. Ferrer and V. Gregori, *A sequentially compact non-compact quasi-pseudometric space*, Monatsh. Math. **96** (1983), 269–270.

[76] ———, *Completeness and Baire spaces*, Math. Chronicle **14** (1985), 39–42.

[77] J. Ferrer, V. Gregori, and C. Alegre, *Quasi-uniform structures in linear lattices*, Rocky Mountain J. Math. **23** (1993), no. 3, 877–884.

[78] ———, *A note on pairwise normal spaces*, Indian J. Pure Appl. Math. **24** (1993), 595–601.

[79] J. Ferrer, V. Gregori, and I.L. Reilly, *Some properties of semi-continuous functions and quasi-uniform spaces*, Mat. Vesnik **47** (1995), no. 1–2, 11–18.

[80] P. Fletcher and W.F. Lindgren, *Quasi-uniform spaces,* M. Dekker, New York 1982.

[81] C. Franchetti and I. Singer, *Best approximation by elements of caverns in normed linear spaces*, Boll. Un. Mat. Ital. B (5) **17** (1980), no. 1, 33–43.

[82] B. Fuchssteiner and H. König, *New versions of the Hahn-Banach theorem*, General inequalities, 2 (Proc. Second Internat. Conf., Oberwolfach, 1978), pp. 255–266, Birkhäuser, Basel-Boston, Mass., 1980.

[83] B. Fuchssteiner and W. Lusky, *Convex cones,* North-Holland Mathematics Studies, 56. Notas de Matemática [Mathematical Notes], 82. North-Holland Publishing Co., Amsterdam-New York, 1981.

[84] T. Fukutake, *On (Ti,Tj)-Baire spaces,* Bull. Fukuoka Univ. Ed. III **41** (1992), 35–40.

[85] G.N Galanis, *Differentiability on semilinear spaces*, Nonlin. Anal. **71** (2009), 4732–4738.

[86] L.M. García-Raffi, *Compactness and finite dimension in asymmetric normed linear spaces*, Topology Appl. **153** (2005), no. 5–6, 844–853.

[87] L.M. García-Raffi, S. Romaguera, and E.A. Sánchez Pérez, *Extensions of asymmetric norms to linear spaces*, Rend. Istit. Mat. Univ. Trieste **33** (2001), no. 1–2, 113–125.

[88] L.M. García-Raffi, S. Romaguera, and E.A. Sánchez-Pérez, *The bicompletion of an asymmetric normed linear space*, Acta Math. Hungar. **97** (2002), no. 3, 183–191.

[89] ———, *Sequence spaces and asymmetric norms in the theory of computational complexity*, Math. Comput. Modelling **36** (2002), no. 1–2, 1–11.

[90] ———, *The dual space of an asymmetric normed linear space*, Quaest. Math. **26** (2003), no. 1, 83–96.

[91] ———, *On Hausdorff asymmetric normed linear spaces*, Houston J. Math. **29** (2003), no. 3, 717–728 (electronic).

[92] ———, *Weak topologies on asymmetric normed linear spaces and non-asymptotic criteria in the theory of complexity analysis of algorithms*, J. Anal. Appl. **2** (2004), no. 3, 125–138.

[93] ———, *The Goldstine Theorem for asymmetric normed linear spaces*, Topology Appl. **156** (2009), 2284–2291.

[94] L.M. García-Raffi, S. Romaguera, E.A. Sánchez-Pérez, and O. Valero, *Metrizability of the unit ball of the dual of a quasi-normed cone*, Boll. Unione Mat. Ital. Sez. B Artic. Ric. Mat. (8) **7** (2004), no. 2, 483–492.

[95] L.M. García Raffi and E.A. Sánchez Pérez, *Asymmetric norms and optimal distance points in linear spaces*, Topology Appl. **155** (2008), no. 13, 1410–1419.

[96] A.L. Garkavi, *Duality theorems for approximation by elements of convex sets*, Uspekhi Mat. Nauk. **16** (1961), 141–145.

[97] ———, *On a criterion for best approximation*, Sibirskii Mat. Z. **5** (1964), 472–476.

[98] ———, *Minimax balayage theorem and an inscribed ball problem*, Math. Notes **30** (1981), 542–548 (the translation of Mat. Zametki **30** (1981), 109–121).

[99] E.G. Golshtein, *Duality Theory in Mathematical Programming and its Applications*, Izdat. Nauka, Moscow, 1971 (Russian).

[100] R.L. Graham, B.L. Rothschild and J.H. Spencer, *Ramsey Theory,* Second edition, Wiley-Interscience, New York, 1990.

[101] V. Gregori, J. Marín, and S. Romaguera, *Some remarks on right K-completion of quasi-uniform spaces*, Acta Math. Hungar. **88** (2000), no. 1–2, 139–146, Third Iberoamerican Conference on Topology and its Applications (Valencia, 1999).

[102] R.C. Haworth and R.A. McCoy, *Baire spaces,* Dissertationes Math. (Rozprawy Mat.) **141** (1977), 73 pp.

[103] T.L. Hicks, *Fixed point theorems for quasi-metric spaces*, Math. Japonica **33** (1988), 231–236.

[104] S.M. Huffman, T.L. Hicks, and J.W. Carlson, *Complete quasi-uniform spaces*, Canad. Math. Bull. **23** (1980), no. 4, 497–498.

[105] E. Hille, *Methods in classical and functional analysis*, Addison-Wesley Publishing Co., Reading, Mass.-London-Don Mills, Ont. 1972

[106] N. Kalton, *Quasi-Banach spaces*, in Handbook of the Geometry of Banach Spaces, Edited by W.B. Johnson and J. Lindenstrauss, Vol. 2, 1099–1130, North-Holland, Amsterdam, 2003.

[107] M. Katětov, *On real-valued functions in topological spaces*, Fund. Math. **38** (1951), 85–91.

[108] ———, *Correction to "On real-valued functions in topological spaces"*, Fund. Math. **40** (1953), 203–205.

[109] K. Keimel and W. Roth, *Ordered cones and approximation*, Lecture Notes in Mathematics, vol. 1517, Springer-Verlag, Berlin, 1992.

[110] J.L. Kelley, *General topology*, Graduate Texts in Mathematics, No. 27. Springer-Verlag, New York-Berlin, 1975.

[111] J.C. Kelly, *Bitopological spaces*, Proc. London Math. Soc. **13** (1963), 71–89.

[112] V. Klee, *Convexity of Chebyshev sets*, Math. Ann. **142** (1960/1961), 292–304.

[113] ———, *Do infinite-dimensional Banach spaces admit nice tilings?*, Studia Sci. Math. Hungar. **21** (1986), no. 3–4, 415–427.

[114] V. Klee, E. Maluta and C. Zanco, *Tiling with smooth and rotund tiles*, Fund. Math. **126** (1986), no. 3, 269–290.

[115] ———, *Uniform properties of collections of convex bodies*, Math. Ann. **291** (1991), no. 1, 153–177.

[116] V. Klee, L. Veselý, and C. Zanco, *Rotundity and smoothness of convex bodies in reflexive and nonreflexive spaces*, Studia Math. **120** (1996), no. 3, 191–204.

[117] M. Kleiber and W.J. Pervin, *A generalized Banach-Mazur theorem*, Bull. Austral. Math. Soc. **1** (1969), 169–173.

[118] D. König, *Sur les correspondances multivoques des ensembles,* Fund. Math. **8** (1926), 114–134,

[119] H. König, *Sublineare Funktionale*, Arch. Math. (Basel) **23** (1972), 500–508.

[120] ———, *Sublineare Funktionale*, Papers from the "Open House for Functional Analysts" (Aarhus Univ., Aarhus, 1972), Paper No. 10, Mat. Inst., Aarhus Univ., Aarhus, 1972, 10 pp. Various Publications Ser., No. 23.

[121] ———, *On some basic theorems in convex analysis*, Modern applied mathematics (Bonn, 1979), North-Holland, Amsterdam, 1982, pp. 107–144.

[122] ______, *On the abstract Hahn-Banach theorem due to Rodé*, Aequationes Math. **34** (1987), no. 1, 8995.

[123] ______, *Sublinear functionals and conical measures*, Arch. Math. (Basel) **77** (2001), no. 1, 56–64, Festschrift: Erich Lamprecht.

[124] Ch. Konstadilaki-Savopoulou and I.L. Reilly, *On Datta's bitopological paracompactness*, Indian J. Pure Appl. Math. **12** (1981), no. 7, 799–803.

[125] A.I. Kozko, *Analogues of Jackson-Nikolskiĭ inequalities for trigonometric polynomials in spaces with a nonsymmetric norm*, Mat. Zametki **61** (1997), no. 5, 687–699.

[126] ______, *Fractional derivatives and inequalities for trigonometric polynomials in spaces with a nonsymmetric norm*, Izv. Ross. Akad. Nauk Ser. Mat. **62** (1998), no. 6, 125–142.

[127] ______, *Multidimensional inequalities of different metrics in spaces with an asymmetric norm*, Mat. Sb. **189** (1998), no. 9, 85–106.

[128] ______, *On the order of best approximation in spaces with an asymmetric norm and a sign-sensitive weight in classes of differentiable functions*, Izv. Ross. Akad. Nauk Ser. Mat. **66** (2002), no. 1, 103–132.

[129] M.G. Krein and A.A. Nudel′man, *The Markov moment problem and extremum problems*, Nauka, Moscow 1973 (in Russian). English translation: American Mathematical Society, Providence, R.I. 1977.

[130] H.-P.A. Künzi, *A note on sequentially compact quasi-pseudometric spaces*, Monatsh. Math. **95** (1983), 219–220.

[131] ______, *Complete quasi-pseudo-metric spaces*, Acta Math. Hungar. **59** (1992), no. 1–2, 121–146.

[132] ______, *Quasi-uniform spaces–eleven years later*, Topology Proc. **18** (1993), 143–171.

[133] ______, *Nonsymmetric topology*, Bolyai Soc. Math. Studies, Vol. **4**, Topology, Szekszárd 1993, Budapest 1995, pp. 303–338.

[134] ______, *Nonsymmetric distances and their associated topologies: about the origin of basic ideas in the area of asymmetric topology*, in: C.E. Aull and R. Lowen (editors), Handbook of the History of General Topology, Kluwer Acad. Publ., Dordrecht 2001, pp. 853–868.

[135] ______, *Quasi-uniform spaces in the year 2001*, in: Recent progress in general topology, II, (North-Holland, Amsterdam, 2002), pp. 313–344.

[136] ______, *Uniform structures in the beginning of the third millenium*, Topology Appl. **154** (2007), 2745–2756.

[137] ______, *An introduction to quasi-uniform spaces*, Beyond topology, 239–304, Contemp. Math., 486, Amer. Math. Soc., Providence, RI, 2009

[138] H.-P.A. Künzi, M. Mršević, I.L. Reilly, and M.K. Vamanamurthy, *Convergence, precompactness and symmetry in quasi-uniform spaces*, Math. Japon. **38** (1993), 239–253.

[139] H.-P.A. Künzi and S. Romaguera, *Completeness of the quasi-uniformity of quasi-uniform convergence*, Papers on general topology and applications (Gorham, ME, 1995), Ann. New York Acad. Sci., vol. 806, New York Acad. Sci., New York, 1996, pp. 231–237.

[140] P.Th. Lambrinos, *A topological notion of boundedness*, Manuscripta Math. **10** (1973), 289–276.

[141] ______, *A note on quasi-uniform continuity*, Bull. Austral. Math. Soc. **8** (1973), 389–392.

[142] ______, *Quasi-uniform characterizations of (weak) boundedness and (weak) compactness*, Ann. Soc. Sci. Bruxelles Sr. I **90** (1976), no. 4, 307–316.

[143] ______, *On precompact quasi-uniform structures*, Proc. Amer. Math. Soc. **62** (1977), 365–366.

[144] E.P Lane, *Bitopological spaces and quasi-uniform spaces*, Proc. London Math. Soc. **17** (1967), 241–256.

[145] ______, *On quasi-metric spaces*, Duke Math. J. **36** (1969), 65–71.

[146] C. Li and R. Ni, *Derivatives of generalized distance functions and existence of generalized nearest points*, J. Approx. Theory **115** (2002), no. 1, 44–55.

[147] J. Marín and S. Romaguera, *On quasi-uniformly continuous functions and Lebesgue spaces*, Publ. Math. Debrecen **48** (1996), no. 3–4, 347–355.

[148] E. Matoušková, *Extensions of continuous and Lipschitz functions*, Canad. Math. Bull. **43** (2000), no. 2, 208–217.

[149] R.E. Megginson, *An introduction to Banach space theory*, Springer, Berlin-Heidelberg-New York, 1998.

[150] I. Meghea, *Ekeland Variational Principle with generalizations and variants*, Old City Publishing, Philadelphia, and Éditions des Archives Contemporaines, Paris, 2009.

[151] A.C.G. Mennucci, *On asymmetric distances*, preprint, 2nd version 2007 (1st version 2004), both available at http://cvgmt.sns.it/people/mennucci

[152] M. Mršević, I.L. Reilly and M.K. Vamanamurthy, *Lebesgue sets in quasi-pseudo-metric spaces*, Math. Japon. **37** (1992), no. 1, 189–194.

[153] M.G. Murdeshwar and S.A. Naimpally, *Quasi-Uniform Topological Spaces*, Nordhoff, Groningen, 1966.

[154] C. Mustăţa, *On the best approximation in metric spaces*, Rev. Anal. Numér. Théor. Approx. **4** (1975), no. 1, 45–50.

[155] ______, *A characterization of semi-Chebyshevian sets in a metric space*, Anal. Numér. Théor. Approx. **7** (1978), no. 2, 169–170.

[156] ______, *Some remarks concerning norm-preserving extensions and best approximation*, Rev. Anal. Numér. Théor. Approx. **29** (2000), no. 2, 173–180.

[157] ______, *Uniqueness of the extension of semi-Lipschitz functions on quasi-metric spaces*, Bul. Ştiinţ. Univ. Baia Mare Ser. B Fasc. Mat.-Inform. **16** (2000), no. 2, 207–212.

[158] ———, *Extensions of semi-Lipschitz functions on quasi-metric spaces*, Rev. Anal. Numér. Théor. Approx. **30** (2001), no. 1, 61–67.

[159] ———, *Extension and approximation of semi-Lipschitz functions on a quasi-metric space*, Numerical analysis and approximation theory (Cluj-Napoca, 2002), Cluj Univ. Press, Cluj-Napoca, 2002, pp. 362–385.

[160] ———, *On the extremal semi-Lipschitz functions*, Rev. Anal. Numér. Théor. Approx. **31** (2002), no. 1, 103–108.

[161] ———, *A Phelps type theorem for spaces with asymmetric norms*, (Proc. 3rd International Conf. on Applied Mathematics, Borşa, 2002), Bul. Ştiinţ. Univ. Baia Mare, Ser. B, Matematică-Informatică **18** (2002), no. 2, 275–280.

[162] ———, *On the uniqueness of the extension and unique best approximation in the dual of an asymmetric linear space*, Rev. Anal. Numér. Théor. Approx. **32** (2003), no. 2, 187–192.

[163] ———, *Characterization of ε-nearest points in spaces with asymmetric seminorm*, Rev. Anal. Numér. Théor. Approx. **33** (2004), no. 2, 203–208.

[164] ———, *On the extension of semi-Lipschitz functions on asymmetric normed spaces*, Rev. Anal. Numér. Théor. Approx. **34** (2005), no. 2, 139–150.

[165] ———, *Best uniform approximation of semi-Lipschitz functions by extensions*, Rev. Anal. Numér. Théor. Approx. **36** (2007), no. 2, 161–171.

[166] ———, *On the approximation of the global extremum of a semi-Lipschitz function*, Mediterr. J. Math. **6** (2009), no. 2, 169–180.

[167] R. Ni, *Existence of generalized nearest points*, Taiwanese J. Math. **7** (2003), no. 1, 115–128.

[168] V. Niemytzki, *Über die Axiome des metrischen Raumes* (German), Math. Ann. **104** (1931), no. 1, 666–671.

[169] V.N. Nikolski, *Best approximation by elements of convex sets in normed linear spaces*, Uchenye Zapiski Kalinin Gos. Ped. Inst. **29** (1963), 85–119.

[170] S. Oltra and O. Valero, *Isometries on quasi-normed cones and bicompletion*, New Zealand J. Math. **33** (2004), no. 1, 83–90.

[171] M. Pajoohesh and M.P. Schellekens, *A survey of topological work at* CEOL, Topology Atlas Invited Contributions **9** (2004), no. 2.

[172] G.K. Pedersen, *Analysis now*, Graduate Texts in Mathematics vol. 118, Springer-Verlag, New York, 1989.

[173] M.J. Pérez-Peñalver and S. Romaguera, *On right K-completeness of the fine quasi-uniformity*, Questions Answers Gen. Topology **14** (1996), no. 2, 245–249.

[174] ———, *Weakly Cauchy filters and quasi-uniform completeness*, Acta Math. Hungar. **82** (1999), no. 3, 217–228.

[175] W.J. Pervin, *Quasi-uniformization of topological spaces*, Math. Ann. **147** (1962), 316–317.

[176] M. Pfannkuche-Winkler, *Beste Φ-Approximanten im nicht-symmetrischen Fall*, Skripten zur Mathematischen Statistik [Lecture Notes on Mathematical Statistics], vol. 15, Westfälische Wilhelms-Universität Münster Institut für Mathematische Statistik, Münster, 1988.

[177] R.R. Phelps, *Uniqueness of Hahn-Banach extensions and best approximations*, Trans. Amer. Math Soc. **95**(1960), 238–255.

[178] T.G. Raghavan and I.L. Reilly, *Metrizability of quasi-metric spaces*, J. London Math. Soc. (2) **15** (1977), 169–172.

[179] ______, *A new bitopological paracompactness*, J. Austral. Math. Soc. Ser. A **41** (1986), no. 2, 268–274.

[180] A.-R.K. Ramazanov, *Direct and inverse theorems in approximation theory in the metric of a sign-sensitive weight*, Anal. Math. **21** (1995), no. 3, 191–212.

[181] ______, *Polynomials that are orthogonal with a sign-sensitive weight*, Mat. Zametki **59** (1996), no. 5, 737–752.

[182] ______, *Rational approximation with a sign-sensitive weight*, Mat. Zametki **60** (1996), no. 5, 715–725.

[183] ______, *Sign-sensitive approximations of bounded functions by polynomials*, Izv. Vyssh. Uchebn. Zaved. Mat. (1998), no. 5, 53–58.

[184] A. Ranjbari, H. Saiflu, *Some results on the uniform boundedness theorem in locally convex cones*, Methods Funct. Anal. Topology **15** (2009), no. 4, 361–368.

[185] I.L. Reilly, P.V. Subrahmanyam, M.K. Vamanamurthy, *Cauchy sequences in quasi-pseudo-metric spaces,* Monatsh. Math. **93** (1982), 127–140.

[186] I.L. Reilly and M.K. Vamanamurthy, *On oriented metric spaces*, Math. Slovaca **34** (1984), no. 3, 299–305.

[187] H. Ribeiro, *Sur les espaces à metrique faible* (French), Portug. Math. **4** (1943), 21–40, [correction, the same issue, pp. 65–68].

[188] G. Rodé, *Eine abstrakte Version des Satzes von Hahn-Banach,* Arch. Math. (Basel) **31** (1978/79), no. 5, 474–481.

[189] S. Romaguera, *On equinormal quasi-metrics*, (English) Proc. Edinb. Math. Soc., II. Ser. **32** (1989), No.2, 193–196.

[190] ______, *Left K-completeness in quasi-metric spaces*, Math. Nachr. **157** (1992), 15–23.

[191] ______, *Fixed point theorems for mappings in complete quasi-metric spaces*, An. Ştiinţ. Univ. Al. I. Cuza Iaşi Secţ. I a Mat. **39** (1993), no. 2, 159–164.

[192] ______, *On hereditary precompactness and completeness in quasi-uniform spaces*, Acta Math. Hungar. **73** (1996), 159–178.

[193] ______, *A new class of quasiuniform spaces*, Math. Pannon. **11** (2000), no. 1, 17–28,

[194] S. Romaguera and J.A. Antonino, *On convergence complete strong quasi-metrics*, Acta Math. Hungar. **64** (1994), no. 1, 65–73.

[195] S. Romaguera and E. Checa, *Continuity of contractive mappings on complete quasi-metric spaces*, Math. Japon. **35** (1990), no. 1, 137–139.

[196] S. Romaguera, and A. Gutiérrez, *A note on Cauchy sequences in quasipseudometric spaces*, Glas. Mat. Ser. III **21(41)** (1986), no. 1, 191–200.

[197] S. Romaguera and M. Ruiz-Gómez, *Bitopologies and quasi-uniformities on spaces of continuous functions*, Publ. Math. Debrecen I: **47** (1995), no. 1–2, 81–93, and II: **50** (1997), no. 1–2, 1–15.

[198] S. Romaguera, J.M. Sánchez-Álvarez, and M. Sanchis, *On balancedness and D-completeness of the space of semi-Lipschitz functions*, Acta Math. Hungar. **120** (2008), no. 4, 383–390.

[199] S. Romaguera and M.A. Sánchez-Granero, *Completions and compactifications of quasi-uniform spaces*, Topology Appl. **123** (2002), no. 2, 363–382.

[200] S. Romaguera, E.A. Sánchez Pérez, and O. Valero, *Dominated extensions of functionals and V-convex functions on cancellative cones*, Bull. Austral. Math. Soc. **67** (2003), no. 1, 87–94.

[201] ———, *Quasi-normed monoids and quasi-metrics*, Publ. Math. Debrecen **62** (2003), no. 1–2, 53–69.

[202] ———, *Duality and dominating extension theorems in noncancellative normed cones*, Bol. Soc. Mat. Mexicana (3) **12** (2006), no. 1, 33–42.

[203] ———, *A characterization of generalized monotone normed cones*, Acta Math. Sin. (Engl. Ser.) **23** (2007), no. 6, 1067–1074.

[204] S. Romaguera and M. Sanchis, *Semi-Lipschitz functions and best approximation in quasi-metric spaces*, J. Approx. Theory **103** (2000), no. 2, 292–301.

[205] ———, *Applications of utility functions defined on quasi-metric spaces*, J. Math. Anal. Appl. **283** (2003), no. 1, 219–235.

[206] ———, *Properties of the normed cone of semi-Lipschitz functions*, Acta Math. Hungar. **108** (2005), 55–70.

[207] S. Romaguera and M. Schellekens, *Quasi-metric properties of complexity spaces*, Topology Appl. **98** (1999), no. 1–3, 311–322, II Iberoamerican Conference on Topology and its Applications (Morelia, 1997).

[208] ———, *Cauchy filters and strong completeness of quasi-uniform spaces*, Rostock. Math. Kolloq. (2000), no. 54, 69–79.

[209] ———, *The quasi-metric of complexity convergence*, Quaest. Math. **23** (2000), no. 3, 359–374.

[210] ———, *Duality and quasi-normability for complexity spaces*, Appl. Gen. Topol. **3** (2002), no. 1, 91–112.

[211] W. Roth, *A uniform boundedness principle for locally convex cones*, Proc. Amer. Math. Soc. **126** (1998), no. 2, 1973–1982.

[212] ______, *Operator-valued measures and integrals for cone-valued functions*, Lecture Notes in Mathematics, vol. 1964, Springer-Verlag, Berlin, 2009.

[213] M.J. Saegrove, *Pairwise complete regularity and compactification in bitopological spaces*, J. London Math. Soc. (2) **7** (1973), 286–290.

[214] S. Salbany and S. Romaguera, *On countably compact quasi-pseudometrizable spaces*, J. Austral. Math. Soc. Ser. A **49** (1990), no. 2, 231–240.

[215] J.M. Sánchez-Álvarez, *On semi-Lipschitz functions with values in a quasi-normed linear space*, Appl. Gen. Topol. **6** (2005), no. 2, 217–228.

[216] M.P. Schellekens, *The Smyth completion: a common foundation for denotational semantics and complexity analysis*: in Proc. MFPS 11, Electronic Notes in Theoretical Computer Science **1** (1995), 211–232.

[217] E.A. Sevastyanov, *On the Haar problem for sign-sensitive approximations*, Mat. Sb. **188** (1997), no. 2, 95–128.

[218] J.L. Sieber and W.J. Pervin, *Completeness in quasi-uniform spaces*, Math. Ann. **158** (1965), 79–81.

[219] B.V. Simonov, *On the element of best approximation in spaces with nonsymmetric quasinorm*, Mat. Zametki **74** (2003), no. 6, 902–912.

[220] I.E. Simonova and B.V. Simonov, *On the polynomial of best nonsymmetric approximation in an Orlicz space*, Izv. Vyssh. Uchebn. Zaved. Mat. (1993), no. 11, 50–56.

[221] I. Singer, *Choquet spaces and best approximation*, Math. Ann. **148** (1962), 330–340.

[222] ______, *Best approximation in normed linear spaces by elements of linear subspaces*, Publishing House of the Academy of the Socialist Republic of Romania, Bucharest; Springer-Verlag, New York-Berlin 1970.

[223] ______, *The theory of best approximation and functional analysis*, C. B. M. S. Regional Conference Series in Applied Mathematics, No. 13. SIAM, Philadelphia, Pa., 1974. vii+95 pp.

[224] ______, *Extension with larger norm and separation with double support in normed linear spaces* Bull. Austral. Math. Soc. **21** (1980), no. 1, 93–105.

[225] M.B. Smyth, *Completeness of quasi-uniform and syntopological spaces*, J. London Math. Soc. **49** (1994), 385–400.

[226] R.A. Stoltenberg, *Some properties of quasi-uniform spaces,* Proc. London Math. Soc. **17** (1967), 226–240.

[227] ______, *On quasi-metric spaces,* Duke Math. J. **36** (1969), 65–71.

[228] Ph. Sünderhauf, *The Smyth-completion of a quasi-uniform space*, Semantics of programming languages and model theory (Schloß Dägstuhl, 1991), 189–212, Algebra Logic Appl., 5, Gordon and Breach, Montreux, 1993

[229] ______, *Quasi-uniform completeness in terms of Cauchy nets*, Acta Math. Hungar. **69** (1995), no. 1–2, 47–54.

[230] ———, *Smyth completeness in terms of nets: the general case*, Quaest. Math. **20** (1997), no. 4, 715–720.

[231] J. Swart, *Total disconnectedness in bitopological spaces and product bitopological spaces*, Nederl. Akad. Wetensch. Proc. Ser. A **74**=Indag. Math. **33** (1971), 135–145.

[232] H. Triebel and D. Yang, *Spectral theory of Riesz potentials on quasi-metric spaces*, Math. Nachr. **238** (2002), 160–184.

[233] J.Sh. Ume, *A minimization theorem in quasi-metric spaces and its applications*, Int. J. Math. Math. Sci. 31 (2002), no. 7, 443–447.

[234] O. Valero, *Quotient normed cones*, Proc. Indian Acad. Sci. Math. Sci. **116** (2003), no. 2, 175–191.

[235] ———, *Closed graph and open mapping theorems for normed cones*, Proc. Indian Acad. Sci. Math. Sci. **118** (2008), no. 2, 245–254.

[236] N. Weaver, *Lipschitz algebras*, World Scientific Publishing Co. Inc., River Edge, NJ, 1999.

[237] J.H. Wells and L.R. Williams, *Embeddings and extensions in analysis*, Ergebnisse der Mathematik und ihrer Grenzgebiete, Band 84, Springer-Verlag, New York-Heidelberg, 1975.

[238] D. Werner, *Funktionalanalysis*, 6., korr. Aufl., Springer-Lehrbuch, Berlin, 2007.

[239] W.A. Wilson, *On quasi-metric spaces,* Amer. J. Math. **53** (1931), no. 3, 675–684.

[240] C. Zanco and A. Zucchi, *Moduli of rotundity and smoothness for convex bodies,* Bolletino U. M. I. (7) **7-B** (1993), no. 4, 833–855.

Index

Frontiers in Mathematics

This series is designed to be a repository for up-to-date research results which have been prepared for a wider audience. Graduates and postgraduates as well as scientists will benefit from the latest developments at the research frontiers in mathematics and at the "frontiers" between mathematics and other fields like computer science, physics, biology, economics, finance, etc.

■ **Qin, Y. / Huang, L.**, Global Well-posedness of Nonlinear Parabolic-Hyperbolic Coupled Systems (2012).
ISBN 978-3-0348-0279-6

This book presents recent results on nonlinear parabolic-hyperbolic coupled systems such as the compressible Navier-Stokes equations, and liquid crystal system. It summarizes recently published research by the authors and their collaborators, but also includes new and unpublished material. All models under consideration are built on compressible equations and liquid crystal systems. This type of partial differential equations arises not only in many fields of mathematics, but also in other branches of science such as physics, fluid dynamics and material science.

■ **Bustamante, J.**, Algebraic Approximation: A Guide to Past and Current Solutions (2012).
ISBN 978-3-0348-0193-5

This book presents a unique survey of solutions in algebraic approximation. Several results related with direct and converse theorems in the theory of approximation by algebraic polynomials in a finite interval are discussed. Some of these results are not collected in any other book. In addition, facts concerning trigonometric approximation that are necessary for motivation and comparisons are included. The selection of papers that are referenced and discussed document trends in polynomial approximation from the 1950s to the present day.

The book will be invaluable to anyone seeking to understand the evolution of ideas in algebraic approximation. Its extensive bibliographic character will help finding the correct references for a specific result.

■ **Picado, J. / Pultr, A.**, Frames and Locales (2011).
ISBN 978-3-0348-0153-9

Until the mid-twentieth century, topological studies were focused on the theory of suitable structures on sets of points. The concept of open set exploited since the 1920s offered an expression of the geometric intuition of a "realistic" place (spot, grain) of non-trivial extent.

Imitating the behaviour of open sets and their relations led to a new approach to topology flourishing since the end of the 1950s. It has proved to be beneficial in many respects. Neglecting points, only little information was lost, while deeper insights have been gained; moreover, many results previously dependent on choice principles became constructive. The result is often a smoother, rather than a more entangled, theory.

No monograph of this nature has appeared since Johnstone's celebrated Stone Spaces in 1983. The present book is intended as a bridge from that time to the present. Most of the material appears here in book form for the first time or is presented from new points of view. Two appendices provide an introduction to some requisite concepts from order and category theories.

■ **Gavrilyuk, I. / Makarov, V. / Vasylyk, V.**, Exponentially Convergent Algorithms for Abstract Differential Equations (2011).
ISBN 978-3-0348-0118-8

This book presents new accurate and efficient exponentially convergent methods for abstract differential equations with unbounded operator coefficients in Banach space. These methods are highly relevant for practical scientific computing since the equations under consideration can be seen as the meta-models of systems of ordinary differential equations (ODE) as well as of partial differential equations (PDEs) describing various applied problems.

■ **Dragović, V. / Radnović, M.**, Poncelet Porisms and Beyond. Integrable Billiards, Hyperelliptic Jacobians and Pencils of Quadrics (2011).
ISBN 978-3-0348-0014-3

■ **Elworthy, K.D. / Le Jan, Y. / Li, X.-M.**, The Geometry of Filtering (2010).
ISBN 978-3-0346-0175-7

■ **Østvær, P. A.**, Homotopy Theory of C*-Algebras (2010).
ISBN 978-3-0346-0564-9

■ **Borsuk, M.**, Transmission Problems for Elliptic Second-Order Equations in Non-Smooth Domains (2010).
ISBN 978-3-0346-0476-5

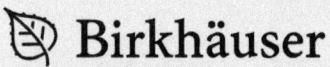